AGRICULTURE AND RURAL DEVELOPMENT

(Emerging Trends and Right Approach to Development)

PROF. K. VENKATA REDDY
M.A., Ph.D., Dip. in Statistics
Professor of Rural Development (Retd.)
and
Formerly Vice-Chancellor
Sri Krishnadevaraya University, Anantapur
Andhra Pradesh.

ISO 9001:2008 CERTIFIED

First Edition : 2012
Reprint : 2018

Published by	:	Mrs. Meena Pandey for **Himalaya Publishing House Pvt. Ltd.**, Ramdoot, Dr. Bhalerao Marg, Girgaon, Mumbai - 400 004 **Phone:** 022-23860170/23863863; **Fax:** 022-23877178 **E-mail:** himpub@vsnl.com; **Website:** www.himpub.com
Branch Offices	:	
New Delhi	:	Pooja Apartments, 4-B, Murari Lal Street, Ansari Road, Darya Ganj, New Delhi - 110 002. Phone: 011-23270392, 23278631; Fax: 011-23256286
Nagpur	:	Kundanlal Chandak Industrial Estate, Ghat Road, Nagpur - 440 018. Phone: 0712-2738731, 3296733; Telefax: 0712-2721216
Bengaluru	:	Plot No. 91-33, 2nd Main Road, Seshadripuram, Behind Nataraja Theatre, Bengaluru - 560 020. Phone: 080-41138821; Mobile: 09379847017, 09379847005
Hyderabad	:	No. 3-4-184, Lingampally, Besides Raghavendra Swamy Matham, Kachiguda, Hyderabad - 500 027. Phone: 040-27560041, 27550139
Chennai	:	New No. 48/2, Old No. 28/2, Ground Floor, Sarangapani Street, T. Nagar, Chennai-600 012. Mobile: 09380460419
Pune	:	First Floor, Laksha Apartment, No. 527, Mehunpura, Shaniwarpeth (Near Prabhat Theatre), Pune - 411 030. Phone: 020-24496323, 24496333; Mobile: 09370579333
Lucknow	:	House No. 731, Shekhupura Colony, Near B.D. Convent School, Aliganj, Lucknow - 226 022. Phone: 0522-4012353; Mobile: 09307501549
Ahmedabad	:	114, SHAIL, 1st Floor, Opp. Madhu Sudan House, C.G. Road, Navrang Pura, Ahmedabad - 380 009. Phone: 079-26560126; Mobile: 09377088847
Ernakulam	:	39/176 (New No. 60/251), 1st Floor, Karikkamuri Road, Ernakulam, Kochi - 682011. Phone: 0484-2378012, 2378016; Mobile: 09387122121
Bhubaneswar	:	Plot No. 214/1342, Budheswari Colony, Behind Durga Mandap, Bhubaneswar - 751 006. Phone: 0674-2575129; Mobile: 09338746007
Kolkata	:	108/4, Beliaghata Main Road, Near ID Hospital, Opp. SBI Bank, Kolkata - 700 010, Phone: 033-32449649; Mobile: 07439040301
DTP by	:	HPH, Editorial Office, Bhandup **(Krunali)**
Printed at	:	M/s. Aditya Offset Process (I) Pvt. Ltd., Hyderabad. On behalf of HPH.

Dedicated to

My Father

Late K. SUBBA REDDY

Dhanojivaripalle
Ganugapenta Village
Pakala Mandal
Chittoor District
Andhra Pradesh

FOREWORD

After the introduction of economic reforms in India in the early 1990s, its GDP growth rate has accelerated and it is now the second fastest growing economy in the world. But rural India continues to lag behind the rest of the economy. The food security achieved as a result of green revolution is now called into question because of slow growth of farm output in the face of rising demand. The growth of output and employment in the rural non-farm sector has also been slow. With nearly 70 per cent of the country's population still living in rural areas and much of the poverty concentrated there, the rural-urban disparities in incomes are increasing. Degradation of natural resources is undermining the sustainability of development while causing deterioration in the quality of life and threatening the security of rural livelihoods. Now, over 80 per cent of our farmers are small and marginal. The out-migration of the young and the educated males has led to the rise in the proportion of women farmers with inadequate skills and property rights in land so that the much talked about demographic dividend still eludes rural India.

The challenges faced in the development of rural India have thus increased manifold in the post-reform era. This book by Prof. K. Venkata Reddy brings out these emerging challenges and discusses policies and programmes for holistic and sustainable, development of rural India. He stresses the need for stepping up investment in agriculture and rural infrastructure, technological upgradation and appropriate institutional arrangements for self-sustaining development. He makes extensive use of the latest information and recommendations of the high-level committees like the National Commission on Farmers headed by Dr. M.S. Swaminathan and the model for urbanization of rural areas advocated by Dr. A.P.J. Abdul Kalam, the former President of India. For imparting human face as well as sustainability to rural development, he invokes Mahatma Gandhi's ideals, in particular, the need for decentralisation of development through peoples' participation.

Prof. K. Venkata Reddy has devoted his whole life to the studies and teaching on rural development. The present book, which has already gone into several editions, is the product of his lifetime endeavour. His approach to rural development in the book is holistic, human and balanced. It is holistic, as it traces the development in its historical perspective and by examining various theories of development bringing out their relevance to the Indian situation. His concept of development itself is comprehensive covering growth in output and income; improvements in social indicators including education and health; changes in the levels of living of different classes of people; the state of environment; and peoples' participation in development and their self-esteem. His approach is humanistic as the focus is on the poor and the vulnerable. It is also balanced because of his objectivity and drawing attention to the achievements made in various spheres of development since, independence even as he brings out the gaps and shortfalls.

Having been a teacher all his life, Prof. Reddy is well aware of the intellectual needs of the students of economics in the country, their academic preparedness and how to relate with them. This is amply reflected in his style of writing. It is simple, devoid of unnecessary jargon, straight and clear. I am confident that the book will be received by the wider sections of graduate and postgraduate students in economics in the country with the same expectations and enthusiasm as before.

Centre for Economic and Social Studies,
Hyderabad.

C.H. Hanumantha Rao
Former Member, Planning Commission, Govt. of India
Former Chairman, Institute of Economic Growth, Delhi
Chairman, Centre for Economic and
Social Studies (CESS), Hyderabad.

PREFACE

Rural India is emerging and fast moving in the direction of a developed nation. During six decades of planned development, there has been significant progress in different sectors of economy. From the stage of food shortage, India has emerged to the level of self-sufficiency in foodgrains production and India is now in a position to export certain items of farm production and earn foreign exchange. Agricultural exports constituted 12.2 per cent of the total national exports in 2007-08. Agriculture accounts for 19 per cent of GDP and this sector continues to be a major employer in the country.

Despite significant economic growth, however, occupational structure in India has not undergone a radical change. Still, some 65 per cent of the population continues to live in villages, seek employment and living in the agricultural sector. The rural sector, thus, assumes critical importance in the economy of India. The rural sector predominantly influences the process and pace of economic development. Growth trends indicate that for some more years to come, rural sector would be of critical importance in the process of development.

At this juncture, some realities have to be noticed and policy measures need to be so designed as to address the basic issues. Agriculture, as a major component of rural economy is overcrowded and it is no longer in a position to support any more population. That apart, Indian agriculture suffers from (1) Low yield per hectare of different crops, (2) Fast depleting water table causing inadequate irrigation, (3) Escalating input costs, (4) Inadequacy of institutional credit, (5) A distorted market, (6) Increased role of intermediaries, denying remunerative price to the producer, (7) Gradual decline in the public investment in agriculture, (8) Poor infrastructure in rural areas, particularly transport and storage facilities, (9) Lack of incentives to attract private investments in agriculture, (10) Inadequate extension service support, (11) Exposure to yield and price risks and (12) Low coverage of crop and livestock insurance.

The cumulative effect of all these different factors, is one of agrarian crisis, leading to rural crisis. Again, the matters of concern today at the end of Eleventh Five Year Plan are: Environmental degradation; regional imbalances in the process of development and consequent social unrest; poverty of high magnitude; lack of food security and social security to considerable number in the society; poor quality of education in the rural areas; the adverse effect of policy of globalization on agriculture and women continue to be a neglected segment of the Indian society. A recent study by United Nations identified that more than 410 million people live in poverty in Indian States. The poverty is found to be concentrated in Bihar, Madhya Pradesh, Uttar Pradesh and West Bengal. The study also pointed out that inspite of India's economic growth, the growth is found to be unequal. Inequalities in-between different states are identified. The study concluded, "Low per capita GDP income does not necessarily mean high poverty. Poverty is linked to low level of social investment in terms of literacy, health, proper human relations etc. Social capital of ancient days needs to be revived and developed as a resource of development.

The Eleventh Five Year Plan speaks of "inclusive growth", clearly indicating that some segments of the society are by-passed, may be some areas or some individuals. This publication examines some of these issues of serious concern. It is suggested that the Gandhian approach is the right approach in the development of India. The Gandhian ideology of rural development with decentralized administration and planning with village as the growth centre; priority to short gestation and eco-friendly schemes like minor irrigation, cottage and small-scale enterprises etc.; value-based and skill building basic education need to be well respected and implemented. The rural areas must become centres of small enterprises. It means creation of non agricultural employment opportunities in the rural sector that would prevent migration of rural labour to urban centres, apart from reducing the pressure on agriculture. Of all the resources, the human resource is most crucial. The need of the hour is one of expanding and strengthening of educational facilities, ensuring quality education particularly in the rural areas. The rural areas must necessarily be provided with all those facilities the urban people are enjoying on the lines indicated by A.P.J. Abdul Kalam. The human resource in terms of literacy and motivation for development need to be developed. The non governmental organizations have a crucial role to play in the betterment of rural areas and rural people and that way pave the way for overall development of the country. The inequalities in the incomes and dichotomy between rural and urban need to be addressed with realistic plan. As Gandhiji indicated, the strategy of development should be one of "production by masses" and not "mass production". By promoting the self-help groups and setting up of micro enterprises in the rural areas employment can be generated without giving room for wide inequalities of income in-between different sections of the society.

In this book an attempt is made, to discuss different issues concerning the growth of the economy with particularly reference to rural sector. Areas of current relevance such as Environment and Climate Change, Ethics, Environment and Sustainable Development, Forest Resources and Joint-forest Management; Natural Resource Management and Livelihood, Participatory Watershed Management for Sustainable Rural Livelihoods; Women Empowerment and Micro Finance; Appropriate Technology for Rural Development, Social Security Schemes, Education for all; PURA Mission for Rural Reconstruction; Regional Imbalances and Social Tensions; Agrarian Crisis and Rural Distress; Policy of Liberalization and its impact on rural economy and "inclusive growth" are briefly discussed in this book. Mainly, the book covers broad guidelines of the strategy of development on the lines of the Gandhian ideology. It is hoped that the book will meet the requirements of both graduates and postgraduate students pursuing courses in different branches of economics and rural development. Some of the topics covered here, I believe are of immense use for competitive examinations at the All-India and State levels. For such of those readers, interested in knowing the growth trends in India, this book could be of some use.

The author has collected required data and information from different sources — Government reports, particularly, Five Year Plan Reports, Census Reports, Research Studies, reports of different committees appointed by the government etc., which are duly acknowledged. In the preparation of this book I received assistance from a good number of scholars, Dr. K. Vijaya, K. Sivaprasad and Prof. A. Ranga Reddy, made available to me useful material. Dr. A. Appasami Pillai has been helpful to me throughout the preparation of this book. Prof. K. Nagi Reddy offered helpful suggestions in designing this book. K. Gokul and K. Nanditha have been highly helpful to me in the proofreading of the book. All my former colleagues of the Department of Rural Development S.K. University have been a source of inspiration in the preparation of this book. Smt. K. Hemalatha, K. Suresh Babu and Smt. K. Sumalatha have been of immense assistance to me during the preparation of this work.

I express my sincere thanks to Dr. Hanumantha Rao, a reputed Economist and Researcher for sparing his valuable time and writing Foreword to my book highlighting the merits of publication. The DTP work is well carried out by Student Xerox, Prakashan Road, Tirupati.

While I place on record the useful suggestions from a good number of well informed persons in the preparation of this book, the opinions expressed here are my own and I only take the responsibility for deficiencies, if any. Suggestions for improvement of this book are always welcome.

I express my sincere thanks to the publishers of this book, Himalaya Publishing House Pvt. Ltd., Mumbai and their efficient editorial staff who helped in publishing this book.

Prof. K. Venkata Reddy

Saraswati Nilayam, 11-219,
S.V. Nagar,
Tirupati - 517 502.

CONTENTS

PART - IV: SCHEMES IN DEVELOPMENT

LIST OF TABLES

CURRICULUM VITAE OF PROF. K. VENKATA REDDY

Sri K. Venkata Reddy hails from an agricultural family: born on 15th August, 1931, at Dhanojivarialler of Ganugapenta Panchayat of Palika Mandal, Chief district, Andhra Pradesh. He had his schooling at Board High School, Pakala. During his school days, in response to Gandhiji's call, he participated in the Independence Movement: Courted arrest twice in 1945 and 1946 for his active participation in the non cooperation activities against the British rule. Sri Vangipuram Rangaswamy Iyangar, a reputed and committed teacher of Board High School pakala, was his mentor and was well trained under the committed teachers of the day. He had the good fortune of seeing Mahatma Gandhi twice in 1944 and 1945. He was the recipient of a good number of prizes in Telugu and English elocution and essay writing competitions and for Historic talents.

He had his Intermediate course in S.V. Arts College, Tirupati during 1950-52 and he secured 1st rank in the Intermediate Examination. Then he moved to Madras University for higher education; he did B.A. (honours) economics in Economics by efflux of time in 1956.

He entered the teaching profession 1955 and served in Tirumala-Tirupati Devasthanam Colleges, S.V. Arts College, S.G.S. College, affiliated to S.V. University for 20 long years, i.e., 1955 to 1975. While working in T.T.D. College, he qualified himself for Ph.D., in Economics, from S.V. University. He also obtained Diploma in Statistics from S.V. University. He moved to Andhra University in 1975 and from there he got into S.V. University postgraduate centre, Anantapur as Reader in Rural Development in 1977 and very soon he was elevated as professor of Rural Development.

In 1981, the Centre was upgraded as a full-fledged university, Sri Krishnadevaraya University. Dr. K. Venkata Reddy served as the Registrar of the University for three years during 1982-1985 when Prof. M. Abel was the Vice Chancellor Dr. K. Venkata Reddy succeeded Prof. M. Abul as the Vice-Chancellor and he served as Vice-Chancellor for three years from 1988-1989 to 1990-1991.

Prof. K. Venkata Reddy had to his credit total teaching of 36 years from 1955 to 1991. He retired as professor of rural development from S.K. University on 30th April, 1991 and simultaneously completed his three years of office a Vice-Chancellor of S.K. University on 11th May, 1991.

Prof. K. Venkata Reddy visited Bangkok, Philippines and Singapore in connection with his participation in an international seminar held at the International Institute of Rural Reconstruction (IIRR), Manila, Philippines during October, 1987.

He had to his credit two major research projects sponsored by the UGC namely;

1. Fertility and Family Planning Attitudes in a Poor Community — A Case Study of Jalaries of Vishakapatnam during 1976-1977 and;
2. Growth and Instability in Crop Production in Drought-prone Rayalaseema Region of Andhra Pradesh, 1980-81.

Under his guidance 12 candidates got Ph.D. and 15 got M.Phil. degree. He is author of 15 books and more than 100 papers published in reputed journals. The popular books of authored by him are (1) Agricultural Economics, (2) Agricultural Productivity in Andhra Pradesh, (3) Green Revolution in India – A Monograph, (4) Rural Indebtedness in India, (5) Reflections on Socially-relevant Education, (6) Education and Socio-economic Development, (7) Rural Development (Poverty and Development), (8) Rural Development in India – A Gandhian Perspective, (9) Mahathmula Sukthi Sudha.

He held different academic and administrative positions and served on different official committees as listed below:

1. He served as Vice-Chancellor of S.K. University, Anantapur during 1988-1991.
2. Member of State Council of Higher Education during 1988 and 1991 under the category of Eminent Educationists.
3. Member of committee setup by the Government of Andhra Pradesh to draft proposals for the proposed rural University 1986-1987.
4. Member of the Committee set up the Government of Andhra Pradesh to draft proposals for the scheme of "EARN WHILE YOU LEARN" 1987-1988.
5. Member of the Committee set by the Government of Andhra Pradesh relating to the Mass Programme for Functional Literacy in 1988-1989.
6. Member of Three-men Committee set up by the Government of Andhra Pradesh to prepare glossary in Social Sciences for Telugu Academy.
7. Member of Senate, Academic Council and Board of Studies of S.V. University Gulbarga University, Bangalore University and S.K. University.
8. Convener of the Expert Committee set up by the University Grants Commission to review VII Plan Development schemes of South Gujarat University and Jodhpur University.
9. Member of the Committee set by the State Council of Higher Education to examine disparities in the level of development among different universities in Andhra Pradesh, 1989-1990,
10. Member of the Planning and Monitory Board of Nagarjuna University, 1978-1988.
11. Member of the Committee setup by the U.G.C. for examining proposals for Travel Grant 1991-1992.
12. Member of the Expert Committee to finalise VIII Plan development proposals of Andhra Pradesh University, Kurukshetra University, Rajasthan State and Sambalpur University.
13. Served as a Member of Andhra Pradesh Pollution Council Board of examine environmental problem in connection with Aswini Bio-Pharmacy Limited, Gajulamandyam, Chittoor district.
14. Served as a Member of the Committee set by the Dravidian University to suggest future plans for development of University, 2002-2003.
15. Served as a member of the Academic and Administrating Audit Committee constituted by S.V. University during 2003-2004.
16. Served as a member of a committee set up to draft rules for Yogi Vermana University, Kadapa, 2008-2009.

Academic Distinctions, Awards and Honours:

1. He was awarded a prize in the All-India Essay Writing Competition conducted by Andhra Pradesh Productivity Council, Hyderabad 1966.
2. Recipient of "Best Teacher Award" from the Government of Andhra Pradesh in 1996.
3. Recipient of Kavikokila Duvvuri Rami Reddy Award on the occasion of 95th Jayanti Celebrations held at Nellore on 9th November, 1990.
4. Honoured as freedom fighter by the Aurabindo Society, Tirupati on the occasion of Jayanti celebration of Aurabindo on 15th August, 2004.
5. Recipient of "Lifetime Achievement Ward" from S.V. University from the hands of the Chancellor on 21st March, 2005.

Currently:

- Member of Balaji Sath Sangh, Tirupati
- Member of Elders Forum, Tirupati
- Member of Aurabindo Society, Tirupati
- Member of Ramakrishna Samithi, Tirupati
- Member of Intellectual Forum, Tirupati
- Member of Friends on SVIMS,, Tirupati

He always keeps himself fully well engaged devoting his time in reading, writing and social service activities.

PART I

INTRODUCTION

INDIA: LAND OF RICH CULTURE WITH POTENTIAL RESOURCES

1. Introduction

India is a country with a glorious past. This sacred land has been nurturing a civilization of high order spanning over 5000 years with rich culture and heritage handed down to us by the native pre-Aryans and Aryans. The Vedas and the Upanishads of this sacred land reflect our wisdom of ancient days. India is the abode of infinite values. Tolerance, compassion, non-violence and truth continue to be the hallmarks of Sanathana Dharma of this sacred land. It is this Dharma of infinite values that is responsible for multi-religious and multi-lingual sections of the society living together. A distinct feature of Indian Culture is well depicted by way of "unity in diversity".

2. Size of the Country

India is the seventh largest country in the world and stands apart from the rest of Asia, marked off as it is by mountains and the sea which give the country a distinct geographical identity. India[1] has a land frontier of about 15,200 kms. The total length of the coastline of the mainland, Lakshadweep Islands and Andaman & Nicobar Islands is 7516.6 kms. The arable land of 160 million hectares (in 2006) in India constituted 11.2% of the World's arable land. India is next only to USA in the extent of arable land. The very size of the land and the extent of coastline clearly indicate, the country's potential resources.

3. Flora and Fauna[2]

India is rich in flora and fauna indicating country's potential in bio-diversity. The Botanical Survey of India estimated that there are more than 46,000 species of plants spread over the country.

Again India has a great variety of fauna. The Zoological Survey of India estimates that there are more than 89,000 species of animals. These resources have to be well preserved to ensure the environmental safety, very much needed for sustainable development.

4. Sacred Land of Sages and Scientists

This sacred land has given birth to a good number of sages, scientists and innovative thinkers. The teachings of different sages are endowed with eternal truths of universal sanctity and validity. The moral sayings and teachings of the Buddha, for example, enjoy universal respect. Similarly, the divine voices of Sri Ramakrishna Paramahamsa, Swami Vivekananda, Aurabindo, Ramana Maharshi, Malayalaswami, Jiddu Krishnamurthy and Mahatma Gandhi convey universal truths, well received all over the globe. Gandhiji's thoughts and his life are truly an expression of the philosophy of our ancient civilization known for peace, non-violence, truth and compassion. These are eternal values which have sustained our society and our nation.

5. Ancient India's Contribution to Different Branches of Knowledge

Ancient India laid foundation for Mathematics and different branches of scientific knowledge. They measured both time and space. They conceived and developed the sciences of logic and grammar and made advances in the fields of Astronomy, Philosophy and Medicine. Long back, the ancient Indians, developed a system, known as Nature-cure. It is not merely a cure of a disease, after it has occurred, but an attempt to prevent a disease altogether by living according to the Laws of Nature. They very rightly, identified that disease of a body is the outcome of physical and mental aspects of man's life and therefore, an integrated approach is required to cure the disease.

6. Different Arts and Architecture

Different Arts, Sculpture and Architectural designs etc., of ancient India constitute a storehouse of ideas and practices of universal validity for all times, very much needed now for a civilized life. In ancient India, music, art, and paintings were well nurtured. The Ajanta and Ellora caves, Taj Mahal in Agra, the Gopuram of the Hindu temples and their sculpture are a living testimony of extraordinary engineering skills of architecture and paintings of those days.

7. Indigenous Technical Knowledge (ITK)

The ITK covers a wide spectrum with a focus on agricultural growth. Agriculture in India has a long history behind spreading over 5000 years. Ploughing of land for sowing the seed was the earliest of ITK. Soon the technology of rainwater harvesting, and well preserving the water in Tanks or Ponds was developed. Also, soil conservation technology; farm equipment, technology, seedling technology, post-harvest preservation, multiple cropping system, pest control etc., were known in the past. The diseases of both human beings and animals were cured by the indigenous methods. Another important technology of the ancient times was one of designing bullock-cart, quite appropriate to the rural areas and it was a pioneering work of our ancients.

India has been reputed in Textile technology in terms of producing a refined variety of fabric with export value. Indian artisans have been well known for their skills in goldsmithy, blacksmithy, carpentry, pottery and a multitude of handicrafts. A reference to 'Pushpaka Vimanam' used by Sri Rama to cross over the ocean and reach Lanka indicates that a technology to fly was known in those days.

Ancient Indians had their own methods of spreading knowledge. Folklore, songs, poetry, village community centres etc., served as media and centre of spreading knowledge. Also knowledge was transmitted by word of mouth and that way knowledge was widely spread. Further village elders had acted as agents in spreading the knowledge.

8. Tribute to Indian Culture

Arnold Joynbee, after surveying the story of entire human race, observed, "It is already becoming clear that a chapter which had a western beginning will have to have an Indian ending, if it is not to end in the self-destruction of the human race.... At this supremely dangerous moment in human history, the only way of salvation for mankind is the Indian-way — Emperor Ashoka's and Mahatma Gandhi's principle of non-violence and Sri Ramakrishna's testimony to the harmony of religions."[3].

N.A. Palkivala makes a fitting observation of our culture as 'Priceless Heritage'. Indian culture which is primarily concerned with spiritual development is of special significance in our age which is marked by the obsolescence of the materialistic civilization. In the words of Sri Aurobindo,[4] "India of the ages is not dead, nor has she spoken her last creative word; she lives and has still something to do for herself and human race".

Peace and non-violence are the hallmarks of the Indian Culture. India never attempted military aggression on countries outside her borders. Friedrich Max Muller[5] in his work, "India, what can it Teach us?" observes, "If I were to look over the whole world to find out the country most richly endowed with all the wealth, power and beauty that nature can bestow, the paradise on the earth, I should point to India. And if I were to ask myself from what literature we, here in Europe, may draw corrective which is the most wanted in order to make our inner life more perfect, more comprehensive, more universal, infact more truly eternal life, again I shall point to India." Such of those who seek solace and mental peace, drawn from different parts of the globe regularly visit places like Aurabindo Ashram at Puducherry, Ramana Ashram at Tiruvanamalai and Ramakrishna Ashram at Bengaluru, etc.

9. India's Scientific Progress

In Science and Technology, India is strong in its own way as an international player. India can be proud of the role of visionary scientists, like Homi Bhaba, Vikram Sarabai, Sahani Randhawa who had glorified plans for the eradication of poverty through scientific knowledge. Pandit Nehru, during his tenure as Prime Minister, laid down strong foundation for scientific and technological infrastructure through a network of public sector research laboratories. India's contribution for promoting biotechnology is really significant. Pharma companies like Cipla, Dr. Reddy's Labs, Ranbaxy have gained global reputation. In the preparation of herbal drugs and Ayurvedic formulations, India has been reputed all through.

India has made a name in information technology. It is the science and technology advancement in our own way that is responsible for evolving new variety of seeds. Under the aegis of Indian Council of Agricultural Research (ICAR) more than 2000 high yielding hybrid varieties of foodgrains and cash crops have been developed. Renowned scientist, M.S. Swaminathan's contribution for promotion of Green Revolution, particularly in the field of rice research, is worth remembering. Evolving hybrid cotton, hybrid sorghum, hybrid castor and hybrid mango are some of the significant achievements of Indian Agricultural Research.

Besides the Green revolution, White revolution in milk production, the Yellow revolution in oil seeds, Blue revolution in fish production and Golden revolution in Horticulture bear ample testimony to the contribution of our agricultural scientists in making the country self-sustained in terms of food production. We are proud of the great scientist of India A.P.J. Abdul Kalam for his contribution in the field of Satellite Launch. The Indian Satellite first launch in August, 1979 was the outcome of efforts of a good number of scientists under the leadership of A.P.J. Abdul Kalam.

10. Indicators to Assess Development

India has never been a backward country and to depict India so is most uncharitable. Gandhiji opined that development of a country need not be assessed from the angle of Gross National Product (GNP) and per capita income alone. On the other hand, the family ties, good relationship and understanding among different sections of the society, development of different Arts and Architecture etc., need to be taken into consideration in assessing the development of a country. The per capita income can indicate the extent of wealth in the hands of the people as a whole. But per capita income does not indicate that they all have the same amount of money. It is the average of the income of both rich and poor. The same per capita figure, as A.P.J. Abdul Kalam[6] rightly points out does not indicate the amount of well-being within a country or within a state or region.

Many indicators can be used to assess the degree of development of a country. There are both economic and non-economic indicators. The economic indicators such as Gross National Product, the Balance of Payment, Foreign Exchange Reserves, rate of economic growth, the extent of international trade, are no doubt, important. Along with these, food security, social security, the average life expectancy, the infant mortality rate, health and sanitation facilities etc., become more important.

Gandhiji indicates another method of assessing the development of a country. Gandhiji's[7] strikingly simple criterion was that "every action proposed or contemplated should in its implementation wipe the tears of the poor and the downtrodden persons" Pandit Nehru's vision of development was one of elimination of ignorance, illiteracy, poverty, disease and inequalities in the opportunities. During the planning era, very rightly, series of schemes have been initiated to address the severity of poverty and poverty related problems.

11. Agriculture as a Major Economic Activity and a Way of Life

Agriculture has been in vogue since a long time; In fact agriculture is a way of life and part of Indian culture. Agriculture has been a major employer providing employment to a vast majority of people in India. The growth models of development emphasize the significant role of agricultural sector in economic development. Hence some details about agriculture in India merit consideration. India has been raising a variety of crops, both foodgrains and horticultural items. B.S. Raghavan very rightly opines, "No other country grows such a wide range of fruits, vegetables and flowers and in such abundance, as India.... The rich variety of medicinal plants and herbs when processed and marketed can help India, take care of the health needs of its population besides coming into its own as the World's Pharmaceutical Giant" (*The Hindu Survey,* Indian Agriculture, 1994, p. 13).

After the monumental studies of Asian countries, Gunnar Mydral concludes: "It is in the agricultural sector that the battle for long-term economic development in South Asia will be won or lost"[8] With reference to India, Coale and Hoover stress the same point: "Very substantial progress in that most backward part of the economy (agriculture) was a prerequisite to successful development of the Indian economy as a whole."[9] In emphasizing the role of agriculture in the development of the Indian economy, it is observed that "If one sector limits the growth of the other, it is more likely to be a case of agricultural growth, limiting non-agricultural sector than *vice versa.*"[10]

The trend of economic growth and occupational structure, clearly indicates that for some more years to come agriculture will continue to be a major source of employment and living to a vast majority of population in the country. The population of India was 361.1 million in 1951 and in 2007-08, population was estimated at 1130 million. This resource need to be well tapped for productive purpose. With increase in population in the absence of employment opportunities from the non-agricultural sector, there is growing pressure on land. Land utilization trend is shown in Table 1.1.

Table 1.1

Land Utilization Trend in India, 1950-51, and 2005-06

(Million Hectares)

S.No.	Classification	1950-51	2005-2006
1	2	3	4
I	Geographical Area	328.73	328.73
II	Reporting Area for Land Utilization Statistics	284.32	305.27
	1. Forest %	40.48 14.2	69.79 22.90
	2. Not available for cultivation (A + B)		
	(A) Area under non-agricultural uses %	9.36 3.3	25.03 8.2
	(B) Barren and Unculturable Land %	38.16 13.4	17.48 5.70
	3. Other cultivable land excluding fallow land (A + B + C)		
	(A) Permanent pastures and other grazing lands %	6.68 2.3	10.42 3.4
	(B) Land under Misc. Tree crops and Groves not included in Net Sown area %	19.83 7.0	3.38 1.10
	(C) Culturable Wasteland %	22.94 8.1	13.12 4.3
	4. Fallow Lands (A + B)		
	(A) Fallow lands other than current fallows %	17.45 6.1	10.50 3.4
	(B) Current fallows %	18.48 3.8	13.67 4.5
	5. Net Area sown (6 - 7) %	118.75 41.8	141.89 46.5
	6. Total cropped area (Gross cropped area)	131.89	192.80
	7. Area sown more than once	13.15	50.90
	8. Cropping Intensity	111.1	135.9
III	Net irrigated Area	20.85	60.20
IV	Gross irrigated Area	22.56	82.63

Source: *Agricultural Statistics at a Glance 2008,* Directorate of Economic and Statistics, Ministry of Agriculture, Government of India, pp. 260 to 262.

Cropping intensity is percentage of the gross cropped area to the net area sown.

From Table 1.1 it is clear that in 2005-06, the forest area was 22.9% of the total geographical area as against 14.2% in 1950-51. But yet, even the increased percentage is less than the optimum level of one-third of geographical area as recommended by the experts for environmental safety. That there has been a growing pressure on land is evident. The barren and uncultivable land; land under misc. trees, crops and cultivable wasteland declined considerably. The cultivable waste land has declined by 50%, fall in the fallow lands is again to the extent of about 50%; Net area sown increased from 118.75 to 141.89 million hectares; total cropped area increased from 131.89 to 192.80 million hectares; net irrigated area increased from 20.85 to 60.20 million hectares; the gross irrigated area increased from 22.56 to 82.63 million hectares. Area sown, more than once stepped up from 13.15 million hectares to 50.90 million hectares. The cropping intensity increased from 111.1% in 1950-51 to 135.9% in 2005-06.

The percentage of net irrigated area to net sown area in 2005-06 works out to 42.43. It means nearly 60 per cent of the cropped area is still under the rain-fed cultivation.

It is evident that 'Land' as a cultivable asset may not be able to support any more ensuring livelihood. There are many signals to that effect. Since 1990s, growth rates of production of different crops have been declining. We also notice gradual increase in the number of agricultural labour during this period. Research studies reveal that in recent years, there has been migration of labour from the rural areas to urban centres in search of livelihood.

With migration of labour class, labour scarcity in rural areas creates a problem. The cost of different inputs, including labour, is soaring up and the farm producer is not able to get remunerative price to his produce. So some landowners also have chosen to migrate to the urban centres, resulting in fall in the foodgrain production.

12. Demand for Foodgrains

India's population is projected at 1.3 billion by the year 2020 and 1.7 billion by 2050. As the economy grows, people earn more and consumption level and pattern changes. There is bound to be a change in the lifestyle indicating growing demand not only for cereals, but also for different items of non-cereals. A.P.J. Abdul Kalam[11] basing on TIFAC study estimates at 7 per cent income growth, projected household demand for foodgrains in India by 2020, would be 343.0 million metric tones. Similar estimates are also made for non-cereals. This calls for greater importance to agriculture with diversified production.

While over 65 per cent of the population was dependant on agriculture, the sector's contribution to GDP has come down to about 18 per cent in 2010. Addressing a policy forum at the International Conference on "Eliminating hunger and poverty", hosted by the M.S. Swaminathan Research Foundation it is observed "Unless we do something quickly to alleviate the conditions of those who depend on agriculture, we will not be able to see inclusive growth" (*The Hindu,* Monday, August 9, 2010). In this connection, among other steps it is suggested, remunerative price for the farm produce, reduce proportion of people dependant on agriculture for livelihood and incentives to the small and marginal farmers by way of direct subsidies.

13. Water Resource Management

Water continues to be an essential input that promotes or restricts growth. As per the study of M.A. Chitale the total water resources available for India are as follows:

1. Rainfall 4000 km^3
2. Average Annual flow in the rivers 1880 km^3
3. Utilizable water 1140 km^3
   (a) Surface water 690 km^3
   (b) Groundwater 450 km^3

The requirements of water for various sectors upto the year 2025 is expected to be as shown in Table 1.2.

Table 1.2

Projected Requirements of Water for Various Sectors upto 2025

S.No.	Sector	1990 %	2000 %	2025 %
1	Domestic	25 (4.5)	33 (4.4)	52 (5.0)
2	Irrigation	460 (83.3)	630 (84.0)	770 (73.3)
3	Energy	19 (3.5)	27 (3.6)	71 (6.7)
4	Industry	15 (2.7)	30 (4.0)	120 (11.5)
5	Others	33 (6.0)	30 (4.0)	37 (3.5)
	Total	**552 (100.0)**	**750 (100.0)**	**1050 (100.0)**

Source: M.A. Chitale, *Science, Population and Development,* Publication and Information Directorate, CSIR New Delhi, 1992, pp. 132 & 133.

From Table 1.2 it is evident, that in the year 2025, the total requirement of water for various sectors is estimated to be 1050 km^3 which is nearly equal to the utilizable water available in India. From the Table 1.2 it is clear that water for domestic requirement, water for energy, and industry will increase. The decline in the consumption of water for irrigation is also marginal. The increased use of water for domestic and industrial purposes will give rise to further deteriorating quality of water. M.A. Chitale rightly points out that "in a country which is getting progressively urbanized and industrialized, it will not be enough to plan for water only in terms of quantity. It is necessary to plan for the desired quality of water as well. Good quality of water is essential for the health of the people and for the cattle population".

Since water potential from rainfall is estimated to be of high order, effective steps are required for harvesting of rainwater. As indicated by Mahatma Gandhi, rural people must view the work of harvesting and well preserving the rain water as a social responsibility and collective action is required in this regard. At the same time, steps must be taken in storing the excess monsoon flood water for use during the dry part of the year and hence a need for developing a large number of storage structures on the rivers and their tributaries. Water requirements for various sectors are presented in Table 1.3.

Table 1.3

Water Requirements for Various Sectors

S. No	Sector	Water Demand in Km (or to Cm)		
		2010	2025	2050
1	Irrigation	557	611	807
2	Drinking water	43	62	111
3	Industry	37	67	81
4	Energy	19	33	70
5	Others	51	70	111
	Total	**710**	**843**	**1180**

Source: Eleventh Five Year Plan Report, p. 46.

It is evident from the Table 1.3 that demand for water has been on the high from all sectors. Apart from demand for water for Irrigation purpose, water demand for industry and energy has also been significant. This situation makes out a case for giving importance to irrigation sector in the planned development.

It is also stated in the Eleventh Plan that even after considering the average irrigation intensity of 140%, the ultimate irrigated area in the country would be only 70% of the net sown area. This reality calls for concerted effort for fully utilizing the irrigation potential, apart from fully tapping development of rain-fed cultivation with appropriate technology.

14. Irrigation Potential

Creation of irrigation potential in the country under the major and minor irrigation projects received a fill up after the commencement of AIBP from the year 1996-97. Yet, there is a gap between ultimate potential and potential created. Attempts were made to reduce the gap by Command Area Development Programme. Major schemes need clearance from different Departments fulfilling conditions with regards to ecology and environmental aspects, inter-state issues etc. As per Eleventh Five Year Plan report as many as 300 remain unapproved projects. The ultimate potential is not fully utilized as is evident from the following Table 1.4.

Table 1.4

Irrigation Potential

	Number of Unapproved Projects	Ultimate Potential (Thousand Hectares)	Potential created upto Tenth Plan (Thousand Hectares)
Major Projects	90	5960.58	930.85
Medium Projects	136	809.82	153.16
Extension, Renovation and Modernization Projects (ERM)	74	1177.07	135.10
Total	**300**	**7947.47**	**1219.11**

Source: The Eleventh Five Year Plan, p. 49.

The gap between ultimate potential and potential created is wide in the case of all projects. As many as 300 projects are yet to get approval.

15. Holistic River Water Management

A major problem of serious concern in the coming years would be one centred round water and water-sharing. Appropriate steps both short-term and long-term are required. The World Commission on Dams (WCD) in its report, "Dams and Development" has suggested an integrated approach with focus on flood management in the direction of meeting growing requirements of water. According to the International Water Management Institute (IWMI), the total demand for water in India will increase by 22 per cent by 2025 and a further eight per cent more by 2050 due to population increase and changes in lifestyle choices (*The Hindu,* Thursday, August 26, 2010). It will be fuelled by industrial and domestic use. This challenge has to be met by exploring ways and means of stepping up of water supply. In this context, putting a halt to the volume of run off water caused by floods becomes necessary. There is a strong case for increasing storage in the projects so as to regulate the vast amount of run-off water. Linking of different major rivers facilitates prevention of run-off water ensuring at the same time water supply to water scarce areas.

The hurdle to an Integrated Water Management strategy is inter-state dispute in the water sharing. Some 90 per cent of rivers in India flow through different states and problems crop up in water sharing. It is necessary that legislation is enacted declaring all major rivers as of National Property and at the same time, protecting the rights of all stakeholders of rivers, guided by the principles of equitable and reasonable utilization of the National property, by all concerned. Also, there is a need for a setting up of desalination plants along the coastline, ensuring supply of fresh water.

Along with these long-term measures, steps, should also focus on minor irrigation in terms of rainwater harvesting and steps to improve groundwater table through a scheme like watershed development programme.

16. Minor Irrigation (MI) Potential

Even with regard to MI, there is a wide gap between potential created and potential utilized during the Tenth Plan as shown in Table 1.5.

Table 1.5

Performance of Minor Irrigation

(Physical in MH)

S. No.	Year	Potential Created	Potential Utilized
1	2002-03	0.687	0.548
2	2003-04	0.628	0.502
3	2004-05	0.740	0.592
4	2005-06	0.545	0.440
5	2006-07	0.918	0.734
	Total	**3,518**	**2,816**

Source: *The Eleventh Plan Report,* p. 50.

That potential created is not fully utilized is evident from the Table 1.5. The reasons for such situation are said to be the following:

1. Poor economic status of small and marginal farmers
2. Non-availability of assured water supply
3. Absence of subsidy for development of groundwater
4. In hard rock areas, groundwater resource is low
5. Overexploitation of ground water in certain areas causing depletion of water tables resulting in failure of wells.

These issues have to be addressed immediately, keeping in view the crucial role of irrigation aiming at the welfare of the small and the marginal farmers. The productivity level of different crops must be stepped up with the application of bio-technology, keeping in view, the quality expected in the international markets for different farm products.

17. Mineral Resources

India is a land well endowed with mineral resources. There are some resources, which are said to be fairly sufficient such as iron ore, manganese, mica, thorium, limonite and chromate. Apart from these, we have 'bauxite, non cooking coal, limestone, dolomite, gypsum, silica, copper, zinc, lead, gold, silver, phosphates and crude oil. We have useful minerals fairly sufficient for industrial development. Without causing environmental degradation, and without adversely affecting livelihood of the poor these minerals have to be tapped. The government must necessarily declare that mineral resources of all kinds are of national property. Private individuals should not be allowed to exploit the resources for their personal gains. Right policy would be one, where only Public Sector should have a role in mining, eliminating the role of private sector in tapping mineral resources as a national policy.

18. Appropriate Industrial Policy

Industrial Policy must be such as to encourage the growth of industries only in rural areas. The policy should be one of establishing eco-friendly, micro enterprises in rural areas. SEZs policy should be such as to take small enterprises to rural areas only. That way pressure on agriculture can be reduced and employment generated by way of non-agricultural operations within the rural sector.

Conclusion

India has been well endowed with rich culture and potential resources of varying nature. Extensive geographical area of the country with lengthy coastline serves as a potential resource. However, increasing population, growing urbanization and rapid industrialization combined with need for raising agricultural production naturally lead to complex challenges. Over a period of time, the occupational structure in India has not undergone a radical change. Still a vast majority continue to depend upon agriculture and agriculture continues to be a source of employment and living. In view of growing demand for foodgrains, foodgrain production must necessarily be stepped up. Productivity levels of different crops continued to be lower than that in countries like Japan and China, though in irrigation and other resources, India is well gifted. The reasons for low productivity need to be well analyzed and appropriate steps be taken in the direction of stepping up of productivity. Along with this, in the direction of reducing the pressure on agriculture, non-agricultural employment opportunities have to be created in the rural areas. Now with rapid industrialization, combined with competing claims for water, the water management assumes crucial importance. The gross irrigated area does not seem to be rising in a manner that it should be, given the investment in irrigation. The difference between the potential created and the area actually irrigated, remains, large and this gap needs to be filled up. Unless this gap is bridged, significant increase in agricultural production will be difficult to realize. Along with major and medium irrigation projects, minor irrigation should receive necessary attention. Rainfall must be fully harvested and well preserved by the collective effort of the local people, considering this task as a social responsibility. Also, for each rural household one supplementary source of income becomes necessary. Mineral resources need to be tapped without degradation of environment and without adversely affecting livelihood of the poor.

Appropriate policy is required in the utilization of different resources, keeping in view environmental safety and social justice. Unmindful exploitation of resources at the cost of ecological equilibrium and lives of primitive tribes denying social justice, is only economic 'growth' and not "development". The Industrial Policy should be such as to encourage the spread of Micro enterprises in the rural areas with due care of environment. All potential resources in the country need to be fully tapped. More than anything else, human resource, which is in abundance should emerge as a major player in the development process. The teachings of the sages of this sacred land, if honoured in practice, are bound to improve the quality of human resource, paving the way for overall development of the country on the right lines.

REFERENCES

1. *India 2009,* Ministry of Information and Broadcasting, Govt. of India, p. 1.
2. *Ibid.,* p. 2.
3. N.A. Palkivala, *India's Priceless Heritage,* Bharatiya Vidya Bhavan, Mumbai, 2003.
4. Nirodbaram, *Sri Aurabindo for All Ages,* Sri Aurabindo's Ashram, Puducherry.
5. N.A. Palkivala, *op. cit.*
6. A.P.J. Abdul Kalam, *India 2020,* Pengiun Books, Chennai, 2002.
7. *Ibid.,* p. 3.
8. Gunnar Myrdel, *Asian Drama,* Vol. II, p. 21.
9. Coale Ansley J. and Hoover, Edger M., *Population Growth and Economic Development in Low Income Countries,* Princeton Universal Press, 1958, p. 120.
10. *Ibid.*
11. A.P.J. Abdul Kalam, *op. cit.*

2 SOCIAL CAPITAL

1. Introduction

Social capital is not an alien concept to India. It has been a part of the Indian culture. Rural India of ancient days was well gifted and endowed with social capital. The concept of social capital has roots in our age old practice of voluntary and collective approach of a group of people towards a goal of mutual benefit. The word 'capital' here is not to be viewed in monetary or physical terms. Social capital is different from the concept of social wealth. Social capital is basically a well honoured human behaviour of desirable human relationship, a voluntary behaviour of a group of people based on mutual trust and for mutual benefit.

It is interesting to note that even ants and honeybee exhibit, orderly social relationship which is inborn and spontaneous. Such social relationship can be observed in the behavior of crows, animals etc., perhaps, man has drawn a lesson from these creatures a wise behaviour very much needed for an orderly society.

2. Definition of Social Capital

The concept of social capital has been defined in different ways. Different scholars looked at the concept from different angles, guided by their respective disciplines and the theme of their investigation. Not surprisingly, there is considerable disagreement and even contradiction in looking at and defining social capital. However, it should be remembered that social capital is primarily a sort of human behaviour and in all the definitions of social capital, the focus is on social relations that have productive benefits. Let us examine selected definitions on social capital.

Social capital is defined by its functions. It is not a single entity, but a variety of different entities having some characteristics in common: They all consist of some aspect of social structure and they facilitate certain actions of individuals, who are within the structure (Coleman, 1990, p. 302).

Widely quoted definition is by Putnam. Social capital is defined by Putnam, "as features of social organization such as networks, norms and social trust that facilitate coordination and cooperation for mutual benefit (Putnam, 1995, p. 67), Putnam, further states, that social capital is an asset, a functioning propensity for mutually beneficial collective action with which communities are endowed to diverse extents. It is further observed that "Communities possessed of large amounts of social capital are able to engage in mutually beneficial cooperation over a wide front. Communities that have low levels of social capital are less capable of organizing themselves effectively".

In the words of Fukuyama, "Social capital can be defined simply as the existence of a certain set of informal values or norms shared among members of a group that permit cooperation among them" (Fukuyama, 1997).

Again social capital is defined as "a culture of trust and tolerance in which extensive networks of voluntary associations emerge" (Inglehart, 1997, p. 188).

It is said "Social capital is multidimensional and must be conceptualized as such to have any explanatory value", (Eastis, 1998).

"Social capital is about the value of social networks bonding similar people and bridging between diverse people with norms of reciprocity (Dekkar and Uslamer, 2001). Uslamer opines that "Social capital is fundamentally about how people interact with each other".

There are thus innumerable definitions on social capital. The common feature in all the definitions is one of focus on "social relations that have productive benefits".

3. India and Social Capital

In India there are records to indicate the existence of a good number of community actions akin to the concept of social capital. For example the farmers joining together for works like plantation; harvesting of a crop etc., villagers together working for harvesting of rain water; if a house is on fire trap all the neighbours spontaneously jump into action and undertake rescue operations. This sort of community effort has been in practice in different parts of India since a long time. Even now, to some degree or other, this sort of united, voluntary effort, particularly in rural areas, can be noticed. Apart from informal community approach, some organized institutions also came into operation which are of the nature of social capital. Cooperative society formation; formation of participatory water shed committees, Self-help Groups (SHG) formation, women participation in decision making process etc. come under the social capital.

However with the technological advancement, growth of industrialization and urbanization of Indian economy social relations are undergoing changes. Again, the traditional labour class category of people are emerging powerful under the conditions of equal voting rights, occupational mobility, access to higher education to all sections of society and consequent development opportunities. There is a keen inter-caste and intra-caste competition in social, economic and political life. We can notice, growing social awareness among the rural masses. Rural artisans like carpenters barbers, cobblers and other service persons are gradually leaving their traditional occupations. The 'Jajman system', where some caste-groups provided specialized services for the cultivators and received payment in kind is also disappearing. In the name of development, natural resources, including forest resources are overexploited. In rural areas, agriculture is a major activity providing employment to a vast majority of people. In the agricultural operations, women are a much exploited segment of the society. Encouragingly enough, now rural people have access to educational facilities. The literacy rate of both men and women is slowly increasing. But most disturbing factor is with regards to social relationship.

4. Society as a Relationship

What is society? Society is not merely an association of people, it is primarily a relationship. "Society by itself is non-existent," as Jiddu Krishnamurthy (J.K.) points out. He further observes, "Society is what you and I, in our relationship have created; it is the outward projection of all our own inward psychological states… there can be no significant change whatsoever alteration or modification in society so long as I do not understand myself in relationship with you." J.K. rightly again observes, "There can be true relationship when there is love, there is complete communion and this can only take place when the self is forgotten". Good relationship required between man and man, man and nature, man and biodiversity and man and different services. Human relationship, understandably, is unpredictable, fluctuating and it varies from time to time. The relationship varies from one social group to another and from one set of relationship to another. Under these conditions, formation of social capital in the real sense may be difficult. But it is necessary. Economic development, community peace and democratic participation can all be promoted, by simply investing in the stocks of social capital. In this direction, what is required is a good social relationship.

5. Investigation into Social Capital

Anirudh Krishna investigates what social capital is, how it operates in practice and what results it can be expected to produce. Theoretical premises relating to social capital are tested with the help of evidence collected in some 69 village communities located in two states of Rajasthan and Madhya Pradesh. Very rightly, it is observed, "Social capital is not directly observable, people carry it inside their heads". What one can observe and measure are some manifestations, or behavioural consequences given rise by social capital. Social capital is essentially a psychological problem, that is, a problem of the individual in relationship with society.

Social capital is possible when there is perfect understanding and good relationship among people in the society. For this, what is required is inward revolution on the part of each through self-knowledge.

6. Survey on Social Capital Across the Globe

Basing on World Values Survey for 1991, it is noted that even in developed countries, there is a low degree of association activity and even that being confined only to towns, leaving huge mass of rural residents unaffiliated with any formal organization. In developing countries, the number of operating associations is found to be very low, indicating poor social capital.

An empirical study is attempted covering some 2500 village residents. The analysis is attempted in terms of economic development, community peace and democratic participation in relation to social capital. Social capital was found to be significant for explaining differences among the village communities. Also, along with social capital, the role of younger and educated village leaders is found to be significant in matters concerning development. Educated rural youth are able to be in touch with concerned officials and get necessary information and derive benefits, while illiterate people are not able to enjoy such advantage.

7. Middle Level Institutions in Social Capital

Middle level institutions such as political parties, unions, interest groups, trade associations, etc., help citizens to develop contacts with the state and market organizations. When these organizations are weak, social capital and institutional incentives remain disconnected and social performance suffers. Putman, who makes a comparative study concludes "Middle-level institutions are weak in India and that situation came in the way of effective functioning of social capital". However, the empirical study undertaken reveals that radical changes are taking place in villages. "Participation in politics is no longer a preserve of rich and high caste groups in the villages". Political parties, which earlier banked upon feudal selected strong men for organizing vote banks in villages are increasingly turning towards the newly emerged networks of poorer and lower-caste villagers. In the villages, under study, education is spreading rapidly among youth and mass media has started carrying images of progress, change and equality. The earlier practice of submitting to the traditional village upper-caste heads or rich no longer exists. With growth of literacy, there has been growing social and economic consciousness resulting in substantial changes in social relations. Younger villagers belonging to Scheduled and Backward caste are almost on par with upper caste villagers in terms of basic education, and they are no longer held back by the drag of ignorance and illiteracy.

With their attainment in education and better information on hand, younger persons of lower caste groups could directly deal with the State and Market agencies without seeking the support of the upper caste patrons.

With all these changes, in rural life, the so called "Social Capital" has not resulted in improving the lot of the rural life. The empirical study of Anirudh Krishna[2] indicates the need for improvement in the quality of relationship for orderly society.

Different organizations and schemes initiated from time to time in India, such as cooperatives and Community Development Programme could not deliver better results because of poor quality of human relations among members. Self-help group scheme is able to emerge as an effective social capital in a situation of better quality of human relationship among members. In certain areas, the scheme proved to be a failure because of poor human understanding and poor mutual trust. Gandhiji very rightly desired that every village must have a community centre, where villagers can meet and discuss their problems and resolve petty differences, if any. Gandhiji was keen that all communities in the village must live with good relationship and he believed that community centers would serve as a platform to unite villagers of varying interests. He also stressed the need for strengthening the Gram Panchayats which can play an effective role in ensuring good relationship among different sections. Gram panchayats, if well strengthened, can play an effective role in the formation and effective functioning of social capital in every village. The Non-governmental Organization (NGOs) can play an active role in organizing the rural people for cooperative effort in forming self-help groups, participatory watershed programme, women's participation in decision-making etc. and thus pave the way for active role of social capital in the rural sector.

Conclusion

Social capital as a resource, has a critical role to play in rural reconstruction. Social capital of the ancient days must be revived and it is possible when people with mutual trust begin functioning unitedly for a common good. In harvesting rain water, preserving the water in the irrigation tanks and in protecting the forest resources, villagers must necessarily work together. Cooperative effort in different development activities and mobilization of resources by way of self help groups become necessary. Thus, there is a strong case for revival of the social capital in different forms for the good of the society as a whole.

REFERENCES

1. Jiddu Krishnamurthy, *Freedom First and Last,* Krishnamurthy Foundation, Adayar, Chennai.
2. Anirudh Krishna, *Active Social Capital,* Oxford University Press, 2003, New Delhi, 110001.

PART II

RURAL DEVELOPMENT : THEORETICAL BACKGROUND

3. **Socio-economic Structure of Rural India**
4. **Rural Development: Early Approaches, Experiments and Programmes**
5. **Rural Development: Theoretical Background**
6. **Agriculture in Economic Development**

3 SOCIO-ECONOMIC STRUCTURE OF RURAL INDIA

1. Introduction

Rural India, though emerges as a developed nation, still a vast majority of people in the country depend upon agriculture and allied activities for their living and most of them live in rural areas. Success in achieving the set development objectives in the country rests primarily on the development of rural areas. Rural development in India is the key to National development.

2. Rural-Urban Differences

Rural societies differ from urban societies in many ways. Generally, the rural societies are said to be those in which the predominant human activities are involved in the production of food, raw materials, fibers milk etc. These field activities require a relatively high ratio of land per person. To the extent that a population is engaged in such field activities, it can be considered rural A population is considered urban, on the other hand, if it is engaged in activities by which food, fiber, minerals and raw materials from the field are processed and distributed to their users.[1] The places where these activities are undertaken are usually thickly settled areas and there areas are known as Urban. In these urban areas, generally, there are better education, health etc., facilities. There are other differences between the rural and the urban. In the rural areas, there is an environment with predominance of and direct relationship with Nature. In contrast to this, there is great isolation from nature and predominance of man-made environment in urban areas. Rural people have normally the benefit of fresh air, both at work and at home while the urban residents are denied of it.

Further, in rural areas, social relations or contacts are highly personal and relatively durable, but, in urban areas, the relations are predominantly impersonal, very casual and relatively short-lived. However, there is no absolute boundary line which would show a clearly cut clearage between the rural and the urban community.[2] It is really difficult to present the exact number of rural people and the urban people. There are conceptual problems in estimating the population in different zones. That apart, some people live in a nearby small town, but attend to the farm operations in a village. In the absence of minimum facilities like primary education and health etc., some people have chosen to live in the nearby town or city but attend to agricultural operations in nearby villages where they own lands. In countries like United States and Canada, some of the farmers have individual dwellings on the land they cultivate but yet have distinctive relationship with the urban sector. Such type of population which is neither strictly rural nor urban is classified under the category of "Rurban" Population.

The economic history of different countries indicates that almost every country, to begin with, primarily depended upon agriculture and agriculture played an important role in the development of the economy. With growth of economy, the importance of agriculture declined; declined very fast in countries like U.K. and other western countries. Further growth of industrialization, unsuitability of climate conditions for farm operation and less of population pressure accounted for sharp decline in the number depending on agriculture for their employment and living. In India, agriculture is a source of living of a vast majority of people and agriculture is in fact a way of life. Industrialization is not to that degree as to draw away population dependent upon agriculture as a source of living. Depending upon the trend of development process so far we can certainly expect that for some more period agriculture continues to be a major employer. Agriculture development should therefore receive adequate priority in the planning and developing process.

So long as agriculture continues to be a major employer in the country, the agriculture sector should necessarily receive adequate attention. The percentage of agricultural population to the total population may be taken as an indication of the extent of rural population. In villages there are some who are engaged in the non-agricultural operations like village artisans, traders, etc. With the economic progress of India we notice a gradual increase in the ratio of urban population and gradual decrease in the ratio of rural population. In spite of significant growth of industrialization and urbanization, rural population is still a sizable segment of total population.

3. India as Land of Villages

India is a land of villages. Village is an important social and economic unit in Indian society. Gandhiji long back said "India lives in villages and village constitutes to be the very heart of India". From the time of the Vedas, there had been references to village communities in India who attached great importance to agriculture. In the Buddhist literature, there are references to villages. From Kautilya's, 'Arthasasthra' we get authentic information about village communities. During the Mauranyan and Gupta periods, the village communities were governed by their own Assemblies. Since the ancient times, the villages have been self governing bodies. It is rightly observed that "In an age when communications were slow and fast means of transport were unknown, the strength from within was more important than the help from outside. During the rise and fall of great empires the villages stood like rocks and preserved much that we value today as our cultural heritage."[3] Indian villages are scattered over a vast area and some time they are named after either the places where they are situated or occupation they follow or the name of the dominant family or caste of the local people.

Some villages in India have a long history. They existed on the same site for over 1,000 years. The people developed sentimental attachment to the place where their ancestors lived or they were born and brought up. Some villagers have proud attachment – "My village where I was born" and so on feeling. Some villagers are not normally willing to migrate to some other better place. But some are enterprising and move to places where there is assured irrigation facility and other advantages for better living.

As per the Censes of India 2001, there were 5,93,616 villages in India each with a population ranging between less than 200 and more than 10,000. The areas covered by all the villages are considered as rural areas. Table 3.1 presents population size in the inhabited villages as per census 2001.

Table 3.1

Distributors of Villages as per Census, 2001

Population Size	Inhabited Villages	Percentage to Total Inhabited Villages
Less than 200	91555	15.42
200-499	127510	21.47
500-999	145408	24.50
1000-1999	129976	21.90
2000-4999	80407	13.55
5000-9999	14798	2.49
10000 and above	3962	0.67
Total	**593616**	**100.00**

Source: *India 2009,* publication division Ministry of Information and Broadcasting, Govt. of India, New Delhi.

It is clear from Table 3.1 that some 36.89 per cent of villages are with a population below 500. Some 39.61 per cent of villages are with a population of 1000 and above.

4. Predominance of Rural Population in India

From the Census Reports, it is clear that a vast majority continues to live in villages. However, the percentage of urban population has been gradually increasing, See Table 3.2.

Table 3.2

Relative Growth of Urban and Rural Population

Year	Population in Millions		Percentage of Total Population	
	Rural	Urban	Rural	Urban
1901	270.3	25.6	89.0	11.0
1911	220.4	25.6	89.6	10.4
1921	216.6	27.7	88.7	11.3
1931	237.8	33.0	87.8	12.2
1941	265.5	43.6	85.9	14.1
1951	288.2	41.6	82.4	17.6
1961	347.2	77.6	81.7	18.3
1971	439.0	109.0	80.1	19.9
1981	525.0	160.0	76.6	23.4
1991	628.0	217.0	94.3	25.7
2001	741.6	286.1	72.2	27.8

Source: 1. *Census of India series*

2. *Agricultural Statistics at a Glance,* 2008, Government of India.

From the Table 3.2 it is evident that a vast majority still continues to be rural, although the urban percentage increased from 11.0 per cent in 1901 to 27.8 per cent in 2001. Since rural areas continue to be centres of habitation of a vast majority, the strategy of development in India should necessarily focus on development of rural areas. Non-agricultural employment through setting up of small size enterprises in rural areas is necessary to generate employment and reduce pressure on agriculture land and thus, prevent migration of rural labour to the urban centres. That way rural poverty can be tackled effectively.

5. Living Conditions of Rural People

In all India in 2000-05, some 27.50% of people were estimated to be Below Poverty Line (BPL). The BPL of rural population in the same period was 28.30%, while the BPL percentage of urban population was 25.70%. The degree of poverty varies in between different states. In States of Bihar, Chhattisgarh, Jharkhand, Orissa, Uttaranchal, the BPL percentage was above 40%.[4] Poverty, illiteracy and ill health normally go together. The literacy rate in rural areas continues to be lower. The rich normally have access to information about the health related issues in many ways: through journals and magazines, discussions with others and visits to doctors and medical specialists. That is not the case with many low income groups and poorer people.[5] The literacy level of rural poor is miserably low. The sanitation and drinking water facilities continue to be poor. Dr. A.P.J. Abdul Kalam very rightly speaks of "The two Indians". "A few tens of millions of people in the country have lifestyles equivalent to or even more luxurious than upper strata of the developed world. They enjoy the material wealth and the facilities offered by modern technology, and simultaneously enjoy the benefits of cheap labour. Another 200 to 300 million Indians, the so called middle class have a varied lifestyles, often aspiring to copy the developed world. But having only limited resources. They face the stress of modern life but often do not have the facilities for good living. The rest of the population is engaged in jobs with constant insecurity. This majority does not have a economic surplus and has just enough for covering its bare necessities. Investment in health care is an impossible luxury."[6] A.P.J. Abdul Kalam is outspoken when he states, "We have many experts in remote sensing to map out areas where mosquitoes breed or such areas from which other diseases can spread... Many entrepreneurial scientists and technologies have started small companies and provide services even to foreign clients. Why don't we deploy these talents to benefit the country as a whole, in the big battle ahead to combat disease."[7]

6. Rural Structure and its Components

Rural structure can be analysed under two main classifications (a) social structure and (b) occupational structure. The social structure is composed of different interrelated institutions. The important components of social structure are: (i) Marriage, (ii) Family, (iii) Caste (iv) Kinship, (v) Religious Customs, (vi) Social Hierarchy and (vii) Power Structure.

Society is not merely an association of people. Society is primary a relationship between people. As people begin to live together, various types of human relationships are established; societies are formed and gradually many social and economic institutions come into existence. Marriage is an important social institution. Marriage, to begin with, fulfills a biological need of both man and a woman. But in Indian society marriage is a sacred bond. Spiritual significance is attached to marriages which aim at sacred and permanent association between man and woman. It is through marriage that nuclear, compound or joint families are formed. Marriage facilitates welfare of the family members and ensures adequate security particularly for women. It preserves certain customs and traditions and perpetuates family descendants. In agrarian societies, mostly patriarchal, joint family has been found to be highly useful. Joint families are advantageous for agricultural operations.

The family members themselves normally attend to different farm operations without depending on hired labour. The role of women is more significant in agricultural operations. Apart from attending to routine household work of cleaning the house, vessels, fetching of drinking water, cooking and child-rearing, rural women perform innumerable other functions in agricultural and related operations. Taking care of the livestock, milching, storage, food-processing, marketing of produce, etc., are some of the tasks performed by the rural women.

7. Caste in the Indian Society

Every person becomes a member of a caste by birth and thus, marriage, family and caste are interrelated. Each caste has its own occupation. There are villages with a single dominant caste or with multiple castes. Villages having only a single caste group, generally maintain some understanding and unity in social relations and they are all governed by their own caste norms. Inter-caste differences are bound to exist but by and large it is noticed that a village with a single caste has less problems than a multiple caste village. If there are more caste groups, the complexities of social relations and social organisation naturally increase. Each caste has its own caste panchayat which regulates the behaviour of its members. Sometimes the interests of one caste group may come into conflict with those of another thereby adversely affecting the normal life in the village. With increase in literacy and higher level of education in all caste groups and with mobility of labour and employment opportunities, wide open for all, the caste rigidity is gradually disappearing.

8. Untouchability

The practice of untouchability is an outcome of caste system in India and it is an evil of serious nature. The so-called unclean persons of scheduled castes suffer from discrimination of varying nature. It is a sad state of affair that even now, in some villages untouchability is practiced. The scheduled caste people are not allowed to step into the houses of upper-caste groups. They have no freedom to draw drinking water from the wells along with upper caste people in certain villages only. Mahatma Gandhi long back described untouchability as the sin of Hinduism. Gandhi, B.R. Ambedkar and other social reformers made sincere efforts to abolish many disabilities of the untouchables. But yet still in rural areas the evil of untouchability continues though in the urban areas, the untouchability is not practiced. The practice of untouchability is now unlawful. Several legislations enacted and different developmental programmes initiated in recent years have helped the untouchables in matters of education, occupational mobility, health and sanitation etc., and paved the way for them to join the ranks of privileged groups in society. From time to time, the village society has been undergoing radical changes. All the components of rural caste structure have to be considered as the most important, since they determine individual's social status, occupation, marriage affiliation, economic dependence political power and so on. Various caste groups in a particular village are interdependent in day to day matters as well as rituals. This interdependence is not confined only to social relationships but also to economic relationships wherein each caste group has certain well defined functions to perform in the overall network of economic activities in a village.[8]

Periodically elections to different administrative units are conducted. The contesting candidate needs support from all caste people and caste becomes a crucial factor either being favourable or unfavourable to the contesting candidate.

S.C. Dube observes, "the different castes are linked and kept together by some well defined expectations and obligations which integrate them into the social system. The village, as a unit of social structure cuts across the boundaries of kin and caste and unites a number of unrelated families within an integrated multi-caste community." [9]

The traditional functionaries like carpenter, goldsmith, potter etc., have almost given up their traditional caste occupations unable to face the competition from the industrial products and became the agricultural labourers. However, all the caste groups living in a village are mutually interdependent. Each caste group irrespective of its traditional occupation has to depend on other caste groups for many of its needs. There are several relationships which cut across caste barriers like master and servant, landlord and tenant, creditor and debtor patron etc. On occasions like marriage and death ceremony, servicing castes are important and they perform their special duties on such occasions and receive payment in kind. This type of relationships providing certain services is found in all parts of rural India and this system is known are as "Jajman system."[10] The Jajman system bound together the different castes in a village or a group of neighbouring villages. However, the "Jajman System is gradually disappearing from the rural scene.

The village panchayat, existed even in the ancient days. The village panchayat of ancient days was an informal gathering of village leaders from different caste groups for the purpose of resolving certain issues of the village. Aged persons of noble nature, well-respected and above suspicion, constituted the panchayat. Even the lower-caste groups were invited to attend the meetings, if issues directly related to them came up for consideration. To some extent the panchayats, with traditional nature still exist in some villages with elders above suspicion settling some minor disputes at the village level. Perhaps some sort of co-ordination of the traditional panchayat with present statutory panchayat would help to make rural life more congenial. The 73rd Amendment to the Panchayat Act with the provision of Gram Sabha, attempts such co-ordination with the traditional panchayat.

9. Occupational Structure

The occupational structure in rural India consists of 3 sectors. They are: (a) Agriculture and Allied Sector, (b) Industrial Sector and (c) Service Sector.

10. Agriculture and Allied Sector

Agriculture, in its broad sense includes not only crop production of various types but also plantation, horticulture, sericulture, livestock, dairying, poultry, forestry, fishing etc. Normally, caste decided the occupation of a person in the village. However, most of the people in villages practice agriculture, in addition to their respective hereditary occupations followed by their caste. For example, a potter (Kummar) in a village apart from making his living through his caste occupation of making pots and selling them, also owns land and acts as an agriculturist. Almost all the artisans and servicing castes do not have adequate work and income from their caste occupations. Hence, they do farming either as landowners or tenant cultivators or farm workers.

Intelligent and educated rural people in different parts of India are moving towards different directions within agricultural sector and again non-agricultural occupations without caste barriers. The tendency within the occupation of agriculture is towards growing more of cash crops and cereal production is limited to the extent of domestic requirements only. It is because, the net income per hectare from non-food crops is found to be higher. Some of the rural people taking advantage of institutional credit facilities now available and forming self-help groups and mobilizing resources are taking up dairying, poultry etc., besides crop production. There is a gradual occupational mobility in the rural areas. The villager today by and large is aware of ways and means of improving the income from his farm.

11. Industrial Sector

The rural industrial sector consists of

(a) Small scale industries like rice mill, flour mill etc., located in rural areas.

(b) Cottage industries like khadi industry, ghani, handloom, coir, carpet-making, toy-making, pottery, basket-making, shoe-making etc. In many rural areas, these activities are undertaken as their traditional and caste occupation. Besides, their respective caste occupations, agricultural operations are taken by different people in a village. Sometimes, agriculture may be the main occupation and in some other cases it may be a secondary, taking up non-agricultural activates as the main occupation. Since the beginning of planned development in India, a policy has been followed for the development of industrial areas and industrial estates within such areas to facilitate the establishment of small and medium industrial units.

A recent study on India's pattern of development by Kalpana Kochhi *et.al.,* has noted that "the paradox of Indian manufacturing in the early 1980s, is that a labour-rich, capital-poor economy using too little of the former, and using the latter inefficiently". It is further observed that little has changed since the economic reforms an account of the fact that the labour markets have not been touched and the education expenditure continues to be skewed towards tertiary education. If the structure of Indian industry is to be adopted to the factor endowments of the country, it is evident that the impediments in the way of labour-intensive industries must be removed. Emphasis must also be put on skill development for making the work force employable in such industries."[11]

12. Service Sector

The service sector in rural areas is composed mainly of (1) village trader, (2) transport operator, (3) technical service and (4) professional or functionary service. The village trader may be a wholesale trader or a retail trader. By availing the institutional trade facilities, are found to be opening some petty shops and selling certain domestic provisions. This type of retail trade by some is only secondary occupation, the main occupation being in many cases agriculture.

In recent years, with formation of roads to different villages trucks, are a source of transport. However, bullock-cart transport is still a popular type of transport especially in remote villages with Kutcha roads. The bullock transport is a very flexible type of transport as it can collect smaller quantities of produce from a large number of producers or production centres. Gradually tractors are being used extensively both for transporting of marketable goods as well as for ploughing operations.

The professional service personnel such as carpenters, goldsmiths and other service persons are gradually withdrawing from their traditional occupations. Most of them have converted themselves as agricultural labourers leading to greater pressure on land and large scale unemployment or under employment in rural areas. Also, there are others like dhobis, barbers, cobblers etc., providing professional services to the rural community. However, with the gradual advance of literacy and occupation mobility these categories are no longer engaged in their traditional service activities. The importance of traditional professional services in the rural areas gets diminished.

13. Rural Society in Transition

Gradually rural society is changing. The traditional professional service persons generally of castes, categorised of lower caste are emerging powerful under the conditions of equal voting rights, occupational mobility, increased access to higher education and consequent employment opportunities. There is keen inter-caste and intra-caste competition in social economic and political life. There is now a growing a social awareness among all sections of rural society. Under these circumstances it is not easy for exploitation of one class of people by another. At one time in the past occupation was decided by the caste and there was no freedom to change it. In view of growing mobility of occupation, persons with higher levels of educational attainments, are able to secure occupations of superior type of jobs either in the country or outside the country, unhindered by caste barriers.

Rural artisans like carpenters, barbers, cobblers and other service persons are gradually leaving their traditional occupation. The most disturbing factor in recent years has been the faster increase of agricultural labourers and gradual decline of cultivators. In 1951, the cultivators constituted 71.9 per cent of the total agriculture workers and this percentage declined to 54.4% in 2001. Agricultural labourers were 27.3 million in 1951, which accounted for 28.1%. The agricultural labourers increased from time to time; the agricultural labourers were 106.8 million in 2001 and they constituted 45.6% of the total agricultural workers.

With growing literacy and access of all sections of society to higher and technical education facilities, employment opportunities outside the agricultural sector increased and as a result of this development, there is a gradual disintegration of joint family system in the villages.

Rural poor with growing awareness of their rights have tended to break the age old relationship with the local rich. Master-servant relationship has changed, breaking the traditional attachment or bond. The tendency for the rural poor is to migrate to the urban areas for seeking employment and living outside the village. They think that a town or a city is a better place for living with education and medical facilities, apart from recreation facilities of their choice. Due to increase in the tempo of industrialization and urbanization, there is mobility of labour from the agricultural sector to the industrial sector. This trend has its own adverse effect on the rural economy. Major problem as a result of this migration of labour to the urban centres is one of shortage of labour in villages. This situation forces some landowners to go in for mechanization of agriculture or take to horticulture or commercial crops which normally require less labour. This trend is dangerous in the long run as it adversely affects the foodgrain production. Again, because of industrialization and urbanization there is a growing urban impact on the rural people. The tendency on the part of the rural people now is to imitate the urban life due to "demonstration effect". This is leading to alcoholism, drug addiction, etc. Also, imitation of the urban life by the rural poor is landing them in heavy debts.

Conclusion

We notice a rapid transformation of subsistence-based traditional agriculture into market-based and profit-oriented commercial farming. In many cases, food crops are raised to the extent of domestic requirements only and the farmer finds it profitable to raise commercial crops. Also, the "Jajman" system, where the various non-cultivating castes provided specialized services for the cultivators and received payment in kind is disappearing. The literacy rate of both privileged and non-privileged castes is slowly increasing. The educated youth in the rural areas are seeking employment outside the agricultural sector. The need of the hour is to plan for creation of non-agricultural employment opportunities in the rural areas. Rural areas must also become industrial centres. Small size enterprises on a large scale need to be planned, aiming at the creation of more non-agricultural employment in rural areas.

REFERENCES

1. *Encyclopaedia of Brittannica,* Chicago, Helen Hemingway Benton, 1981, Vol. 15, p. 25.
2. Pritirim Sorokin and Carle C. Zimmerman, *Principles of Rural-Urban Society,* New York, H. Holt & Co., 1929, p. 14.
3. Sushila Mehta, *A Study of Rural Society in India,* S. Chand & Company Ltd., New Delhi, 1980, p. 20.
4. *Agricultural Statistics 2008,* Ministry of Agriculture, Govt. of India, pp. 54-55.
5. A.P.J. Abdul Kalam, *India 2020,* Penguin Books, p. 218.
6. *Ibid.,* pp. 221-222.
7. *Ibid.,* p. 227.
8. S.P. Jain, *"The Social Structure in Rural India",* in Rural Development in India – Some Facets, National Institute in Rural Development, Rajendra Nagar, Hyderabad, Revised Edition, 1987, p. 63.
9. S.C. Dube, "Social Structure and Change in Indian Peasant Communities" in *"Rural Society in India",* Popular Publications, 5th Edition, 1978, p. 203
10. M.N. Srinivas, *Indian Social Structure,* Delhi, Hindustan Publishing Corporation, 1982, p. 14.
11. *The Eleventh Five Year Plan* Report, Vol. III on Industry, p. 149.

4 RURAL DEVELOPMENT: EARLY APPROACHES, EXPERIMENTS AND PROGRAMMES

1. Introduction

The importance of Rural development in India was recognized long back. The origins of schemes in this regard may be traced to several experiments pioneered by different individuals and organizations well committed to the cause of rural reconstruction. The early pioneering efforts include: (i) The Gandhian approach of village upliftment; (ii) The Rural Reconstruction of Rabindranath Tagore; (iii) Martandam Experiment by Spencer Hatch; (iv) The Gurgaon district development work of P.L. Brayne; (v) The Rural Reconstruction in Baroda; (vi) The Firka Development scheme of Tanguturi Prakasam; (vii) The pilot project of Albert Mayer in Etawah and, (viii) The Nilokheri Experiment by S.K. Dey. Besides these, some other programmes were also launched addressing specific areas of development.

2. Gandhian Approach to Rural Development (Please see chapter on Rural Development – Theoretical Background)

3. Srinikethan Experiment of Rabindranath Tagore

As early as 1908, Rabindranath Tagore set up Rural Reconstruction Centres in eight villages in the Kalingram Pargama of West Bengal. Later, he founded Shantiniketan (1921) with a view to bringing about an all round transformation of village life. This experiment is popularly known as Srinikethan Rural Reconstruction Programme. Tagore's Experiment in a small area of few villages was aimed at both economic as well as social development of the rural community. The objective of the programme was to study the rural problem and to help villages to develop agriculture, livestock, formation of cooperatives and improving village sanitation etc.

Attempts were made to develop village crafts. Schools for boys and girls with boarding facilities were established. The programme included running of night school also. Training facilities were arranged for the boys and girls in kitchen garden, poultry, dairy, carpentry and other crafts. Further, games and sports were also encouraged in the schools so that children drawn from different social groups could mingle together forgetting different social barriers. Tagore emphasized the importance of rural reconstruction and wished to draw the attention of the people towards the poor and backward masses.[1] Rabindranath Tagore desired that the local government should take up some responsibility in rural development programmes leading to professional guidance and rural welfare. This idea in those days appeared to be strange and was not well received by the Government. Tagore planned increase of material wealth through cooperative effort and increase of cultural wealth through music, drama and dance at Shantiniketan. His ideas spread and influenced people all round shantiniketan. He was aware that several rural problems remained uncovered by his schemes and they needed understanding and appropriate action by the governmental and other agencies. Nevertheless, Tagore's contribution to all-round improvement of the villages covered by his experiment is commendable.[2]

4. Martandam Experiment by Spencer Hatch

Rural reconstruction programme was initiated by Spencer Hatch of Y.M.C.A. at Martandam (South Travancore) in 1921. The programme aimed at the betterment of living conditions of the rural poor through constructive schemes of the Y.M.C.A. The Martandam rural reconstruction programme was based on certain principles known as "Pillars of Policy". The important principles were: (i) the programme of rural reconstruction must be people's own and the personnel associated with the programmes would only guide the people to help themselves, (ii) people of all communities must be included in the development programmes and focus of help must be on the poorest of the poor, (iii) it must be a comprehensive programme covering development of cottage industries like mats and basket-making, palmyra sugar, hand-oven cloth, poultry-keeping, bee-keeping, etc., (iv) spirituality should be the basis of every programme, (v) simplicity at all levels should be the keynote so as to achieve the results with less cost, benefiting the rural poor and (vi) honorary extension services must be tapped wherever possible. In this programme, demonstration centres were established at selected places to convince the people to take to bee-keeping, poultry-keeping, basket-weaving, etc.

Attempts were made to carry out the different developmental works through well-organized "Clubs" like egg clubs, honey clubs, bull clubs, weavers' club; etc. Through these different clubs, villagers were trained to produce and sell vegetables, pickles cashew nuts, jaggery, peanuts, baskets, mats, etc. From headquarters, experts were sent to the villages for proper guidance but it was made clear that the responsibility for carrying out different developmental activities was of people themselves. Cooperation in all activities of life was emphasized. Some social activities like sports, scouts, folk dances and folk singing etc., were also initiated. It is said that this experiment had emphasized cooperation in work, play and in all activities of life.[3]

5. The Gurgaon Experiment by F.L. Brayne

F.L. Brayne, a British civilian, initiated a village development programme in Gurgaon district in Punjab and the scheme is popularly known as the "Gurgaon Scheme" This scheme also known as the Martandam scheme, emphasized that the villager himself must be made to take necessary interest in himself and in his village for his own betterment and the betterment of his village. The governmental agencies would only coordinate and properly guide the individual villagers in different developmental schemes. The different development works were initiated to convince the villager that improvement is possible and to make him adopt better ways of farming and living. Under the Gurgaon scheme, a school of rural economy was set up to train the village guides for rural uplift work. The curriculum of the school covered cooperation, agriculture, public health, domestic and village hygiene and sanitation, livestock breeding and elementary veterinary training, scouting, first aid, infant welfare, etc. The functions, of the village guides were development of co-operatives, preparing people as vaccinators, cleaning of villages by digging of manure pits, agricultural demonstration and sale of improved seeds and implements, etc.

Further, a Domestic School of Economics was also set up to uplift the village women. The curriculum in this school covered reading and writing up to the primary standard for illiterate women, instructions in sewing and embroidery works, toy-making, cooking, hygiene, sanitation, first-aid, child welfare, etc.

Rural sanitation works were also taken up under the scheme by way of using manure pits as latrines, preserving rubbish and dung in properly dug pits, vaccination against smallpox, inoculation against plague, etc.

Agricultural development programmes were also initiated by way of encouraging the farmers to use improved seeds and implements, use of preventive measures against crop pests, killing of field rats, and monkeys, and consolidation of landholdings on cooperative basis. Much emphasis was laid on cooperative efforts in different lines of development.

Importance was given to the village school teacher under the Gurgaon experiment. In the words of Brayney, "The village school teacher with his school library, his night school and his scouts must be the centre of uplift and culture and he must be so trained that he can solve all the simple problems of the villager, whether they are agricultural, social or moral or related to public health."[4] under the experiment, emphasis was also laid on social reforms like prohibition of child marriages, development of co-education, abolition of 'purdah' etc. The scheme was gradually extended all over the province of Punjab, though it did not last long. But yet, this experiment did make a significant contribution to the cause of rural reconstruction.

6. Rural Reconstruction Programme in Baroda

State-initiated rural reconstruction programme was commenced in Baroda province as early as 1890, by Maharaja Sayaji Rao. First the programme was confined to a group of villages. The programme took interest in the establishment of village panchayats, taluk and district boards. Focus was laid on making education compulsory to all children in the age group of 6 to 11 years. Village libraries were set up and people were encouraged to make use of the library. This initial step paved the way for a comprehensive programme of rural reconstruction spread to other parts of Baroda state due to the missionary zeal of V.T. Krishnamachari. In 1932, the rural reconstruction centre was set up and the personnel of the programme were initially trained by Y.M.C.A., volunteers of the Martandam project. The Rural Reconstruction Movement in Baroda had the main objective of improvement in all aspects of rural life through changing the outlook of the agriculturists towards higher standards of living and to develop the best type of village leadership.

In the economic field, the programme laid emphasis on agricultural production. Through expansion of irrigation facilities, conservation of soils, setting up nucleus seed farms and multiplication and distribution of improved seeds, etc. Further, supplementary occupations like kitchen gardening, weaving, poultry farming, silk production, bee-keeping etc., were also encouraged.

Before the commencement of the rural reconstruction programme, a survey was made to identify the minimum needs of the villages. Local works such as construction of feeder roads, connecting villages with the nearest railway station, digging of village wells, formation of grazing fields, etc., were taken up with 50 per cent contribution by the villages in the form of money and/or labour except in the most backward areas.

Under this scheme, adult education was encouraged and wide propaganda was made against the evils of child marriages and many other unreasonable customs and practices. The scout movement was also encouraged. The Debt Regulation Act of 1935 and the Debt Conciliation Act of 1938, were some of the important steps taken up for the welfare of rural poor. In the Baroda experiment, the village panchayat, cooperatives, village school and library served as centers for development activities. Through these institutions, the officials and non-officials worked together aiming at rural reconstruction.

7. Firka Development Scheme

The Firka Development Scheme (FDS) was first launched in 34 Firkas in Madras State in 1946, by Tanguturi Prakasam and later extended to other Firkas from April 1, 1950. The Scheme was primary based on the Gandhian idea of "Village Swaraj" aiming at self-sufficiency of village through the collective efforts of the villagers themselves. The object of the scheme was to address the different rural problems through certain short-term as well as long-term plans. The important short-term plans were development of rural water supply and communications, formation of panchayats, organization of cooperatives covering every village, improvement of sanitation, etc. The long-term schemes covered the attainment of village self-sufficiency through agricultural, irrigational and livestock improvements and development of khadi and cottage industries. Rural water supply and health improvement schemes had first priority, followed by better communications and improved agricultural practices. The different development schemes were implemented with the close coordination of the different government departments like agriculture, irrigation, veterinary, medical and public health industries and communication, etc. The different developmental schemes were implemented through a Firka development committee under the direct supervision of a Firka development officer with reputation as a distinguished social worker.

8. Evaluation of Early Reconstruction Schemes

The different attempts at rural reconstruction during the pre-independence days, no doubt, did something really good in spite of many hurdles. But some of the schemes gradually disappeared from the scene and some were merged later with government-sponsored schemes. The early attempts could not last long or yield the desired results for various reasons like lack of encouragement from the government, lack of financial support, inadequate, inexperienced and untrained staff, lopsided approach to different aspects of development, absence of needed supplies and services, inadequate cooperation and coordination from other departments and agencies, etc. Among others, according to the First Five Year Plan, the experiences of the early rural reconstruction schemes reveal that different schemes were forced on the villagers and as such there was no enthusiasm amongst them, and there was lack of initiative from the people which is essential for the success of the schemes[5].

Post-independence schemes of rural reconstruction: Immediately after India became independent and before the commencement of planned economic development, through series of five year plans, two important schemes of rural reconstruction were introduced in India.

9. Etawah Pilot Project

In 1948, a development project confined to 64 villages in Etawah district of U.P., came into existence. This project was first conceived by M.R. Abert Mayer motivated by a spiritual sense of a mission and service of humanity. It is said that the project was sponsored by Pandit Jawaharlal Nehru. It had no foreign collaboration or external financial aid. It was primarily financed and administered by the government of U.P. The purpose of the Etawah project was to see what degree of production and social improvement as well as of initiative, self-confidence and cooperation could be achieved in the villages of a district which was not the beneficiary of any programme so far.

There were two important functionaries in the project: (i) the rural life analyst, and (ii) the village participation officer who gave an applied social science orientation to the project.

In this project, training of workers received a very high priority. The approach to training was more informal and essentially pragmatic and functional. The project trained village level workers not only for Etawah, but also for other districts where pilot projects were initiated after Etawah. It provided in-service training to village level workers, outside job training, training to village leaders, social service cadets, adults literacy, literacy for panchayat secretaries and functionaries of other associated departments. Active and enthusiastic participation of the people in activities related to the overall development of rural areas was the main objective of the pilot project. Among others, the important functions of the project were creating awareness in the villagers, communicating with them, helping them, listening to them and exploring their needs. These activities helped establish close ties of confidence between the people and the project workers. Through adult literacy programmes, question and answer sessions, kisan melas, fairs, camps, individual and group approaches, the workers tried to gain the villagers' confidence and ensure their participation in several activities of socio-economic development.

The programme of work of the project relating to agricultural development covered the use of improved varieties of seeds, chemical fertilizers, irrigation, improved implements, plant protection measures, horticultural development, soil conservation, improvement of animal husbandry, provision of cooperative credit, marketing, etc. Other developmental schemes consisted of better sanitation and health services, maternity and child welfare services; improvement of roads, water supply, drainage and other public utility works; improvement of housing; field demonstrations, provision of reading room and library service. It was reported that significant results could be achieved in the project, mainly centered around agriculture.

The important development in the project was the growth of village organizations and institutions such as panchayats, schools and cooperatives. It is reported that the project attracted world wide attention and many visitors from other parts of India as well as from other countries visited the project area. Its success paved the way for the establishment of community development projects on a large scale.[6]

10. The Nilokheri Experiment

In 1948, some steps were taken by S.K. Dey to rehabilitate the displaced persons coming from Pakistan due to partition of the country, by developing a new township at Nilokheri. S.K. Dey wanted the refugees to be actively engaged in constructive programmes assuring them the right to live, the right to work for a living and the right to receive what is earned. These three rights formed the basis of a new scheme known as Mazdoor Manzil. This scheme aimed at not only preventing one-way traffic of labour material, skill and culture from villages to towns, but also to develop a decentralized, agro-based economy, forming a nucleus township with a population of about 5,000. The township was intended to include institution for medical relief and sanitation, middle and high school education, technical and vocational training, agricultural extension covering crop production, horticulture, poultry, piggery, fishery, sheep-breeding, etc. The township would also provide recreation facilities through reading rooms, drama, music and other cultural activities. Vocational centres and work centres for different crafts were started. Weaving, calico printing, soap making, tin and blacksmith, leather tannery and many other crafts and trades began to flourish at Nilokheri. Cooperatives were also started to look after the problems of marketing. Efforts made under this scheme greatly influenced the future course of formulating community development programme in the country.

11. Area Approach to Rural Development

The important programmes of area approach are:

(i) Backward Area Development Programme, (ii) Hill Area Development Programme, (iii) Tribal Area Development Programme, (iv) Drought Prone Area Development Programme, (v) Desert Development Programme, (vi) Command Area Development Programme, (vii) Intensive Agricultural Development Programme followed by Intensive District Development Programme.

The area approach programmes, no doubt have helped the development of specified areas. But it has been noticed that this approach mostly benefited persons with certain assets, like land owners, etc., and not the rural poor. In the context of rural development, the need for development of "backward", "hill" and "desert" arising arises from several factors. They are: (i) growing population all over and the decreasing man-land ratio, particularly in developed pockets necessitates the development of backward tracts too, (ii) the backward regions have to necessarily contribute to farm production. The drought prone areas are major contributors in the production of groundnut, bajra, jowar and pulses. These crops account for almost one-fourth to one-third of the total national farm production, (iii) the drought prone areas are major livestock tracts providing the country's draught animals and also accounting for about 60 % of the total sheep population of the country benefiting the poor too, (iv) development of backward areas helps in reducing the glaring disparities in income and standard of living between dry and backward areas on one side and irrigated and prosperous regions on the other, (v) large scale exodus of poverty-stricken people from backward regions to cities can be checked if backward areas are developed and (vi) the overall growth of economy depends on the development of different constituents of the country.

In due course of time these different schemes are either merged with other programmes or separate programmes aiming at development of less developed areas are designed. There are specific programmes now directly addressing poverty of varying nature in the country.

12. Programmes of Rural Development

In order to ensure active participation of people rural reconstruction, a scheme known as Community Development (CD) programme was designed. The CD programme was launched on October 2nd 1952. The scheme aimed at building up of community asserts like village roads, school building, wells etc. by involving the rural people. It may be said that the CD programme which helps assert building is based on Nurkse's theory of Capital formation with surplus labour in the agricultural sector of under developed countries.

Basing on the recommendations of Balavantrai Mehta study team, the three-tier system of panchayat raj was introduced in 1959. Under this arrangements, there is Zilla Parishad at district level, Panchayat Samithi at taluk level and Gram Panchayat at village level. After the establishment of Panchayat Raj, the Zilla Parishad took the overall responsibility of implementing the different programmes. At the Block level/Taluk level the Panchayat Samiti implements different programmes. The Block development officer with the assistance of different extension officers implements the different schemes. At the village level, there is a Village Development Officer in charge of implementing different schemes. Different Study Teams on the working of CD programme made certain observation worth noting: (i) Neglect of development schemes and greater stress on welfare programmes (ii) Rural poor are not benefited by the programme, (iii) Absence of clearly defined priorities, (iv) Tempo of programme far exceeded the resources, (v) Lack of required technical knowledge for agricultural development and (vi) The programme has not been people's programme, but only government programme.

Conclusion

With the launching of the Integrated Rural Development Programme (IRDP) different programmes are promulgated and they are covered by IRDP. The CD programme has undergone many changes in different States and CD programme continues to function with different nomenclature.

REFERENCES

1. Singh, Prabhakar, *Community Development Programmes in India,* New Delhi, Deep & Deep Publications, 1982, p. 33.
2. Ramabhai, B., *The Silent Revolution,* Jiwan Prakashan, Delhi, 1959, p. 10.
3. Memoria, C.B., *Agricultural Problems of India,* Delhi, Kitab Mahal, Ninth Edition, 1979, p. 26.
4. Quoted by Memoria, C.B., *op. cit.,* p. 827.
5. Govt. of India, *First Five Year Plan,* pp. 223-24.
6. Pareek, Uday, *Education and Rural Development, op. cit.,* pp. 120-23.

5 RURAL DEVELOPMENT: THEORETICAL BACKGROUND

1. Introduction

In the analysis of "Rural Development", the concepts of "Rural", "Urban", "Development" and "Integrated Development" figure in. Understanding of the theoretical background becomes necessary for clarity of these different concepts. The word "rural" has much wider implication than the spatial references suggested in the word, as distinguished from the term "urban". The classical economists did not focus their attention on "Development" or 'Rural Development' *per se*. They only referred to the concept of growth. They perhaps assumed that economic growth would naturally lead to 'development'. It was towards the end of World War II around 1945, that "development" became an important field of study and attracted attention of the scholars.

Development is a complex and integrated process which is affected by both economic and non economic factors. The importance of non-economic factors in development was duly recognized by the classical school. John Stuart Mill thought that non-economic factors like beliefs, habits of thought, customs and institutions play an important role in economic development. He attributed the backwardness of underdeveloped countries to the outdated, anti progressive character of their customs, institutions and beliefs.

Boeke attempted to explain underdevelopment in terms of sociological dualism. He defines sociological dualism as "the clashing of an imported social system with an indigenous social system of another style". On the basis of his understanding, largely based on the Indonesian experience, he concludes that "the kindest thing that the western world can do for developing countries is to leave them alone. Any effort, to develop them along the western lines can only hasten their retrogression and decay."[1] But it should be remembered that developing countries are benefited through technical and capital assistance from the western countries. For example, Green Revolution in India is partly the outcome of technical assistance from the United States Agency for International Development (USAID). In some fields, societies have grown by imbibing outside cultures, not by isolation.

One may, therefore, not agree with the theory of sociological dualism advanced by Boeke. But at the same time we have to consider sociological, cultural and psychological factors that play a role in economic development. The psychological and sociological requirements of development are as important as the economic requirements. We can notice the role of non-economic factors in the development process.

2. Concept of Rural Development

Anker[2] gives the following definition: "Strategies, policies and programmes for the development of rural areas and the promotion of activities carried out in such areas, (Agriculture, Forestry, Fishery, Rural Crafts and Industries, The Building of Social and Economic Infrastructure) with the ultimate aim of achieving a fuller utilization of available physical and human resources and thus, achieve higher incomes and better living conditions for the rural population as a whole, particularly the rural poor". A World Bank Publication[3] defined rural development as "Improving living standards of the masses of the low income population residing in rural areas and making the process of rural development self-sustaining". Michael P. Todaro[4] examines the concept of rural development from three aspects: (a) Improvements in levels of living, including employment, education, health and nutrition, housing and variety of social services, (b) Decreasing inequality in the distribution of rural incomes and in rural-urban imbalances in income and economic opportunities and (c) The capacity of rural sector to sustain and accelerate the pace of these improvements. The equity aspect of development is crucial. A study[5] of the United Nations Asian Development Institute considers the essence of the philosophy of rural development centered around five: (i) man as the end of development, which is, therefore, to be judged by what it does to him; (ii) delineation of man in the sense that he feels at home with the process of development in which he becomes the subject and object; (iii) development of collective personality of man in which he finds his richest expression; (iv) participation as the true form of democracy and (v) self-reliance as the expression of man's faith in his own abilities.

3. Concept of Integrated Rural Development

The concept of integrated rural development has acquired a distinct connotation in economic literature. C. Subramanyam, a former Finance Minister has put forward the thesis of integrated rural development at the all India Science Congress, Waltair, 1976. He views integrated rural development as a "Systematic, scientific and integrated use of all natural resources as part of this process enabling every person to engage himself in a productive and socially useful occupation and earn an income that would meet atleast the basic needs.[6] In this definition essential aspects of integrated rural development can be identified as: (i) an application of science and technology for the use of natural resources, (ii) functional linkages between the use of these resources, (iii) the use of natural resources should be socially useful and should lead to full employment, (iv) it should serve the basic needs of society and (v) science and technology should be geared to the production of basic needs.

Lalit Sen discusses the concept of integrated area development in much greater detail and its twin aspects, namely functional and spatial. Integration according to him refers to the appropriate location of social and economic activities over a physical space for the balanced development of a region.

From this brief review of literature on the concepts of rural development, we can identify critical elements. A shift of resources from urban to rural areas, development of infrastructure application of science and technology for better utilization of resources. All these must be geared to the basic needs of the rural masses. The process of their use must generate full employment ensuring forward and backward linkages.

There are a good number of growth models. Growth models relevant for rural development are worth mentioning here.

4. Lewis Model of Economic Development

In 1954, W. Arther Lewis has put forward a growth model. He rightly identifies that in many developing countries, there is abundant of labour whose marginal productivity is negligible, zero or even negative. This labour is available in unlimited quantities at a wage equal not only to the subsistence level of living but also 'subsistence plus wage'. As the supply of labour is unlimited he believes that new industries can be set up and the existing ones can be expanded without limit at the ruling wage rate. Lewis is aware of the need for skilled workers. However, he maintains that the skilled labour is only a temporary bottleneck and can be overcome by providing training facilities to unskilled workers. Lewis believes that the marginal productivity of labour in the capitalistic sector is higher than the ruling wage rate and hence, there results a capitalistic surplus. This surplus can be used for capital formation, which makes possible employment of more people from the subsistence sector. The increase in investment by the capitalists raises the marginal productivity of labour which induces capitalist employers to increase their labour force till the marginal productivity of labour falls to a level equivalent to the ruling wage rate. This process goes on till the capital-labour ratio rises to the point where the supply of labour becomes inelastic

Technical progress in the capitalist sector may also increase the share of profits in the national income as long as there is surplus labour. The share of profits may increase through innovation or because the capitalist sector itself grows. According to Lewis, there is possibility of rise in the capital formation from 4% to 5% to about 12% to 15% of national income.

Capital is created not only out of profits but also out of bank credit. In a developing country, characterized by unemployed resources and scarcity of capital, credit creation will expand output and employment in the same way as profits. Credit financed capital formation, however, results in a temporary rise in prices. The inflationary forces come to an end when voluntary savings from increased profits are large enough to finance new investment, without resort to bank credit.

Lewis Model seems to provide a good framework to understand the process of economic development in labour-surplus developing countries like India. Its basic premise is that labour productivity in agriculture must increase substantially in order to generate surplus in the form of food to be used for development of the non-farm sector and to release the surplus labour from agriculture for meeting the growing needs of the non-farm sector. However, the Lewis Model has certain limitations such as: (i) labour unions may push the wage rate up as labour productivity increases, and keep the rate of profit and rate of capital formation lower than expected, (ii) The capitalist employers may use the surplus for speculative or no-productive purposes instead of ploughing back for development purpose, (iii) to meet their rising expectation, rural people may consume more and save less than predicted by the Model and thereby dampen the pace of development. It is to be noted that in Lewis Model there is no satisfactory answer for the growth of agricultural sector in terms of higher productivity. In this Model there is no indication of appropriate population control measures. No strategy of agricultural and national development would ever succeed in the absence of appropriate population control measures. Lewis Model however focusing stress on utilization of surplus labour has its own relevance in rural development strategy.

5. Institutional Approach of Gunnar Myrdal

The central idea in the institutional approach is one of integrated approach to development. Gunnar Myrdal[7] in his monumental work, Asian Drama, observes "History and politics, theories and ideologies, economic structures and levels, social stratification, agricultural and industry population growth, health and education and so on must be studied not in isolation but in their mutual relationships". Gunnar Myrdal rightly identifies differences in socio-economic conditions in between the rich western countries and the poor South Asian counties. Under South Asia conditions of socio-economic institutional structure, the problem of development here is one of calling for induced changes in the social and economic structure.[8]

The "Social System" consists of a good number of "conditions" that are casually interrelated in such situation, a change in one will cause a change in the other. Myrdal classifies these "conditions" as: (a) output and income, (b) conditions of production, (c) levels of living, (d) attitudes towards life and work, and (e) institutions and policies. Myrdal emphasizes the interrelationship between attitudes and institutions on one side and the economic factors implied in the first three conditions on the other side. Myrdal opines that attitudes and institutions represent heavy elements of social inertia that hamper and slow down the circular causation within the social system among the conditions in these categories.[9] He is critical of the blind application of the concepts of savings and investments to conditions in South Asian countries. In these countries, levels of living, adequate food, better housing, improved facilities for health, education and training, and general improvement of cultural activities are all desirable in their own right and as a means to the future development of the human personality. By itself, he argues, industrialization can do little to raise labour utilization in the non-traditional sectors of the economy in rural areas. These problems must be addressed in their own right by specific policies designed to promote reform.

The institutional approach of Myrdal places emphasis on attitudes and institution which are themselves developed and sustained in the process of development. It appears that this model ignores issues relating to power structure and the underlying class forces behind unfavourable attitudes and institutions. This is to be a major weakness in offering a viable strategy of development that could pave the way to integrated rural development.

It has become clear that problems of poverty cannot be solved without a direct attack on unemployment, poverty and inequality. Growth strategies must necessarily give thought to these problems. Since the target groups have abundance labour resources, but only poor capital and land resources, the labour-intensive technology is needed. The experience of development programmes with target groups has abundantly made clear that such programmes are likely to succeed only when the surplus labour is well utilized. But it is rightly pointed[10] "what is missing even here is the inadequate recognition of the need for a shift of power to the groups who should be beneficiaries of new policies."

6. Rosenstein-Rodan's Theory of Big-push

Rosenstein-Rodan Theory (1970), points out that a minimum quantum of investment is necessary and with that investment, a "Big-push" is required for development to go on. This theory identifies three kinds of indivisibilities, which are main obstacles to the development of developing countries. There are (i) The indivisibility in the supply of social overhead capital (lumpiness of capital), (ii) The indivisibility of demand (complementarity of demand) and (iii) The indivisibility (kink) in the supply of savings. This theory argues that a big-push in terms of minimum quantum of investment is required to scale the economic obstacles to development created by these kinds of indivisibilities, and the external economies to which they give rise. This implies that the development process is a series of discontinuous 'jumps' and each jump requires a 'big-push'. The theory suggests that international trade may reduce the size of the minimum push required to obviate the effect of indivisibility (complementarity) of demand.

Rosenstein-Rodan does not offer any specific effects of and predict suggestions to overcome the adverse effects of indivisibilities. The suggestion that a trust-with capital from outside be established is not the effective step. The services required to give 'big-push' are of a high order in a major country like India. This theory is not able to provide a practicable solution for the development to go on. However, conceptually, this model continues to be of some guidance to planners and administrators.

7. Leibenstein's 'Critical Minimum Effort Thesis'

According to Leibenstein (1957) Model, economic backwardness is characterized by a set of interrelated factors and in order to attain sustained growth, initial stimulant to development of a certain minimum size is essential. The stimulants have a tendency to raise per capita income above the equilibrium level. It is noted that the economy is always being subjected to certain shocks and the actual values are different from the equilibrium values. Hence, some minimum effort is required to overcome the shocks. In backward economies, the magnitude of stimulants is too small. Hence, there is a case for critical minimum effort to ensure sustained growth.

In backward economies the income-depressing factors are more significant than the income-raising forces. Population growth, for example, is on such phenomenon. Population growth may be at a maximum rate between 3 per cent and 4 per cent. The need for a minimum effort arises to overcome internal and external diseconomies of scale, so as to overcome income depressing obstacles, which may be generated by the stimulants to growth so as to generate sufficient momentum in the system to get the economy moving on the right lines.

Leibenstein Model is more realistic than Rosenstein-Rodan's "big-push" theory. Giving a big-push to the programmes of industrialization and other areas of development is not practicable in under developed countries, while the critical minimum effort can be properly programmed into a series of smaller efforts to put the economy on the path of sustained development. It is clear that this theory is consistent with the concept of decentralized democratic planning to which India also is committed. Thus, this theory provides necessary clues to design planned development in India.

8. The Dependency Theory of the Marxist School and Neo-Marxian Approach

Karl Marx, (1818-1883) and Engels (1820-1895), were highly critical of capitalist and free market model of development. They believed that the process of social change was not gradual and evolutionary. Instead, it was characterized by conflict of interests between classes in society or class struggle. They argued that capitalistic system was an exploitative one. The system exploited colonized nations in terms of securing raw materials and other inputs at cheap rate and convert these nations as market centres for the products produced in well developed nations. This was the fundamental criticism.

The Neo-Marxian approach suggests that capitalism is at the root of underdevelopment and it is the peculiar nature of capitalism in underdeveloped countries that led to a few islands of affluence within sea of stagnation. Now what is required is appropriate development strategy favouring the poor. The old structures are to be replaced and the new structures of power favouring the poor are to be created. The Neo-Marxist approach is one of revolutionary steps destroying the old structures and creating new structures favouring changes in society. The Neo-Marxist theory believes that conflicts generated within the society of capitalistic system create revolutionary forces destroying the old structures and create new structures favouring changes in society. They reject gradualism and gradualistic approach. However, India which is committed to democracy, peace and non-violence may not welcome any system involving class conflict.

9. Human Capital Model of Development

This model emphasizes the importance of human capital investment in the process of economic and social development. By human capital, we mean acquired mental and physical ability through education, training, health care and pursuit of some spiritual methods. The acquisition of human capital is largely through the investment on human resource the simplest and effective step would be schooling model, giving priority to literacy and that way create necessary awareness among people. The classical and neo-classical economists did not explicitly include the quality of human resource through education and awareness building. It was Theodore Schultz (1964), who elaborated the concept of human capital and explicitly considered the investment in human capital as an important determinant of economic development. Subsequently, quite a few other scholars got interested in the economics of human capital, especially the economics of education. Very encouragingly, discussion at different levels began and useful material emerged with regards to human capital. This model considers the totality of human potential and emphasizes the need to harness it for the good of the people at large. The merit of this model, is that it respects people's culture and religions, and social values and structures. It is applicable to different countries of varying socio-economic conditions, especially countries like India.

This model is based on certain assumptions which have been ignored in the classical theory of development. These assumptions are: (i) It is to be remembered that human physical and mental capabilities are partly inherited and partly acquired. The assumption of a homogenous labour force is not realistic. (ii) Human capital directly contributes to development through its positive effect on skill building and that way contributes for higher productivity through reduction in resistance to the diffusion of new technologies in the economy, especially in the rural sector. (iii) Human resources are available in plenty in countries like India; If property developed and utilized, human resources can contribute significantly to development.

The merit of this model is in terms of shift from physical capital formation to human capital formation and shift from industrial development to rural development as a basis for overall development. This model seems appropriate for labour surplus and large scale illiterate population countries like India. There is a clear merit of human resources. They are renewable and hence, inexhaustible. The human capital can be substituted for exhaustible non-renewable physical capital in the process of development. The tertiary (service) sector is now the fast growing sector. This sector requires skilled and innovative human resources for its success. Then there is a strong case for human resource development strategy. Human resource development through literary and education; nutrition, health care, appropriate training and empowerment of women deserves the highest priority in the process of development. Here this model is relevant for countries like India.

10. The Gandhian Model of Rural Development

The Gandhian Model, covers a wide range of crucial issues relevant to Indian conditions. In this Model village, village panchayat development, realization of truth, non-violence and religious tolerance, priority to food production, minor irrigation and small scale industries, upliftment of women, Economic equality, untouchability received priority. As one who well understood the socio-economic structure of Indian society needs and aspirations of the people he proposed a very realistic plan of development. This model is centered round the poverty stricken people. It is evident from his Talisman to administrators: "Whenever you are in doubt when the self becomes too much with you apply the following test: Recall the face of the poorest and weakest man whom you may have seen and ask for your self if the step you contemplate is going to be of any use to him". Development, Gandhiji argued need not be assessed only in terms of gross national product and per capita income and similar such material yardsticks. Development, meaningfully, must be judged in term of sound relationship in between different members of a family, sound relationship in between different religious groups; development in different arts and architecture etc. Gandhiji's concern for spiritual and moral aspects of life is well revealed when he speaks of seven social sins, namely, Politics without principles, Wealth without work, Pleasure without conscience, Knowledge without character, Commerce without morality and science without humanity, worship without sacrifice.

As one well understood the Indian situation, Gandhiji rightly states "India lives in villages and if villages perish India too perishes". Gandhiji's Model of development is basically one centered round poor and downtrodden. The goals of development as enunciated by the Gandhiji model are:

Satisfaction of basic needs and not greed and he believes that the earth provides enough to satisfy man's needs but not greed. Every village must strive to emerge as a self-sufficient unit. However, the villagers, if necessary, may secure goods and services from outside the village, which the village could not produce decentralization of administrative machinery so as to enable people to have easy access to administration and pave the way for participation of people in the development programmes. Every farm household must have a supplementary occupation to enable them to get supplementary income. It is suggested that depending upon local resources and market situation, appropriate projects may be identified. Agriculture particularly production of foodgrains must receive priority. Water being a crucial input in farming, in the direction of providing assured irrigation, Gandhiji, advocated collective effort of the villagers in harvesting and conserving the rain water. Gandhiji desired that every village must emerge as a self-sufficient Republic with village panchayat well strengthened with adequate resources and functions covering areas of legislative, executive and judicial. Gandhiji desired that the local problems and minor disputes must be settled and resolved locally at village panchayat level. Gandhiji advocated Gram Nyaya System to settle petty disputes. Gandhiji advocated effective cooperatives system such as cooperative farming, Credit cooperatives, weavers, spinners, and dairy cooperatives. Gandhiji is for economic equality. In his words "Working for economic equality means abolishing eternal conflict between capital and labour. It means levelling down of the few rich in whose hands is concentrated the bulk of the national wealth on the one hand and the leveling up of the semi-starved maked millions on the other. A non-violent system of government is clearly an impossibility so long as the wide gulf between the rich and the hungry millions persist." Gandhiji advocated the concept of Trusteeship. Gandhiji believed that land and other natural resources belong to god or community. Such resources should be collectively owned by the community. He stated that one should own land to the extent of need to support the family. The surplus land must be handed over to the Trust. He saw in the principle of Trusteeship a non-violent method of persuading wealthy persons to voluntarily surrender their surplus wealth to a Trust for community benefit, Gandhiji advocated Nai Taleem or Basic Education. It is a system, where class room teaching is linked with vocational training. In the words of Gandhiji,[15] "The object of basic education is the physical intellectual and moral development of children through the medium of a handicraft. All instruments must be through the medium of provincial language. Gandhiji as one who was fully committed to rural development advocated that during the period of education students must regularly visit a chosen village with a clear-cut plan of educating the rural poor. Gandhiji's thinking was 'rural service' must form an important component of the curriculum of higher education. To minimize the exploitation of poor countries by the rich countries associated with foreign trade and investments, Gandhiji advocated two Doctrines, *viz.,* (a) Swadeshi Doctrine and (b) the Doctrine of International Cooperation. The first was to replace the principle of comparative advantage and the second was to displace the international competition. The Swadeshi Principle stressed self-reliance of the basic unit of Indian society, namely the village. The same principle extended to international economic relations stresses maximum economic self-sufficiently or minimum economic dependence on other countries as a means to minimize exploitation of one nation by another.[16] Gandhiji firmly believed that means are as important as ends. He asserted that "of one takes care of the means, the end will to the care of itself. Gandhiji speaks of Sarvodaya, the greatest good of all creatures – humans, animals and micro-organisms of all kinds and a variety of plants. Sarvodaya principle advocates non-violence to all creatures and welfare of one and all.

Gandhiji advocated labour-intensive method of production only Gandhiji argued what is required for good of the society is not mass production by selected few capitalists, but production well dispersed in small units owned by many. If production is spread over it would be primarily a labour-intensive one. In this system there is no scope for damage of environment and wealth is not concentrated in the hands of limited rich. There is no scope for exploitation of one by another

Gandhiji, argued that a machine that helps an individual has a place, but there should be no place for machines that concentrate power in a few hands and turn the masses into mere machine minders. Gandhiji advocated application of improved indigenous technology without giving room for displacement of labour. The Gandhian Model is quite suitable to conditions prevailing in India.

The Gandhian Model, though has its own critics, under the prevailing socio-cultural and economic conditions in India, the Model is quite suitable. The goal of sustainable development can be reached when the Gandhiji's Model is adopted. In the wake of present economic policy characterized by privatization, liberalization and globalization, there is tendency for mass production, concentrated in the hands of selected capitalists. In such situation there is bound to be damage to the environment in many ways and exploitation of poor. Gandhiji's Model is worth honouring.

11. Schumacher Model of Intermediate Technology

Dr. Schumacher[17] in his thought provoking publication *"Small is Beautiful"*, proposed a new model "Model of Intermediate Technology". Dr. Schumacher maintains that man's current pursuits of profit and progress through increased specialization and giant organizations has resulted in environmental pollution, inequality of incomes of high order and inhuman working conditions. Dr. Schumacher challenges all these developments and proposes a system of Intermediate Technology based on smaller working units, communal ownership and regional work places utilizing local labour and resources. With the emphasis on person not on the product, "Small is Beautiful" points the way in which capital serves the man instead of man remaining a slave to capital.

Dr. Schumacher's model is on the lines of the Gandhian ideology. He has all the appreciation of the Gandhian views, quotes him extensively in his work. He was critical of man's attitude to nature. In his words "Modern man doesn't experience himself as a part of nature, but as an outside force destined to dominate and conquer it. He even talks of a battle with Nature, forgetting that, if he wins the battle, he would find himself on the losing side. Far larger is the capital provided by Nature and not by man and we don't even recognize it as such. This larger part is now being used up at an alarming rate and that is why it is an absurd and suicidal error to believe and act on the belief, that the problem of production has been solved". In this era of "the learning society", we have to live peacefully, not only with our fellow men but also with nature. He very rightly observes "To talk about the future is useful only if it leads to action now." We have science and technology to help as along the road to peace and plenty and all that is needed is we should not behave irrationally and "Kill the goose that will lay golden eggs". This is the message to the rich and they must be intelligent enough to help the poor. He rightly remarks that "the development of science and technology over the last hundred years had been such that the dangers have grown even faster than the opportunities". He rightly observes "Scientific of technological 'solutions' which poison the environment or degrade the social structure and man himself are of no benefit, no matter how brilliantly conceived or how great their superficial attraction. Wisdom demands a new orientation of science and technology towards the organic, the gentle, the non-violent, the elegant and the beautiful".

Schumacher stresses the need for restricting the desires and depending on others for securing needs. Every increase of needs tends to increase one's dependence on outside forces over which one cannot have control and therefore increased existential fear. Only by a reduction of needs can one promote a genuine reduction in those tensions which are the ultimate causes of strife and fear.

Schumacher desires the science and technology must help to produce equipment cheap so that: (1) they are accessible to all, (2) suitable for small-scale application and (3) compatible with man's need for creativity. The cost of the machines must stand in some definable relationship to the level of incomes in the society. That machines must be suitable for small-scale application without giving room for displacement of labour the most important of all – that methods and equipment should be such as to leave ample room for human creativity. More than all these, checking the adverse effects on the environment and the quality of life of industrialism, population growth and urbanism becomes important.

Schumacher attaches importance to spiritual aspect of economics. He speaks of meta-economics and Buddhist economics. These concepts only convey 'right livelihood' with liberation from wants. While the materialist is mainly interested in goods, the Buddhist is mainly interested in liberation. It is not the wealth that stands in the way of liberation, but the attachment to wealth.

In the process of 'development', the question of scale assumes crucial importance. Millions of people start moving about deserting the rural areas. Smaller towns now emerge as big cities. Some selected capitalists with capital intensive technology, begin to own and run large scale production. It means mass production, and not production by masses. With mass production, there is damage to environment and a situation of rich becoming richer and the poor becoming poorer.

Very rightly the role of education as the greatest resource in the process of development is well presented. The essence of education is said to be transmission of values. The training in different branches of knowledge must have metaphysical and ethical roots. Metaphysics is a branch of philosophy that deals with the nature of existence, truth and knowledge. When people ask for education, they normally mean something more than mere knowledge of facts and something more than a mere diversion. Schumacher stresses on moral aspects of personality development. Education, he says, can help us only if it produces 'whole man', man with inner clarity. Such persons will not be in doubt about his basic convictions, about his view on the meaning and purpose of his life. In the words of Schumacher, "The problems of education are merely reflections of the deepest problems of our age. They cannot be solved by organization, administration, or the expenditure of money, even though the importance of all these is not denied. We are suffering from a metaphysical disease and the care must therefore be metaphysical. Education must be recognized as a greatest resource of development and proceed on right lines".

In his model of development, Schumacher attached due importance to the proper use of land. Man, he rightly argues, is only a child of nature and not at all the master of nature. He only stresses on 'rational' and 'sacred' use of land, keeping in view 'ecological problems' and long-term interests of the society at large. His model of development is centered round "Technology with human face". He looks at technology from the angles of employment and environment. Both must be taken care of.

As Gandhiji said, the poor of the world cannot be helped by mass production, only by production by the masses. The system of mass production based on sophisticated machinery is highly capital intensive owned and managed by selected rich. Schumacher observes "The technology of mass production is inherently violent, economically damaging, self-defeating, in terms of non-renewable resources and stultifying for the human person. The technology of production by masses making use of the best of modern knowledge and experience, is conducive to decentralization, compatible with the laws of ecology, gentle in its use of scarce resources and designed to serve the human person instead of making him the servant of machines". He named it Intermediate Technology to signify that it is superior to the primitive technology of bygone ages, but at the same time much simple and cheaper than the super technology of the rich. We can call it self-help technology or democratic or peoples' technology, a technology to which every body can gain admittance".

In his Model, Schumacher points out, mass rural unemployment and mass migration into cities. This leads to 'dual economy' in which there are different patterns of living. In the dual economy of typical developing country, we may find some 15 per cent of the population in the modern sector mainly confined to selected big cities. The rest of 85 per cent is spread over rural areas and small towns. Most of the development effort goes into the big cities with 85 per cent of the rural population are largely by-passed by different development activities. The development experts rarely referred to the dual economy and its twin evils of mass unemployment and mass migration into the cities. They believed "what is best for the rich must be best for the poor". This belief prevented diagnosing the real problem and appropriate solution.

The basic problem in development is poverty. There are many causes for poverty, both material and immaterial. Among the causes of poverty, the material factors such as lack of natural wealth or lack of capital or insufficiency of infrastructure are some. The primary causes of poverty are of immaterial in nature such as deficiencies in education, organization and discipline. Schumacher states that "development does not start with goods; it starts with people and their education, organization and discipline. Without these three, all resources remain latent, untapped potential.

If the primary cause of poverty are deficiencies in these three respects, then the alleviation of poverty depends primarily on the removal of these deficiencies Schumacher states that development is a process of evolution. Education does not 'jump'; it is a gradual process. Organization does not 'jump'; it must gradually evolve to fit changing circumstances. And much the same for discipline. All these three must evolve step by step and the foremost task of development policy must be to speed up this evolution. All three must become the property, not merely of a tiny minority, but of the whole society.

Schumacher categorically states that in many places in the world today the poor are getting poorer, while the rich are getting richer and the established process of development planning appear to be unable to overcome this tendency. "It is always easier to help those who can help themselves than to help the helpless". In most of the developing countries, rural people face the problems of underemployment and unemployment. Rural unemployment becomes urban unemployment. The strategy of development should be to bring health to economic life outside the big cities particularly in villages and small towns. It is necessary therefore that development effort should address directly rural people by way of creation of an agro industrial structure in the rural and small towns. Provision of work opportunities is the primary need and should be the primary objective of economic planning.

Schumacher indicates four propositions to tackle the underemployment and unemployment of the rural poor: (i) work places have to be created in the areas where the people are living now and not in metropolitan areas into which they tend to migrate, (ii) there should be sufficient number of work places accessible to all, (iii) the production methods employed must be relatively simple meeting the needs of skills matters of organization, raw material supply, financing, marketing and so on; (iv) the production should be mainly from local materials and mainly for local use. These four requirements can be met if (a) there is regional approach to development and (b) there is a conscious effort to develop and apply what might be called "Intermediate Technology". The thrust of this model is to help those who need it most". It is necessary that an important part of the development be directly concerned with the creation of an agro industrial structure in the rural and small town areas. All unemployed man being a desperate man, he is forced into migration. The provision of work opportunity is the need of the hour and economic planning should move in this direction. The choice of industry and the choice of technology should be guided by creation of employment. It means labour intensive industries need to be identified and that way create more employment Economic development in poverty-stricken areas can be fruitful on the basis of labour-intensive, "Intermediate Technology".

Schumacher reasoning of Intermediate Technology is as given below: If we define the level of technology in terms of cost per worker we can call the indigenous technology of a typical developing country – symbolically speaking – a £1 technology, while that of the developed countries could be called say of £1000 technology. The gap between these two technologies is so enormous that a transition from one to the other is simply impossible. In fact the current attempt of the developing countries to infiltrate the £1000 in to these economies inevitably kills off the £1 technology at an alarming rate, destroying traditional work places much faster than modern work places can be created and thus leave the poor in a more desperate and helpless position. If effective step is to be brought to those who need it most, a technology is required which would range in some intermediate position between £1 technology and £1000 technology. Let us call it symbolically a £100 technology.

Such an Intermediate Technology would be immensely cheaper than the sophisticated, highly capital-intensive technology of modern industry. The Intermediate Technology would also fit much more smoothly into the relatively unsophisticated environment in which it is utilized. The equipment would be fairly simple and therefore understandably suitable for maintenance and repair on the spot.

There are admirers as well as critics of Schumacher Model of Intermediate Technology. Different issues of development with special reference to India are well highlighted in this Model. Rural unemployment is one which indicates underutilization of labour force. Migration to urban centres continues in India and this particular factor is responsible for population growth of urban centres being 4 per cent with all adverse consequences. Exploitations of natural resources causing damage to environment is well identified mass production confined to selected capitalists leading to inequalities of income in between rural and urban is well identified. The central concern of development policy he says, "must be the creation of work opportunities for those who remain unemployed without contributing anything to the fund of either 'wage goods' or capital. Employment is the very precondition of everything else." This Model is subject to criticism on certain grounds. It is argued that with this type of technology, superior technology is ignored and people would only take to inferior and outdated technology, Schumacher's reply to this line of argument is that this kind of voice is only of those who are not in need and who can help themselves. It is argued that this is not the voice of those with whom we are concerned and the poverty-stricken multitudes who lack basic necessities of life. It is stated that the application of Intermediate Technology is not universal. There are products which are themselves the typical outcome of highly sophisticated modern industry and cannot be produced except by such an industry. The reply for this type of argument is that these products are not normally of the need of the poor. What the poor need mostly are simple things like building materials clothing, household goods, agricultural implements, and better returns for their agriculture products. Schumacher argues that Intermediate Technology is easily capable of producing all those needed by the poor and those who need help.

In this connection, Prof. Godgil's views on Intermediate Technology are worth considering. Prof. Godgil of Gokhale Institute of Politics and Economic at Pune has outlined three possible approaches to the application of Intermediate Technology as follows: The approach may be to start with, existing techniques in traditional industry be transformed by utilizing knowledge of advanced techniques suitably. Transformation implies retaining some elements in existing equipment, skills and procedures. This process of improvement of traditional technology is extremely important for preventing technological unemployment. Another approach would be to start from the end of the most advanced technology and to adopt and adjust so as to meet the requirements of Intermediate Technology. In some cases the process would also involve adjustment to special local circumstances such as type of fuel or power available. A third approach may be to conduct experimentation and reach in a direct effort to establish Intermediate Technology. In this direction National Laboratories, Technical Institutions and different University Departments must give thought in clearly defining Intermediate Technology and identify its relevance and application in the process of development.

Conclusion

This review of theoretical background to development is to be used to identify appropriate elements suitable for the development of India on right lines. India being primarily agriculture-based rural economy with persistent poverty, inequality and regional variations in development and consequent social tensions, a 'Development Model' must necessarily address these burning problems. Apart from some relevant clues from almost every Models, the Gandhian and Schumacher's Models provide necessary inputs for India to emerge as a "Developed Country".

REFERENCES

1. Kapar Singh, *Principles, Policies and Management,* Sage Publications, Third Edition, New Delhi, 2009, pp. 59-73.
2. Desmond L.W. Anker, "Rural Development Problems and Strategies", Vol. 108, *International Labour Review,* 1973.
3. Uma Le Le, *The Design of Rural Development – Lessons from Africa,* John Hopkins Publications, London, 1975, p. 20.
4. Micheal P. Todaro, *Economics for a Developing World;* Longmans, London, 1977, p. 249.
5. Whiden Hague *et. al.*, Towards a Theory of Rural Development, *United Nations Asian Development Institute,* Dec. 1975.
6. C. Subramanyam, "Strategy for Integrated Rural Development" in *Community Development and Panchayat Raj Digest,* Vol. 8, April 1977.
7. Gunnar Myrdal, *An Approach to the Asian Drama,* Methodological and Theoretical Vintage Books, 1970, pp. 1535-1536.
8. *Ibid.*, p. 26.
9. *Ibid.*, p. 228.
10. D.B. Gupta *et. al., Development Planning and Policy – Some Essays in Honour of Professor V.K.R.V. Rao,* Wiley Eastern Ltd., New Delhi, 1982, pp. 96-108.
11. Thus Spoke Gandhi (Quotes of Gandhi) complied by Dr. Ramesh Bharadvaj, p. 155.
12. *Ibid.*, p. 191.
13. Thus Spoke Gandhi, *op. cit.*, p. 84.
14. *Harijan,* February 11, 1939, p. 8.
15. Thus Spoke Gandhi; *op. cit.*, p. 49.
16. Amrutananda Das, *Foundations of Gandhian Economics,* St. Martin Press, New York, 1979, pp. 66-67.
17. Schumacher, *Small is Beautiful,* Penguin Books Ltd., London, 1973.

6 AGRICULTURE IN ECONOMIC DEVELOPMENT

1. Nature and Significance of Agriculture

Agriculture, as a primary industry plays a significant role in the process of economic development of a country. But the contribution of the agricultural sector to the overall economy varies from country to country depending upon the level of economic development. In the early stages of economic development, normally, agriculture is the major contributor to national income and it provides employment to a majority of people. At later stages of a fairly high level of economic progress, the importance of agriculture gradually declines. However, agriculture with its fundamental importance has its own contribution to make throughout the unending process of economic development.

2. Definition of "Economic Development"

Economic development may be defined as transformation of an economy which is predominantly agricultural and traditional into one largely industrial and modern. The agricultural and non-agricultural sectors in the former remain separate, presenting economic dualism, while in the latter they get integrated together. In the overall economic development, the role and functions of agricultural sector have figured prominently in the theories of development, particularly, since World War-II.

Historical records clearly show that no country has moved from the stage of chronic stagnation to the take-off stage of economic development without first achieving a substantial increase in agricultural production. In the earlier stages of economic development of modern advanced state, high rate of agricultural production has played a crucial role in furthering overall economic growth. In Western Europe, revolutionary changes took place during 1850 to 1980. Among other factors, agriculture, on the eve of the Industrial revolution, was favourable for a take-off. In the United Kingdom agriculture did provide cheap and sufficient food in the early decades of the Industrial Revolution and did provide self-finance for more efficient agriculture; but it did not, as agriculture did in Japan in later years, help capital formation and industrial development in general. The take-off periods of France, Belgium, Germany and Sweden also rested upon a firm base of a rising agricultural productivity.[1] Economic development of Russia, China and Japan reveals the basic role played by agriculture in earlier stages.

Russian economic progress was made possible by an initially substantial food surplus. Japan with an agricultural economy, characterized by land shortage and labour surplus adopted policies which created a food surplus, the basis and foundation for its spectacular economic development. Communist China also recognized the importance of achieving an agricultural surplus in the process of economic development.[2] In India, too, during the planning era, some of the State Governments have accorded high priority to agriculture.

After his monumental studies of Asian countries, Gunnar Myrdal concludes; "It is in the agricultural sector that the battle for long-term economic development in South Asia will be won or lost."[3] With reference to India, Coale and Hoover stressed the same point: "Very substantial progress in that most backward part of the economy (agriculture) was a prerequisite to successful development of the Indian economy as a whole."[4] In emphasizing the role of agriculture in the development of Indian Economy, it is observed that, "If one sector limits the growth of the other, it is more likely to be a case of agricultural growth, limiting non-agricultural sector than *vice versa.*"[5]

3. Role and Functions of Agriculture

Agricultural development is important for many reasons. Agriculture must primarily provide adequate food and nutrition supplies to the fast growing population. In the early stages of economic growth, the income elasticity of demand for food in underdeveloped countries is estimated to be 0.6 or higher which is twice or thrice as much as that in Europe, the United States and Japan.[6] In an underdeveloped country like India, with low nutritional standards, any rise in income is most likely to increase the demand for food and more diversified foodstuffs. Moreover less developed countries are experiencing population explosion. In order to supply adequate food and nutrition to the growing population, the rate of agricultural production should be higher than the rate of growth of population, which in these countries varies from 2 per cent to 3 per cent per annum. If domestic production fails to increase adequately, the gap has to be filled up by imports, involving a heavy drain on foreign exchange resources. If the food supplies fail to expand in pace with the growth of demand, the result is likely to be a substantial rise in food prices and consequent inflationary situation both in rural and urban sectors.

The farm sector has to supply the raw materials for growing manufacturing industries. Agricultural raw materials enter many industries. Nearly one-third of India's industrial output till recently depended upon the supply of agricultural commodities such as raw cotton, jute, oilseeds and rice.[7] If the agricultural sector in a country fails to supply adequate quantity of raw material, the gap is to be filled up with imported raw material. Imported raw material is most likely to cost more leading to a higher price of final product with consequent adverse effect on the marketability of products.

In under developed economies, the agricultural sector with surplus labour can supply manpower to the growing manufacturing industry. The agricultural sector, with unlimited supply of manpower, as Arthur Lewis[8] has put it, has to release the labour force for industrialization. For such labour transferred to the industrial sector, agricultural should also provide food, clothing and other essential consumption goods. In the early stages of economic development, manpower for manufacturing sectors can be drawn easily from agriculture. "Agricultural sector can even provide much labour at zero opportunity costs because a considerable part of the labour force in agriculture is redundant in the sense that its marginal productivity is zero."[9] In countries where the amount of labour is unlimited, labour can be transferred to the industrial sector through higher productivity per man in the agricultural sector. On the other hand, in countries where the size of population is large and its rate of growth is also high, transformation of agriculture becomes operative "To prevent an overflow of population to the cities, in a futile search for employment."[10]

The agricultural sector creates demand for more and new industrial goods. As agricultural development takes place, the per capita farm income will improve. Now the farmers would be in a position to buy more of modern agricultural inputs and consumer goods from the industrial sector. Further, with the development of processing, packing and distribution services in the agricultural sector, there is greater scope for marketing of agricultural products. The improved inputs enhancing productive efficiency in agriculture, leading to more of marketable surplus, can be exchanged for goods and services in the industrial sector.

Agriculture also contributes to capital formation which is essential for economic growth. The contribution of agriculture to capital formation can take place in three ways. First, increased agricultural productivity leads to lower food prices which, in turn, raises real income and promotes saving. Second, increased farm output may generate higher levels of farm income and a part of it may be saved. These saving may be utilized for investment purpose for overall economic development, "With the rising of agricultural productivity with comparatively smaller outlays", observe Johnston and Mellor,[11] "Greater is the scope for agricultural sector to make contribution to the capital requirements for overall economic growth." "Thirdly, capital can be derived by taxing the agricultural sector. In China, Russia and Japan, farm tax revenue helped significantly their economic growth. The system of agricultural taxation may be diverse in form such as land taxes, agricultural income tax, export duty and irrigation tax. When agricultural incomes rise, a larger tax revenue can be secured on a rising income base. However, a policy of ruthless or too high taxation as that of taxes on agricultural exports in Argentina could be like "killing the goose that laid the golden eggs which were expected to pay for the programmes of industrialization".[12] If agricultural production is large enough to yield an exportable surplus, it will earn the much-needed foreign exchange for imports of essential equipment, technical know-how and industrial raw materials.

It is hardly a surprise that in the initial stages of growth of many presently developed countries", observes Simon Kuznets, "agriculture was a major source of exports and that the resulting command over the resources of the more developed countries played a strategic role in facilitating modern economic growth."[13] For instance, the post-Meiji industrialization of Japan was crucially assisted by the rapidly expanding agricultural exports.

4. Theoretical Models

The theoretical models of agricultural development also emphasis the significant role of agricultural sector in economic development. According to these theoretical models, the transformation of traditional agriculture into modern one involves several phases of development. The first phase consists of transferring surplus labour (W.A. Lewis) accompanied by transfer of food capital to the industrial sector. If marginal productivity of labour is zero, surplus food will be available (as Fei and Ranis Model Postulates).[14]

If marginal productivity is positive, surplus food for transfer has to be secured by stepping up productivity (as per the hypothesis of Jorgenson Model).[15] In the second phase, as agricultural productivity is raised, surplus income as well as output emerges and the strategy of development called for mobilization of the rural surpluses and their transfer to the industrial sector for capital investment. The third phase is characterized by a mature agricultural sector and a fairly developed industrial sector helping each other.

5. Interdependence between Agriculture and Industry

The agricultural sector contributes to industrial development by way of: (a) providing food to the growing industrial workers, (b) providing raw material to industrial development of both large-scale as well as small-scale industries, (c) providing labour to the industry, (d) providing demand for industrial products, (e) helping capital formation required for industrial development.

Agriculture production not only accelerates the growth of industries, but also brings stability to the industrial sector. Thus agriculture helps industry and is also helped simultaneously by industry in many ways. There is, thus, a continuous process of feedback in the economic system. Agriculture and industry, therefore, are not conflicting alternatives, but complementary ones. In other words, agriculture, industry and transport, etc., mutually help and fortify one another towards the integrated growth of an economy. To quote Jagdish Bhagwati, "Those who stress the role of industrialization generally underestimate the importance of agricultural sector in developing economies. But the two need not be in conflict. Agriculture and industry compete for national resources. But this does not mean that those who emphasise the need for agricultural expansion should necessarily be opposed to industrialization."[16] In view of diversity of agricultural conditions and of the general economic situation in different underdeveloped countries, it is difficult to state the priority between agriculture and industry as a universally acceptable proposition.

However, it may be taken as the general consensus that "an agrarian revolution, preceeding and running parallel with industrial revolution, is a sound strategy that could take a country along the golden path of economic development."[17]

6. Importance of Agriculture in the Indian Economy

The importance of agriculture in India can be judged mainly from its contribution to national income and employment. Agriculture continues to be a major source of income and employment to a vast majority of (about 65 per cent) people in India. Official estimates of national income and its components are available on a regular basis in India since 1948-49. According to the estimates of individual scholars like Dadabhai Naoroji, Curzon, Vakil and others, the contribution of agriculture to the national income since the latter part to the nineteenth century till about World War I had remained around 66 per cent.[18]

Agriculture continues to be a main source of livelihood for a vast majority of the population in India. According to the decennial census reports, the proportion of cultivating and agricultural labourers to main workers was 72 per cent in 1951, 76.0 per cent in 1961, 62.2 per cent in 1971, 62.5 per cent in 1981, 59.7 per cent in 1991 and 54.4 per cent in 2001. It is clear that the dependence of the workers on agriculture, though slightly decreasing continues to be significantly higher.

The agricultural sector in India supplies food to the fast growing population and raw material to the manufacturing industry. The agricultural sector with surplus labour is in a position to supply manpower required for the industrial sector in urban areas. However, there seems to be no dearth of manpower in the urban areas also. Further, with nearly 65 per cent of the people depending on agriculture for their living, the agricultural sector creates demand for industrial products. With the advent of the Green revolution, there has been a considerable increase in farm incomes in areas with relatively better irrigation facilities. Farmers in areas like Punjab, Haryana, Coastal Andhra, etc., are able to contribute to capital requirements for overall economic growth in India. Through agricultural taxation, some state governments are able to raise considerable amounts. The contribution of the agricultural sector to foreign exchange resources is significant. The agricultural sector contributed 12.2 per cent to the total exports in 2007-08. The share of agriculture in total exports will be higher if the exports of agro-based manufacturers are taken into account.

In 1991-92 the contribution of agricultural commodities to exports was ₹ 8,228 crore and it increased to ₹ 77,768.7 crore in 2007-08.[19]

There is a close interdependence between agriculture and industry. The interdependence between agriculture and industry clearly show that further growth both in agriculture as well as industrial production in India is materially dependent on the rapid increase in the production of the input supply capacity of both the sectors to each other. It is such an interdependent relationship that will culminate in the fusion of the two sectors and formation of an integrated modern economy.

Conclusion

Agricultural development, however, is not synonymous with rural development. While the development of agriculture is critical for self-sufficiency in food and for the economic development of India, it is equally necessary to develop simultaneously other sectors in rural areas, so as to reduce pressure on land. This step is most likely to contribute to the improvement of living conditions of many. Moreover, the rural population consists, of a large number of landless people, who form the hard core of poverty. In the schemes of agricultural development, the rural poor without any assets like landless labour, village artisans, etc. are not likely to be benefited substantially. Therefore some, radical institutional changes aiming at redistribution of rural wealth in favour of rural poor become necessary as a part of rural development programmes. Land reforms in India, no doubt, help to some extent, the neglected section of rural population lie land labour, etc. At the same time, it should be noted that inspite of land reforms, a large number of people remain without land ownership in rural areas.

Therefore, alternative programmes like rural industrialization, provision of institutional credit with supporting services to rural poor for asset-building, payment of minimum wages to agricultural labour, social security measures benefiting rural poor, etc., should receive greater attention of the government. SHGs must be well promoted in rural areas and that way ensuring asset-building. Paying enterprises like dairying horticulture, sericulture, etc., may be popularized in the village ensuring regular income and necessary purchasing capacity to secure all those inputs needed for agriculture.

REFERENCES

1. Nicholls, William, H., "The Place of Agriculture in Economic Development" in *Agriculture in Economic Development,* Edited by Eicher and Witt, New York, McGraw-Hill Book Company, 1964, p. 19.
2. *Ibid.,* p. 40.
3. Myrdal Gunnar, *Asian Drama,* Vol. II, p. 1, 241.
4. Coale, Ansly J. and Hoover, Edgar M., *Population Growth and Economic Development in Low Income Countries,* Princeton, New Jersey, Princeton University Press, 1958, p. 120.
5. *Ibid.,* p. 139.
6. Mellor, John W., *The Economics of Agricultural Development,* Bombay, Vora & Co. Publishers Pvt. Ltd., First Indian Reprint, 1969, p. 64.
7. Bhagwati, Jagdish, *The Economics of Underdeveloped Countries,* London, World University Library, p. 147.
8. Lewis, W. Arthus, "Economic Development with Unlimited Supplies of Labour", *The Economics of Underdevelopment,* (Eds.) A.N. Agarwala and S.P. Singh, Oxford Press, 1958, pp. 400-49.
9. Schultz, Theodore. W., *Transforming Traditional Agriculture,* Ladhiana, Lyall Book Depot, First Indian Edition, 1970, p. 8.
10. Bhagwati, Jagdish, *op. cit.,* p. 148.
11. Johnston and Mellor, "The Role of Agriculture in Economic Development", In *American Economic Review,* Vol. 511, September 1961, p. 572.
12. Nurkse, Ragner, "Trade Fluctuations and Buffer Policies of Low Income Countries", In *Agriculture in Economic Development,* (Eds.) Eicher and Witt, New York, McGraw-Hill Book Co., 1964, pp. 320-32.
13. Kuznets, Simon, *Economic Growth and Structure – Selected Essays,* Lonon, Heinemann Educational Book Ltd., 1966, p. 249.

14. Fei, C.H. and Ranis, Gustar, *Development of the Labour Surplus Economy: Theory and Policy,* Yale University, Economic Growth Centre, Homewood, III, Irwin, 1964.

15. Jorgenson, Dale W., "The Development of a Dual Economy", In *Economic Journal,* June 1961, Vol. XXXI, pp. 309-34.

16. Bhagwati, Jagdish, *op.cit.,* p. 147.

17. Reddy, K.Venkata, *Agricultural Productivity in Andhra Pradesh,* Tirupati, S.V. University, 1977, p. 22.

18. Govt. of India, *National Commission on Agriculture,* New Delhi, Ministry of Agriculture and Irrigation, 1976, p. 2.

19. *Agricultural Statistics 2008,* Ministry of Agriculture, p. 36.

PART III

RESOURCES AND PLANNING FOR DEVELOPMENT

AGRICULTURAL DEVELOPMENT DURING THE PLANNING ERA

1. Introduction

Since time immemorial, agriculture has been a major source of living and employment to a vast majority of people in India. Even now (in 2010) agriculture accounts for about 19 % of GDP and about two thirds of the population is dependent upon the sector. As per census 2001, as many as 127.3 million or 54.4% were cultivators and 106.8 million or 45.6% composed of agricultural labourers. In 2007-08, agricultural exports could earn ₹ 77769.7 crore and this constituted 12.2% of total national exports. Agriculture is the source of supply to wage goods, namely, food supplies and raw material for the growing manufacturing industry in the country. Agricultural sector continues to create market for industrial products.

Agriculture in India in comparison with world agriculture is presented in Table 7.1.

Table 7.1

India in World Agriculture-2006

Sl. No.		Item	India	World	Percentage of India's Share
I		**Area and population**			
	(a)	Total Population (Million in 2006)	1152	6593	17.5
	(b)	Total Area (Million hectares)	329	13445	2.4
	(c)	Land Area (Million hectares)	207	13015	2.3
	(d)	Arable land (Million hectares)	160	1423	11.2
	(e)	Irrigated area (Million hectares)	51	350	7.0
II		**Crop Production (Million Tonnes)**			
	(a)	Rice (Paddy) " "	137	635	21.6
	(b)	Wheat " "	69	606	11.4
	(c)	Cereals " "	239	2221	10.8
	(d)	Potatoes " "	24	315	7.6
	(e)	Total pulses " "	14	60	23.3
	(f)	Groundnut (in shell) " "	5	48	10.4

(g)	Rape seed (including mustard seed)	8	49	16.3
(h)	Coffee (green) " "	0.27	7.84	3.4
(i)	Sugarcane " "	281	1392	20.2
(j)	Tea " "	0.89	3.65	24.4
(k)	Jute and allied fibres " "	2.04	3.11	65.6
(l)	Cotton (Lint) " "	356	2484	14.3

Source: *Agricultural Statistics at a Glance, 2008,* Ministry of Agriculture, Government of India, New Delhi.

From Table 7.1 it is clear that India with 11.2% world's arable land, has to support 17.5% of world's population. It indicates the need for strengthening agriculture sector. Table 7.2 presents Land Use Classification (2005-06) of India.

Table 7.2

Land Use Classification 2005-06

(Million Hectares)

Sl. No.	Item	2005-06	Percentage to Reporting Area
I	Geographical area	328.73	
II	Reporting area for land utilization	305.57	
	(i) Forest	69.79	22.9
	(ii) Not available for cultivation (A+ B)		
	A. Area under non-agricultural uses	25.03	8.2
	B. Barren and uncultivable land	17.48	5.7
	(iii) Other uncultivated land excluding fallow land (A + B + C)		
	A. Permanent postures and other grazing lands	10.42	3.4
	B. Land under misc. tree crops	3.38	1.1
	C. Cultivable wasteland	13.12	4.3
	(iv) Fallow lands (A + B)		
	A. Fallows land other than current fallows	10.50	3.4
	B. Current fallows	13.67	4.5
	(v) Net area sown	141.89	46.5
III	Gross cropped area	192.80	63.1
IV	Area sown more than once	46.57	15.2
V	Net area irrigated	55.11	18.1
VI	% of net area irrigated to net sown area	38.8	
VII	Grass area irrigated	76.03	24.9
VIII	% of gross are irrigated to gross cropping area	39.4	

Source: *Agricultural Statistics at a Glance 2008,* Ministry of Agriculture, Government of India.

Table 7.3 shows distribution of irrigated area, source-wise.

Table 7.3

Distribution of Irrigated Area (Source-wise) 2000-01

Size Class	Canals	Tanks	Wells	Tube wells	Others	Total
1	**2**	**3**	**4**	**5**	**6**	**7**
Marginal	3405	855	1296	5419	1409	12384
Small	2929	587	1971	4335	1069	10891
Semi-Medium	3219	463	2500	4740	1010	11932
Medium	3447	276	2511	4502	811	11547
Large	1578	77	952	1715	550	4872
Total	**14578**	**2258**	**9231**	**20711**	**4849**	**51627**

Source: *Agricultural Statistics at a Glance, 2008,* p. 308.

2. Features of Indian Agriculture

Indian agriculture demonstrates certain distinguished features. The land-use classification presented in Table 7.2 reveals broad outlines of land-use in India. The chief features of Indian agriculture are as follows:

2.1 It is still predominately a rainfed cultivation

In 2005-06 out of net sown area of 141.89 million hectares, net irrigated area was 55.11 million hectares. The net irrigated area constituted only 38.8% to net sown area. It is thus clear that 61.2% of the cropped area was under rainfed cultivation. Some land though classified under irrigated area such land enjoys irrigation under the conditions of rainfall only. If such land is excluded, the percentage of irrigated area to net sown area would be less than 38.89%.

From Table 7.3, it is clear that minor irrigation (wells, tube wells, irrigation tanks etc.) continues to be a major source of irrigation. In 2000-2001, out of total irrigated area of 51.63 million hectares, irrigated area under minor irrigation was 37.1 million hectares or 71.8%. The canal irrigation constituted only 28.2% (Table 7.3) Minor irrigation sector which is normally eco-friendly needs much more attention from the government as well as private individuals.

2.2 Indian agriculture is characterized by multiplicity of crops

Under the varying climatic and soil conditions in the country, a variety of crops are grown in India. Agricultural production of varying nature meets the need for food items of different consumers and a variety of raw material required by the manufacturing sector. In recent years we notice a change in cropping pattern. There is a tendency to grow commercial crops rather than food grain crops.

We can clearly notice decline in the area under food grains and gradual increase in the area under commercial crops. In certain parts of India, under the rainfed cultivation farmers have chosen to go in for single groundnut crop in the place of multiple cropping pattern.

2.3 Inequalities of higher magnitude in the ownership of land

The distribution of operational holdings, All India, clearly indicate the inequalities of higher magnitude in the ownership of land. Table 7.4 presents distribution of operational holdings, category wise. From the table it is clear that out of total operational holdings, holdings of small and marginal farmers constitute 81.3% of total holdings, but the area operated by these two categories works out only to 38.9% of the total area operated. Again the average of size of holdings of marginal formal category was 0.4 hectares and that of small farmers was 1.42 hectares. On the other hand the average size of operational holdings of large farmers was 17.12 hectares and of the medium farmers 5.81 hectares. Thus the inequality of higher magnitude was well pronounced in the ownership of operational holdings.

Table 7.4

Distribution of Operation Holdings – All India (2000-01)

Category of holdings	No. of co-operational holdings	Area operated	Average size of operational holdings
Marginal (Lessthan1 Hectare)	75408 (62.3)	29811 (18.7)	0.40
Small (1.0 to 2.0 hectares)	22685 (19.0)	32139 (20.2)	1.42
Semi-medium	14021 (11.8)	38193 (24.0)	2.72
Medium (4.0 to 10.0 hectares)	6577 (5.5)	38217 (24.0)	5.81
Large (10.0 hectares and above)	1230 (1.0)	21072 (13.2)	17.12
All holdings	119231 (100.0)	159436 (100.0)	1.33

Note: Figures in the parentheses indicate the percentage of respective column total.
Source: *Agricultural Statistics at a Glance, 2008*, p. 303.

2.4 Indian agriculture towards commercial agriculture

In recent years there has been visible change in the attitude of farm producers. Even the small and marginal farmers are interested in raising cash crops. All categories of farmers, under pressing cash requirements raise cash crops intended for marketing. Indian agriculture is no longer subsistence agriculture.

2.5 Low level of yield per hectare

Table 7.5 presents productivity levels of different crops in India in comparison with that in selected countries.

Table 7.5

Yield Per Hectare of Major Crops in India in Comparison with that of Selected Countries in 2006

Country	Paddy	Wheat	Maize	Groundnut	Sugarcane	Tobacco (Unmanufactured)
India	3124	2619	1938	859	66945	1486
Bangladesh	3904	1605	-	-	40862	1285
Pakistan	3164	2519	2906	-	49229	1996
China	6265	4455	5365	3118	82528	1997
Japan	6336	-	-	-	-	2417
Myanmar	3500	-	-	1400	-	-
Philippines	3684	-	2366	-	62061	1458
Thailand	2906	-	-	1773	50904	1750
World Avg.	**4112**	**2804**	**4815**	**2149**	**68257**	**1724**

Source: *Agricultural Statistics at a Glance, 2008,* Ministry of Agriculture, Govt. of India, pp. 213-215.

The productivity or yield per hectare of land remains low in India as compared to that in Bangladesh, China, Japan, Myanmar, Philippines. With regards to paddy, while the world average was 4,112 kg. per hectare, in India it was only 3,124 kg. per lecture in 2006 as shown in Table 7.5. With regard to wheat, maize, groundnut, sugarcane and tobacco also, yield per hectare in India was less than the world average as shown in Table 7.5.

2.6 Seasonal unemployment

In rural areas, there is both the problem of unemployment as well as underemployment. In the absence of assured irrigation, timely and adequate rainfall and supplementary occupations, all farm producers do not have adequate work throughout the year. The farmers under the conditions of rainfed cultivation face seasonal unemployment. The small and marginal farmers in a situation of unemployment plan often to migrate atleast temporarily to the urban centers seeking livelihood.

Table 7.6

Agriculture and Five Year Plans

Outlay/ Expenditure	First Plan 1951-56	Second Plan 1956-61	Third Plan 1961-66	Annual Plans 1966-69	Fourth Plan 1969-74	Fifth Plan 1974-79	Sixth Plan 1980-85	Seventh Plan 1985-90	Eight Plan 1992-97	Ninth Plan 1997-02	Tenth Plan 2002-07	Eleventh Plan 2007-11
Total Public Outlay	2,378	4.800	8,577	6,665	15,902	39,303	97,500	1,80,000	4,34,100	8,59,200	15,25,639	36,44,718
Agriculture and Community Development	360	565	1,089	1,051	2,385	-	5,695	10,524	22,467	42,462	58,933	1,36,381
Irrigation and Flood control	400	458	664	445	1,087	-	12,160	16,979	32,525	55,420	1,03,315	2,10,326
Total Expenditure on Agriculture	760	1,023	1,753	1,496	3,472	8,253	17,855	27,503	54,992	97,882	1,62,248	3,46,707
Percentage of Agriculture Expenditure in Total Public Outlay	32	21	20	22	22	21	18	15	13	11	10.63	9.5

Source: *(1) Five Year Plan Reports*
(2) Agricultural Statistics at a Glance, 2008, Ministry of Agriculture, Govt. of India.

3. Agriculture During the Planning Era in India

The progress of the Indian economy is crucially linked with the growth of agriculture. Immediately after Independence, the most burning problem was one of shortage of foodstuffs leading to food crisis in the country. From 1951, onwards the government had chosen to develop the economy of the country through series of Five year plans. During the First plan, covering 1951-56, priority was given to the agricultural sector with the goal of achieving self sufficiency in food grain production. Table 7.6, presents total outlay and expenditure during the planning era on agriculture, in the first plan 32 per cent of the total public outlay was spent on items relating to agriculture. The allocation to agriculture in terms of percentage gradually declined from plan to plan. In the Ninth plan, only 11 per cent of the total outlay was incurred on agriculture. This percentage has come down to 10.63 in the Tenth plan. In the Eleventh plan, the allocation on agriculture was only 9.5 percent of the total outlay. It is evident that from plan to plan period the percentage of public expenditure on agriculture has been dwindling.

4. Mid-term Appraisal of Tenth Plan

The mid-term appraisal (MTA) for the Tenth Five Year Plan had pointed out severe challenges faced by the Indian agriculture. The Eleventh plan listed out issues of serious concern relating to agricultural sector. They are (i) Slow down in growth rates of different crops; (ii) Widening economic disparities between irrigation and rainfed areas (iii) Due to trade liberalization policy producers of cotton and oil seeds faced severe competition; (iv) Inadequacy and inefficient use of available technology and inputs; (v) Degradation of natural resource base and (vi) Rapid and widespread decline in groundwater table, adversely affecting mainly small and marginal farmers.

Table 7.7, presents growth rates of agriculture and allied sectors. It is evident from the table that the growth rates of cereals has been gradually declining from plan period to plan period. Regarding other items the growth rates are found to be fluctuating. Similar has been the situation with regards to livestock and fishing growth rates.

From Table 7.8, it is clear that the growth rates of the agriculture and allied sectors have been less than the growth rates of the total economy throughout the period under reference. The growth rates of crafts and live stock put to gather also are found to be less than the growth rates of total economy.

From Table 7.9, it is evident that the percentage share of public sector investment in agriculture has been gradually declining. The private sector investment has been fluctuating from year to year, though it has been, by and large, of higher percentage than the public sector investment.

Table 7.10, presents production of major crops during the planning era. From the angle of food security and estimated demand for growing population, the production of different items falls short of the demand. There is a strong case for higher levels of farm production, which demands priority for development of agriculture sector.

Table 7.7

Annual Growth Rates in Output of Various Sub-sector of Agriculture and Allied Sectors (in Percentages)

Period	Cereals	Pulses and Oilseeds	Fruits and Vegetables	Other Crops	All Crops	Live-stock	Fishery
1951-52 to 1967-68	4.19	2.98	2.67	2.42	3.00	1.02	4.68
1968-69 to 1980-81	3.43	0.97	4.82	2.98	3.00	3.26	3.08
1981-82 to 1996-97	3.52	5.41	2.84	1.71	2.97	4.78	5.74
1991-92 to 1996-97	2.36	2.92	6.07	2.18	3.09	4.00	7.05
Ninth Plan (1997-98 to 2001-02)	1.49	-1.43	4.11	3.82	2.25	3.53	2.63
Tenth Plan (2002-03 to 2006-07)	1.28	4.29	2.97	3.58	2.46	3.69	3.23

Source: *The Eleventh Five Year Plan Report,* Vol. III p. 5.

Table 7.8

Annual Average GDP Growth Rates — Overall and in Agriculture (in Percentage)

S. No.	Period	Total Economy	Agriculture and Allied Sectors	Crops and Livestock
1	Pre-green revolution 1951-52 to 1967-68	3.69	2.54	2.65
2	Green revolution period 1968-69 to 1980-81	3.52	2.44	2.75
3	Wider technology dissemination period (1981-82 to 1990-91)	5.40	3.52	3.65
4	Early reforms period 1991-92 to 1996-97	5.69	3.66	3.68
5	Ninth plan period 1997-98 to 2001-02	5.52	2.50	2.49
6	Tenth plan period 2002-03 to 2006-07	7.77	2.47	2.51

Source: *Eleventh Plan Report,* Vol. III, p. 4.

Table 7.9

Investment Trend in Agriculture (₹ in Crores at 1999-2000 Price)

Year	Total Investment	Public Sector Investment	Share of Public Sector in Total (%)	Private Sector Investment	Share of Private Sector in Total (%)
1980-81 to 1984-85	25139	12007	47.8	13132	52.2
1985-86 to 1989-90	23971	9601	40.1	14370	59.9
1990-91 to 1994-95	27263	7915	29	19348	71
1995-96 to 1999-2000	30354	7724	25.4	22631	74.6
2000-01	39027	7155	18.3	31872	81.7
2001-02	48215	8746	18.1	39468	81.9
2002-03	46823	7962	17	38861	83
2003-04	44833	9376	20.9	35457	79.1
2004-05	49108	12273	24.99	36835	75
2005-06	54905	15006	27.3	39899	72.7
2006-07	60762	17749	29.2	43013	70.8

Source: *The Eleventh Five Year Plan Report,* Vol. III, p. 8.

Table 7.10

Production of Major Crops and Allied Items (1950-51 to 2007-08)

(Million Tonnes)

Crops	1950-51	1970-71	1980-81	1990-91	2000-01	2007-08
Foodgrains	50.82	108.42	129.59	176.39	196.81	230.67
Rice	20.58	42.22	53.63	74.29	84.98	96.43
Wheat	6.46	23.83	36.31	55.14	69.28	78.40
Cereals	15.38	30.55	29.02	32.70	31.08	40.73
Pulses	8.41	11.82	10.63	14.26	11.08	15.12
Sugarcane	57.05	126.37	154.25	241.05	295.96	340.50
Cotton (in bales)	3.04	4.76	7.01	9.84	9.52	25.81
Groundnut	3.48	6.11	5.01	7.51	6.41	9.36
Milk	17.00	21.20	31.60	53.90	80.60	100.0
Fish	7.52	17.56	24.42	38.36	56.56	68.6

Source: *Agricultural Statistics at a Glance, 2008,* Ministry of Agriculture, Govt. of India.

5. Tenth Five Plan Outlays and Expenditure

Against the Tenth Plan outlay of ₹ 58,933 crore on agriculture and allied sectors, the utilization was to the extent of 84 per cent at the 2001-02 price level. The utilization in the case of irrigation and flood control was of the order of 85%. In the Tenth plan,[2] several new programmes introduced aimed at diversification of agriculture, strengthening technology validation, demonstration and dissemination, water saving and development and dissemination, water saving and development of infrastructure, such as: (i) The National Horticulture Mission (NHM) became operational during 2005-06. (ii) Micro-irrigation programme was introduced to help spread the network of water saving implements such as sprinkler and drip irrigation throughout the country. (iii) A National Gender Resource Centre in agriculture was set up as a focal point and coordination of gender-related issues. (iv) The National Fisheries Development Board (NFDB) was set up at Hyderabad with the main objective of bringing major activities relating to fisheries and aquaculture for better management. (v) The government decided to set up a Krishi Vigyan Kendra (KVK) in each rural district in the country. By the end of 2009, over 537 KVKs have been established. (vi) Technology dissemination efforts are being strengthened by establishing Agriculture Technology Management Agencies (ATMAs) for extension planning 252 districts covering mostly rural districts. (vii) The National Rainfed Area Authority (NRAA) was constituted as an expert body to advise on various watershed development schemes. (viii) The government of India announced a rehabilitation package amounting to ₹ 16978.69 crore for the farmers in distress in 31 selected districts of four states namely Andhra Pradesh, Maharashtra, Kerala and Karnataka. The package comprises relief from the Prime Minister Relief Fund, strengthening of institutional credit support, irrigation development, promotion of micro-irrigation, watershed development, extension services and income generation, through horticulture, livestock, and fishery in these districts.

However, there are areas of concern in important existing plan schemes. They are: (i) It is noticed that these schemes initiated did not originate from the grass-root level. (ii) The technology mission on oilseeds has not been effective. (iii) The crop-insurance scheme was limited to a few States and to a few crops only. The Operation Flood Programme has not spread as expected and the growth rate of milk has been less than 3 per cent per annum. It is hoped that these different areas of concern will be effectively addressed in the Eleventh Plan.

6. Eleventh Plan and Agriculture

The planned development is a continuous process. The schemes of the Tenth Plan and earlier plans are continued in the Eleventh plan period also. It was noticed that there were some gaps in the development process and hence the Eleventh Plan proposes to identify gaps and take suitable steps as indicated below.[3]

(i) The productivity achieved on farms has fallen short of those in the field trials. (ii) Some technological backwardness is noticed in overcoming Intellectual Property Rights (IPRs) and this has to be addressed effectively. (iii) Integrating methods of traditional and modern biology giving attention to both yield and quality aspects is a problem. (iv) Public sector research in "Hybrid Development" with commercial viability is a critical research gap. (v) Indigenous plant types responsible for higher nutritive values such as millets need to be identified and used for enriching nutrients in rainfed crops. (vi) The implications of climate change need to be studied in depth to combat global warming. (vii) A major thrust is required in the water management. (viii) Integrated Pest Management (IPM) needs greater emphasis. (ix) In horticulture the research should focus on survey of indigenous bio-diversity aiming at improvement in production, productivity and quality produce. (x) In livestock, there is an urgent need to reorient research and assess the genetic potential of indigenous breeds. (xi) With large quantities of animal-products now being produced, research on process technologies, value addition, packaging, storage, transportation, and marketing should receive high priority. (xii) There is need to identity integrated farming system aiming at effective resources utilization, income and livelihood generation and minimize environmental adverse effects.

7. Extension

Extension is an area which needs serious attention. The challenge before extension agency is how to deliver knowledge and skills to all farmers, especially the small and marginal farmers. The present extension service is found to be inadequate and the gaps in the system have to be strengthened by appropriate steps and effective supervision.

8. Price Policy

Price policy continues to be a burning problem. In fixing the minimum support price the government must necessarily balance the interests of both producers and consumers. Public distribution system and minimum support price-policy are very critical and they have to be carefully handled keeping in view the interests of all stakeholders. Price policy should address the problem of fluctuations in price level particularly due to the impact of international prices. Policy of import tariff, insurance, MSP need to be continuously handled. The recommendations of Dr. M.S. Swaminathan Committee need to be implemented.

9. Soil Health Management

Soil degradation through use of agro-chemicals is a serious issue that needs to be addressed on a priority basis. The nature and extent of problem differs in different parts of the country. As the Plan report observes[4] "lack of knowledge on the part of the farmers about the importance of soil health and the information about the status of the soil of different farms is an important constraint. The imbalance use of nitrogen-phosphorous-potassium (NPK) is creating serious problem of soil degradation and adversely affecting productivity. This calls for effective extension services, covering all categories of farmers.

The present system of fertilize subsidy as Plan observes is irrational and has been counterproductive. Since nitrogenous fertilizers are subsidized more than potassic and phosphatic fertilizers the subsidy tends to benefit more the crops and region which require higher use of nitrogenous fertilizers as compared to the crops and regions which require higher application of potassic and phosphate fertilizers. The imbalance in the use of nitrogen-phosphorous – Potassium (NPK) brought about distortions in price ratio in favour of nitrogenous fertilizer. This is creating serous problem of soil degradation and adversely affecting productivity.[5] Balanced use of fertilizer can be achieved with a policy of nutrient neutral or by increasing subsidy on P and K in such a way that farmers are induced to use NPK in the right proportion.[6]

10. Seeds

Supply of quality seed at right time is crucial for growth of productivity. Some times spurious and adulteration seeds are sold in the market and farmers' agitation against such situation is a clear indication of serious lapses. In recent years, there has been considerable increase in the distribution of certified seeds but the quality of these seeds is often questioned by farmers. Both public and private agencies are involved in the supply of certified seed. During the Tenth Plan period the private sector seed supply had overtaken, the seed supply by government agencies. Yet the situation has not improved. The government must take effective steps to ensure the supply of quality seed to the farmers.

11. Agricultural Credit

Timely and adequate institutional credit to all categories of farmers is essential for smooth growth of farm sector. Despite increase of share of commercial banks in total agricultural credit from 54% in 2001-02, to around 69% in 2005-06, still farmers are forced to borrow from private moneylenders and are being exploited by them. The findings of the National Sample Survey Organisation (NSSO) 59th Round (2003), revealed that only 27% of the total number of cultivator households received credit from formal sources, while 22% received credit from informal sources. The fact that even after government had chosen to write off loans taken from the institutions, there have been instances of suicides in rural sector. Some studies clearly reveal that it is an account of credit taken from money lenders and such type of loans are not written off. Mounting over dues in the institutional credit has become a burning problem. There is therefore strong case for complete elimination of the role of money lender in the rural credit by way of strengthening the institutional credit system ensuring adequate and timely credit to all categories of farmers.

12. Risk Management

The frequency and severity of risks in agriculture have increased in recent years as is evident from increase in the severity of droughts, floods and cyclones. There is clear evidence of climate change leading to rising temperatures, resulting in rising sea levels and submergence of coastal lands erratic rainfall etc. Environmental protection needs urgent attention of the government and the world at large. Apart from care of environment, crop and cattle insurance scheme must be strengthened as a part of risk management.

13. Marketing

The Eleventh Plan observes "The agricultural sector needs well-functioning markets to drive growth, employment and prosperity in rural areas. The markets lack even basic infrastructure at many places. Cold storage units are needed in the markets where perishable commodities are brought for sale."[7] The Plan proposes to strengthen warehousing facilities. It is proposed to encourage private sector, cooperatives and panchayats to set up rural godowns so as to ensure marketing facilities to rural people. It is estimated that at present post-harvest losses are still to the extent of 20 to 30 per cent in different horticulture crops. This happens because of inadequate infrastructure development for post-harvesting. Again it is proposed to develop agro-processing as there is rising demand for new products such as dried powder, banana chips, fruit wines, etc. The consumer friendly products such as frozen green peas, ready-to-use salad mixtures, etc., are proposed to be developed through private corporate sector. The Government of India has proposed to set up 10 food processing units (food parks) in the country. Steps are already under way to start one such unit at Mogili village, in Chittoor District of Andhra Pradesh at a cost of ₹ 125 crore in about 140 acres of land (*The Hindu,* dated 8th July 2010).

14. Mechanization

Machines have also been developed and installed for different specialized uses such as chips-marketing, pickles, nutmeg powder, chilli powder, fruit juices and jam making, packaging of coconut water, etc. To keep pace with improved production and productivity, different machines for tilling the soil, sowing seeds harvesting have been developed for agricultural operations.

15. Animal Husbandry, Dairying and Fisheries

Livestock plays an important role in Indian economy. The livestock sector employs about 5.5. per cent of the workforce and the contribution of the livestock to the GDP is about 4.5. per cent. Livestock provides stability to family income especially in the arid and semi-arid regions of the country. Livestock is an insurance against the vagaries of the nature due to drought, famine and other natural calamities. Livestock assures livelihood particularly to the small and marginal farmers. Growth during the Tenth plan has been at the rate of 3.6 percent. The Eleventh five-year plan proposes to increase the growth of livestock between 6 and 7 percent per annum. The aim primarily is to benefit small and marginal farmers and landless labourers particularly in the drought-prone areas. However, lack of credit for livestock production has been a major problem. In the Eleventh plan period, it is proposed to ensure timely availability of inputs and services including credit to livestock farmers.

Fisheries and aquaculture contribute 1.04 per cent to the National GDP of the country and 5.34 per cent to agriculture and allied activities. The fisheries sector has been providing employment to over one lakh full-time and eleven lakh part-time fishermen. It is proposed to export fisheries by the end of Eleventh Plan to the extent of 1.06 million tonnes of fish with the value of ₹ 15,000 crore.

16. Rainfed Agriculture

Rainfed areas in the country account for 60 per cent of the cultivated area, supporting the livelihood of rural poor. The Eleventh Plan proposes to improve rural livelihoods through Participatory Watershed Development Projects. It is hoped through this programme productivity can be increased, contributing to livelihood security. There are plans to develop the rainfed cultivation in eastern India. Crop loan requirement being crucial, the repayment methods would be designed in such a way as to benefit the rural poor.

In the Eleventh Plan it is proposed to develop degraded lands, desert and drought prone areas, where groundwater depletion is very high. The priority during the plan would be to develop arid, semi-arid and dry sub-humid areas through Integrated Watershed Management Programmes, following scientific methods developed by the International Crop Research In Semi Arid Tracts (ICRISAT) Hyderabad and other institutions.

17. Updating Land Records

Eleventh Five Year Plan rightly points out that the "Correct and updated land records" are crucial for the security of land rights and to encourage investment. The proposal is gradual computerization of land records. However, unless the land-records are accurate and updated mere compilation may not be of much use. The steps proposed for effective land administration are: (i) The concerned officers are given specific instructions to look into records of previous 50 years, ascertain the facts and only then computerization should be taken up. (ii) This method of computerization should cover different areas such as issue of land passbooks, dispersal of credit by institutions such as cooperative banks and other credit institutions. In addition to the above steps it is also proposed to integrate three layers of data namely forest summary maps, cadastral maps and revenue records. By integrating these three, it is expected to provide a comprehensive tool for disaster management or any other crucial area of development.

18. Reforms Relating to Land Ceiling

During the planning era efforts have been initiated for imposition of ceiling on land to bring about equitable distribution of land. A total area declared surplus so far has been 73.5 lakh acres only of which 53.9 lakh acres have been distributed. The distribution of the remaining area of land declared surplus is held up mainly due to litigation. The Eleventh Five Year Plan proposes: (i) To take steps for early disposal of the cases and that way take up distribution of land in litigation. (ii) Even after land is distributed, if there is dispute steps should be taken to restore the land to whom it was given. (iii) A thorough survey of the encroachment of government land would be taken up and such that land could be identified and distributed to the landless. (iv) It may not be possible to provide government land alone for the landless poor, hence the state government are directed to consider purchase from private land owners by the state and distribute to the poor in a group and not individual. In the Eleventh Plan steps would be initiated for the preparation and updating of land records in the tribal areas. The district authorities are expected to take necessary steps for the detection of encroachment of tribal lands, if any and restoration of land belonging to the scheduled tribes.

19. Land for Housing

It is estimated some 13 to 18 million families in rural India remain landless, of which about 8 million lack homes of their own.[8] In the Eleventh plan the following steps are proposed, to provide security in terms of housing: (i) All landless families with no house of their own should be allotted 10-15 cents of land each. Female headed families should have priority. (ii) Required resources in this connection may be mobilized through schemes like Indira Awas Yojana, Rajiv Swagruha Scheme, NREGP etc. (iii) Wherever there is insecurity in the ownership of a house, the issue will be settled, ownership is regularized in the name of the wife. (iv) The land for the homeless could be allotted within one kilometre or less of their existing village habitation. (v) The concerned panchayats are expected to take necessary steps to help the homeless.

20. Strategy for Agricultural Growth

The strategy of the Eleventh Plan is to accelerate agricultural growth to the extent of 4 percent per annum. Towards this end the steps proposed are: (i) Bringing technology to the farmers. (ii) Improving efficiency of investments, increasing systems support and rationalizing subsidies. (iii) Diversifying farm production aiming at food security. (iv) A strategy of inclusiveness by which the poor will get better access to land, credit and skills. (v) Priority to research towards evolving cropping patterns suited to various agro-climate condition. (vi) The ICAR would be strengthened and be made to be more accountable. (vii) Public expenditure (both plan and non-plan) on agriculture research will be increased from the present 0.7% of GDP to 1% by the end of the Eleventh Plan. (viii) High level expert committee at the centre and in states would be formed to oversee progress in agricultural research.

21. Irrigation

In the Eleventh Plan the focus would be on completing ongoing irrigation projects and on modernizing existing ones along with steps for improving the efficiency of water use. It is proposed that the allocation to Accelerate Irrigation Benefit Programme (AIBP) would be increased. Also, it is stated that Participatory Irrigation Management (PIM) approach would be made aiming at reducing the gap between potential created and the actual utilized. Through better regulation of water deliveries and conjunctive use, it is expected that wastage would be reduced. Eleventh plan pointed out that the subsidy structure on micro-irrigation equipment must be restructured to enable promoting of community sprinkle system by water use association. Also it is stated that NREGP fund be used to Supplement Command Area Development. The Plan proposes regular and accurate assessment of groundwater use in both rural and urban areas to correlate this with recharge and extraction. It is also proposed to empower and entrust village communities with the right and responsibility of groundwater use and rainwater harvesting. As a part of strengthening irrigation system, natural resource management and watershed development would be under taken. The Eleventh plan gives importance to soil health and agricultural extension aiming at higher levels of productivity. Diversification towards horticulture and livestock is taken as a major element for achieving 4% agriculture growth in the Eleventh Plan. Available demand projections suggest that the demand for good grounds will grow at 2.25% per annum. During Eleventh Plan, it is expected that diversification offers opportunities for rising farm income with less pressure on natural resources, apart from contributing to higher level of produce. Agricultural marketing is found to be a critical element of diversification strategy. It is proposed that food processing system be strengthened to create demand for agricultural produce, cut down or eliminate post production losses and provide value added products. The Eleventh plan proposes livestock and fishing development. In fishing the proposal is to establish more hatchers and ensure stockable sizes of seed for ponds, tanks and reservoir sites. The State governments are expected to take appropriate steps on the lines indicated by the Eleventh Plan.

22. Gender Equity

With the role of female workers in agriculture being of higher magnitude, it is stated in the plan that there is an urgent need to ensure women's rights to land and infrastructure support by way of: (a) Women's names should be recorded as cultivators in revenue records, when women operate the land having ownership in the name of male members. (b) Again it is stated that the gender bias in functioning of institutions for information, extension, credit, inputs and marketing should be suitably corrected. (c) Women's cooperatives and other forms of group effort should be promoted for the dissemination of agricultural technology. Again, wherever possible, a group approach for investment by women should be promoted.

An important innovation during the Eleventh plan is the new Rashtriya Krishi Vikas Yojana (RKVY) with an outlay of ₹ 25,000 crore. The programme requires the cooperation from the state governments to draw district wise plans with details of proposed activities so as to avail assistance from RKVY. The success of RKVY depends more on the role of state governments in the preparation of appropriate districtwise plan of development.

The Eleventh Plan strategy of inclusive growth rests upon not only substantial increase in public sector outlay but also considerable increase in the private sector. Investments in NREGP, Backward Regions Grant Fund (BRGF) and rural infrastructure under Bharat Nirmaan would also strengthen the growth of agricultural sector. These measures, as stated in the Eleventh plan are expected to create conditions for active role of private sector through private investment as complementary to public investment in agricultural sector.

Conclusion

The policy measures proposed in the Tenth and Eleventh Five Year Plans are very appropriate towards the goal of sustainable agriculture. What is important is effective steps in the implementation of different schemes initiated. Agricultural production primarily depends upon the decisions of individual farmers, which in turn are influenced by the willingness and ability to invest on different inputs. This in turn depends on availability of remunerative price and adequate credit for investment. In this context, assured irrigation, timely and adequate institutional credit, supplementary occupation to each household, guarantee of remunerative price, extension services in areas of new farm technology and rural industries, quality improvement measures etc., become crucial. The existing gap in these areas needs to be plugged effectively through appropriate policy measures. Public sector allocation to agriculture needs to be increased apart from necessary incentives for private investment in agriculture. The farm sector must be well strengthened to meet the challenges arising from the policy of liberalization and globalization. India being a member of WTO, the farm sector must be in a position to meet the challenges of growing competition by way of improvement in farm technology leading to higher productivity and quality upgradation meeting the standards of international market and be in a position to sell agricultural commodities at competitive prices.

REFERENCES

1. *Eleventh Five Year Plan Report,* Vol. III, p. 4.
2. *Ibid.*, pp. 9-10.
3. *Ibid.*, p. 12.
4. *Ibid.*, p. 16.
5. *Ibid.*
6. *Ibid.*
7. *Ibid.*, p. 12.
8. *Ibid.*, p. 31.
9. *Ibid.*, pp. 25-26.

8 NATURAL RESOURCE MANAGEMENT AND LIVELIHOOD

1. Introduction

India is a vast country with a rich bio-diversity endowed with natural resources of varying nature ensuring livelihood for people. Natural resources are gift of Nature, not made or caused by human effort. The natural resources are of varying nature and they are needed for existence and survival of man, animal, plant, insects and all kinds of microorganism. The basic natural resources needed for human survival, serving as life supporting systems are: land, water, air, forest resources, soils, vegetation, climate and rainfall.

2. Concept of Livelihood

The concept of livelihood conveys a means of making a living by earning money or other needs of life. In the words of Robert Chambers and Gordon Convey. "Livelihood comprises the capabilities, assets (including both material and social resources) and activities required for a means of living". To make a living there must be a means in some form or other in physical terms. That apart, in a society one needs the cooperation of another for survival in some form or another and that is what is known as social resource or social capital. If one has ownership right on some resources, the livelihood is well assured. However, one need not necessarily have ownership on some basic resources, it is enough, if one has access to natural resources to make a living.

3. Land Resources and Livelihood

Land resources are the most important of all natural resources. Land resources in a broad sense include, cultivable land, water resources, climate, raw materials, etc. Land may take many physical forms: Plains, swamps, hills, mountains, forest, prairie and different kinds of climate, from hot to cold, from humid to dry.[1] Barlow,[2] (1958), states that the word 'land' has at least seven major meanings:

1. It is space or room and surface, upon which life takes place:
2. It is nature, or natural environment, including access to sunlight, rainfall, wind and other climate conditions including soil and natural vegetation;
3. It is a factor of production in economic process, comparable to labour and capital;
4. It is a consumption good, especially when used as a site for dwellings, parks, etc.
5. It is a situation or location with respect to markets, geographic features, other resources and other countries.
6. It is a property, with legal connotation as to rights of ownership of individuals and rights and responsibilities of ownership of individuals and sovereignty of governments, and
7. It is capital, in a realistic economic sense.

Table 8.1 presents land use pattern.

Table 8.1

Land Use Pattern

(Million Hectares)

Sl. No.	Item	1950-51	2005-06
1	Area under forests	40.48 (14.2)	69.79 (22.9)
2	Area not available for cultivation	47.52 (16.7)	42.51 (13.9)
3	Other uncultivated land (excluding fallow land)	49.45 (17.4)	26.92 (8.8)
4	Fallow land	28.12 (9.9)	24.17 (7.9)
5	Net area sown	118.75 (41.8)	141.89 (46.5)
6	Total cropped area	131.89 (46.4)	192.80 (50.90)
7	Area sown more than once	13.15 (7.1)	50.90 (15.0)
8	Cropping intensity	111	136
9	Total reporting area	284.32	310.62
10	Total geographical area	(100.0)	(100.0)

Source: *Agricultural Statistics at a Glance 2008,* Ministry of Agriculture Government of India, New Delhi.

Figures in the parenthesis indicate percentage to total reporting area.

From Table 8.1 some important inferences may be drawn regarding land use in India: (1) There has been improvement in the reporting area. (2) Area under forest, net area sown and total cropped area have all increased substantially. (3) Fallow land percentage declined from 9.9 to 7.9. (4) Fallow land is negligible indicating that there is very limited scope for extensive cultivation. (5) Area under forest cover is only 22.9% of the reporting area in 2005-06 and this percentage is below the optimum level of 30 per cent.

The scope of shifting some more land to the area sown is limited due to (a) high cost involved in the conversion of current fallows, other fallows and the cultivable waste, (b) need to expand forest cover, (c) pressure on land for housing etc., and (d) increase of industrial activities.

However the area sown more than once is negligible and with concerted efforts, particularly with expansion of irrigation, area own more than once can certainly be increased. This step will pave the way for increasing cropping intensity which is low in India.

In the context of India emerging as a developed country, land resources in the form of cultivable land, raw materials, favourable soils and climate, etc. would be of great significance. The available land resources must be put to the optimum use through proper land use planning.

4. Land and Livestock Resources

The livestock wealth assumes crucial importance in the rural economy of India. The livestock wealth is composed of large number of cattle, buffaloes, sheep and goat. The cattle and buffalo are the main source of draught power in agricultural operations and rural transport. Livestock will provide essential foods of animal origin like milk, and milk products, meat, etc. Further, they provide large quantities of skins, bones, wool etc., which have great commercial value serving as raw materials for some industrial products. Valuable organic manures are provided by the livestock. The farmyard manure is widely used in India.

As cattle and buffalo maintenance involves intensive use of labour on the part of the member of the family, they offer significant employment and income opportunities particularly to small and marginal farmers. Livestock provides main occupation for some and subsidiary occupation for many others. Normally women take care of the livestock and their sweat is very much involved in the maintenance of livestock. The National Commission on Agriculture observes "A very large proportion of the female labour finds scope for fuller utilization in several operations connected with cattle and buffalo rearing."[3]

The total livestock in 1951, was 292.8 million and it increased to 485.0 million in 2003. All India milk production has been increasing from time to time. Milk production in 1985, was 44.0 million tones and it increased to 100.0 million tones in 2006-07. During the same period Eggs production increased from 16.1 Billion to 50.7 Billion. The wool production increased from 39.1 million Kgs to 45.1 million Kg. However, the average annual milk yield per cow is of the order of 160 kg and of buffalo 5000 kg. In counties like Denmark, New Zealand and the USA, the average annual production cow is between 3000-4000 kg. This low milk production is due to several factors and lack of adequate feeding is a major problem.

In recent years, the common property of grazing land is fast declining. Such lands are sometimes encroached and used for purpose other than grazing of the cattle. This has adverse effect on feeding the animals. Some research studies reveal that the environmental problem in India are primarily centred round land; decline in common property like grazing land, cultivation of uncultivable soils, particularly in semi arid areas. This demands better management of land resources leading to restoration of environment ensuring livelihood to the poor.

5. Mono Cropping and its Adverse Effects

In recent years the trend has been towards commercial crop shifting to single groundnut crop in the place of multiple cropping composed of combination of jowar, maize, pulses etc. The fodder is assured under this type of multiple cropping and not under mono cropping.

6. Farming System Approach

It is good old practice in India. It is a mixed farming where livestock is linked with cultivation. This practice is now termed as Farming System Approach. Here one product or byproduct of a system will be used as an input in another system. Farming system approach embraces activities like sericulture etc. along with cultivation and that way make farming a profitable concern. Assured livelihood from agriculture depends on a host of complimentary activities such as improvement of soil; supply of quality inputs like seed, fertilizer and pesticides; adequate and timely institutional credit; cold storage facility; processing and marketing services; potable water and health services; schools with quality education regular electricity, etc.

7. Natural Resources Management: Key Issues

There has been growing demand for land for non-agricultural purposes; problems of soil erosion; mining activities; climate change and its impact on environment and land use. These different issues have to be addressed.

8. Growing demand for Land for Non-agricultural Purposes

Growing demand for land for non agricultural purposes also contribute for land degradation. Historically, common property resources have served as a major source of grazing. With increasing pressure on lands, permanent pastures and grazing lands have declined from 13.26 million hectares in 1970-71, to 10.45 million hectares in 2003-04. In the same way, miscellaneous tree crops and groves which also support livestock declined by 22% during the same period.

India has a long coast line of 7,600 kilometres that supports the marine fisheries. Due to fishing activities on a large scale, marine fishers declined. In this way, livelihood prospects suffered due to land degradation in terms of loss in grazing lands and marine fisheries. Soil fertility issues and soil salinity and water logging create problems detrimental to the livelihood of the poor.

9. Land Use Planning

Owing to increasing pressure of population on land and ever growing demand for food and raw materials, there is a pressing need to use every piece of land properly. It calls for scientific, rational and systematic planning of land use resources without disturbing ecological balance and livelihood of the poor.

The concept of "land use" is a wide and complex one. It is a function of land, water, air and man. It is the man who uses the land to fulfill his changing needs. Land use requires a multi-disciplinary approach to arrive at sustained land use planning.

Our land needs proper planning as there is misuse, under use and overuse of land than the optimum use. Land use planning should take into account (a) Multiplicity of uses, (b) Multiplicity of cropping and (c) Maximization of production assuring livelihood. Land use planning is important since the land is an important basic natural resource for human survival. There are three important stages in the land use programme. They are (i) Survey, (ii) Analysis and (iii) Planning for the optimum use of the land in future.

First of all, a through survey of land use must be undertaken. The reason for some lands remaining fallow and some lands being overcrowded must be carefully examined. In the course of the survey, the necessary data have to be collected. After the survey and collection of necessary data proper analysis of the data, scientific interpretation and drawing appropriate conclusions become important in the second stage.

On the basis of available knowledge of the present use of land, a future plan must be thought of. The demand for land arises from the following sources: (i) Land for Producing food, raw materials and other agricultural products; (ii) Land for housing purpose which means land development of villages, towns and cities; (iii) Land for industrial development. It means the demand for land arises for the purpose of factory buildings, workshops, etc.; (iv) Land for recreational purposes such as clubs, play grounds, national parks and zoos; (v) Land for transport like road transport, railway transport, etc.; and (vi) Land for security purposes like the requirement for training the army, navy and air force personnel and maintaining them.

It is most likely that the present use of land is not optimal. Therefore, it is necessary that we plan for better use of land in future. In land use planning, no doubt, there are certain limitations. The implementation arises from physical factors. If nature is unkind and a particular area suffers from unfavorable weather conditions, rocky soil, etc., proper land use planning in that region is a difficult task. However, it becomes necessary to make use of the knowledge available from the scientific and technological advancement for the purpose of optimum land use planning. In the matter of land use planning, the following principles may be observed (i) The suitability of land for different uses as determined by the agro-climatic conditions should be carefully examined and right decision must be taken; (ii) It is necessary to make optimum use of every piece of land available; (iii) It is appropriate to plan for multiple use of the land; and (iv) In the matter of land utilization, care must be taken to preserve forest area.

For better land utilization appropriate cropping pattern which is more remunerative must be chosen. The land use system must be flexible and adjustable to development in agricultural technology. The expansion of town or village or industry should be done generally on lands not suitable for agricultural purposes. It may not be possible to have rigid principles in the matter of land use planning. However, the guiding principle in land use planning should be one of the optimum use of the available scarce land resources.

10. Natural Resources Management

Soil and water conservation measures become crucial for increasing agricultural output in the country. During the planning era a number of schemes have been initiated as soil and water conservation in the catchments of River Valley Project (RVP) was started in the Third Five Year Plan. Subsequently another scheme of Flood Prone Rivers (FPR) was started in the Sixth Five Year Plan, keeping in view the magnitude of floods in 1978. Both the scheme were clubbed together and the scheme began functioning as Macro Management Mode since November 2000. Under this programme for the catchment Management of River Valley Projects and Flood Prone Rivers, 53 catchments spread over 27 states were covered by 2004-05 Again centrally sponsored scheme of reclamation of Alkali Soil was taken up in Punjab, Haryana and Uttar Pradesh during the seventh Five year plan. During the Ninth Plan the programme was extended to other states. The scheme at present stands subsumed with in the Macro Management Scheme.

The scheme of Watershed Development Project launched during the Eight Plan aims at soil and water conservation at appropriate place. During the Tenth Plan an area of 0.89 lakh he has been treated with an expenditure of ₹ 88.32 crore up to the end of 2005-06.[5] The Government of India has accorded high priority to the holistic and sustainable development of rain fed area through integrated watershed development approach. The key attributes of the watershed approach are conservation of rain water and optimization of soil and water resources in a sustainable and "cast-effective mode". It is hoped that productivity levels in the areas covered by the programme would increase.

11. Water Resource Management and Livelihood

Water is the most important single requirement for the growth of plant, animal and human life. Water resource management, therefore, becomes pre-requisite in all productive activities. The scope for further increase in the cropped area being limited, further growth in farm production can only he hoped through higher productivity per hectare of different crops. A crucial input facilitating higher productivity is assured irrigation.

There are different sources of irrigation – namely canals, Tanks, wells and tube wells. Irrigation sources are broadly classed as major, medium and minor irrigation. Table 8.2 presents distribution of irrigated area according to sources of irrigation and classification of farmers covered by each of these sources of irrigation.

Table 8.2

Distribution of Irrigated Area: Source-wise Category of Farming-wise 2000-01

Category of Farmers	Canals	Tanks	Wells	Tube Wells	Others	Total
1	2	3	4	5	6	7
Marginal	3,405	855	1,296	5,419	1,409	12,384
Small	2,929	587	1,971	4,335	1,069	10,891
Semi-Medium	3,219	463	2,500	4,740	1,010	11,932
Medium	3,447	276	2,511	4,502	811	11,547
Large	1,578	77	952	1,715	550	4,872
Total	**14,578**	**2,258**	**9,231**	**20,711**	**4,849**	**51,627**

Source: *Agricultural Statistics at a Glance, 2008,* Ministry of Agriculture, Government of India, New Delhi, p. 308.

From Table 8.2, it can be seen out of 51,627 thousand hectares, 14,578 thousand hectares or 28.2 per cent of cropped area is covered by canal irrigation. All other sources put together cover 370 49 thousand hectares and this constitutes 71.8 per cent of the area irrigated. It is evident that minor irrigation is the major source of irrigation. In minor irrigation, wells and tube wells are predominant. These two sources cover mostly small and marginal farmers. Under the Tank irrigation, again, small and marginal farmers are benefited. Table 8.3 shows the extent of irrigated area, source-wise covering different Crops.

It is evident from Table 8.3, that the extent of unirrigated area is of higher magnitude. The land of small holdings owned by small and marginal farmers is also of higher magnitude under the cultivation of unirrigated cultivation. As we have seem in the earlier chapters more than 60 per cent of cropped area is under rainfed cultivation only. This calls for provision of assured irrigation to the rain fed cultivation and appropriate technology for rain fed cultivation.

12. Harvesting of Rainfall

Rainfall is like a fountain serving as a source of water to different channels of irrigation. Water resource management programme should necessarily give priority for harvesting of every drop of rain water as well preserving it in the widely scattered irrigation tanks. Also individual farms must evince interest to have small ponds on their field to preserve the rainwater. Harvesting of rainwater and well preserving in the tanks and ponds has been a traditional practice and it is an indigenous technology which needs to be revived and strengthened.

A.P.J. Abdul Kalam[6] rightly observes "There is an urgent need to conserve water in a number of ways ranging from water harvesting to drip irrigation. There are technologies which can contribute a great deed to agriculture. We need to use all available methods because coming years are not going to be easy on the food front. India is strong in the area of remote sensing technologies. We have capabilities in utilizing remotely sensed data for various application; groundwater targeting, soil salinity assessment, crop yield estimates, and so on. He further states "Water should be treated as a national resource and asset". The growing demand for water for domestic, industry, energy etc. purposes necessitates efficiency in water use. Sprinkle and drip irrigation are necessary in many areas, as also recycling of water. The storage of water during floods or heavy rain, including ground storage is a crucial national task. Wherever possible, particularly for horticultural crops, efforts should be made to introduce modern methods of irrigation". In a vast country like India, with varying geo-physical conditions, we find regional variations in tapping water resources.

Some studies reveal in Southern States of India, groundwater is fully exploited, while in the Eastern and North-Eastern parts of the country irrigation potential remain to a greater degree untapped.

13. Forest Resources Management and Livelihood

Forest resources have been a source of livelihood to the rural poor since times immemorial. The poor collect firewood for domestic use as well as for marketing and that way make a living. The poor utilize forests for grazing purpose of their cattle. Building materials, raw material for paper making, medicinal plants etc., are secured from the forest. Some cut the trees in the forest, burn it and convert it into coal for commercial purpose. Some others hunt and through the sale of birds or meat of killed animals make a living. The tribal population have special attachment to the Forest Resources. They collect honey etc., and make their living. (More details of Forest Management are presented in a separate chapter in this book.) It is rightly pointed out,[8] "Loss of forests can make it difficult for developing countries to meet their growing demands for paper, building materials and most serious of all, fuel wood. It could take hundred of years to reverse some of these forms of environmental degradation. And the climate changes global or regional, induced by large scale deforestation could prove to be irreversible. The stock of wildlife is declining under a variety of development related pressures. The best available estimates suggest that if current trends continue some 15 to 20 per cent of the world's million of plant and animal species may become extinct. The implications go beyond ethics and aesthetics".

Table 8.3

Distribution of Irrigated and Unirrigated Area under Major Crops in India : Category of Farming-wise and Operation Holdings-wise 2000-01

(In thousand hectares)

Crops	Irrigated						Unirrigated					
	Marginal	Small	Semi-Medium	Medium	Large	Total	Marginal	Small	Semi-Medium	Medium	Large	Total
1	2	3	4	5	6	7	8	9	10	11	12	13
Paddy	7,547	5,413	5,013	4,069	1,563	23,606	5,515	2,955	4,682	3,316	1,131	17,599
Wheat	5,117	3,952	4,670	5,107	2,336	21,181	575	404	678	166	266	2,089
Total cereals	13,294	10,148	10,667	10,125	4,274	48,507	10,006	11,125	12,385	10,813	5,178	49,507
Pulses	304	431	592	792	358	2,477	2,043	2,917	3,557	3,740	1,894	14,141
Sugarcane	906	1,063	1,124	821	189	4,103	106	93	92	59	13	363
Oil seeds	757	852	1,110	1,304	589	4,612	1,941	3,386	4,477	4,822	1,907	16,533
Fibres	368	429	594	937	583	2,911	600	1,789	2,150	1,935	542	7,016
Total crops	**17,260**	**14,223**	**15,539**	**15,482**	**6,733**	**69,237**	**16,503**	**20,769**	**24,497**	**23,859**	**12,256**	**97,884**

Source: *Agriculture Statistics at a Glance, 2008,* Ministry of Agriculture, Government of India, p. 309.

Conclusion

The objective of sustainable development can be effectively pursed only if natural resources are well protected and preserved. Prevention is more important and less costly than remedial action. It is necessary that the impact of proposed development projects, particularly major irrigation projects, on environment need to be well analyzed to determine what its effects are likely to be on the natural resources and livelihood of the poor. When the effects are seriously adverse, the project should be redesigned aiming at reducing the hazards. Protection of natural resources as Gandhiji stated should be taken as a social responsibility. There must be an integrated approach with active participation of local people in well managing the natural resources ensuring sustainable development and assured livelihood to the poor.

REFERENCES

1. David, L. Sills (ed), *Encyclopedia of Social Sciences,* The Macmillan Company and The Free Press, 1968, pp. 550-551.
2. *Ibid.*, p. 72.
3. Report of the National Commission on Agriculture, Part VII, *Animal Husbandry.*
4. *India, 2009,* Publication Division, Ministry Information Broadcasting Government of India, p. 68.
5. *Ibid.*,
6. A.P.J. Abdul Kalam, *India 2020 – A Vision for New Millennium,* Penguin Books, 2002, pp. 69-70.
7. Warren C. Baum, Slotes M. Tollersh, *Investing in Development,* Oxford University Press, 1985, p. 524.

9 COMMON PROPERTY RESOURCES (CPRs)

1. Concept of Common Property Resources Defined

The common property resources are those resources managed by a community or society rather than by individuals. If a resource is owned by an individual, he alone has the right over it and the resources may not be accessible to community use. If a resource is a under the control of government, it becomes no man's property and the use of such resource may be confined to selected individuals and not by the community at large it is therefore, felt the need for keeping some resources as a common property being accessible to the benefit of the entire community. The common property resources, being the property of the community as a whole, no individual can claim ownership right over it. Legally, the property may be of the government, but no one has any exclusive ownership right over such land or any other resource. The individuals have only 'usufruct rights' that is right to use the yield from the resources but not the ownership right as such. In this way, the concept of common property resource came into popularity. This type of common property has been in vouge since a long time.[1]

Forest resources, fishing, village grazing lands, village irrigation tanks, village common lands, permanent pastures, uncultivable waste lands, community wells, streams, cattle watering points, ground water resource, etc., come under the common property resources. Since times immemorial, common property resources have been in existence contributing to the sustenance of the poor as well as to the benefit of animal husbandry and farming. During the Vijayanagar rule of the 16th century, for every one revenue village one irrigation tank was designed as common property. The irrigation tank was the property of the entire village and it was the responsibility of the villagers to well maintain the common property, harvest the rainwater preserve the water in the tank for the common good of the villagers for farming and livestock use etc.

2. Benefits from Common Property Resources

Almost all the Indian villages even today have some common property resources and the community at large derives benefit from the CPRs in many ways as follows:[2]

(i) More importantly, community grazing lands, wherever they are well preserved, continued to be of immense use for grazing purpose.

(ii) Village irrigation tanks again serve as an important resource for preserving the rain water for the purpose of farming and farming related activities.

(iii) For the purpose of threshing operations, common property is used.

(iv) From the waste and barren lands, of common property, firewood is collected.

(v) For dumping of waste material, common property land is utilized.

(vi) Rocky mountains which serve as community property are useful for stone cutting for the purpose of securing material for house construction purpose.

(vii) Some public places are utilized for community gathering and that way CPRs facilitate community life.

(viii) Groundwater, again, serves as a useful resource for community benefits.

(ix) Sometimes waste and barren land classified as common property is utilized for stock breeding purpose.

3. Forest Resources as CPRs

Forest resources constitute an important component of the CPRs. The concerned people have only 'usufruct rights' to utilize the resources in multiple ways. The villagers use the forest resources by way of collecting fuel wood, securing wood, securing fruits and nuts, grazing of cattle, obtaining rope making material and some leaves for domestic use etc.,

More importantly forests are a source of medicinal plants of varying nature.[3] *Oscimum Sanctum* (Thulasi) is one such item. It is a small herb with an aromatic smell. Many particularly the Hindus consider this plant as a sacred one. This plant is found in waste land and on the tank beds of the rivulets. The thulasi leaves are used as sacred item in daily prayers. They are used as medicinal plants to heal head boils and some other diseases. A good number of medicinal plants are secured from CPRs.

The neem tree plays an important role as a medicinal plant. The neem leaves and seeds are used for medicinal purpose. The neem has been an important resource for a variety of uses, such as preparation of neem cake, neem oil, neem soap, neem based bio fertilizers and bio pesticides etc. The neem plant, thus has direct bearing in promoting eco friendly agriculture. In view of growing demand for neem based products, there is a keen competition among the rural people for collecting the neem seeds and that way make their living.

Forests are a major source of livelihood to many, particularly to Tribal people by way of collecting honey etc., and that way making their living. The rural poor collect leaves of certain trees sell them and make their living. Apart from leaves, fruits are also collected from the CPRs. The popular fruits collected are: Zizyphus Jujuba (Regi), Phoenix Sylvestries (Eethapandu) and Ficus Glomerata (Medical), etc.

Tamarind is a very popular tree raised all around a village. Tamarind tree is found in the forest as well as in the barren wastelands. Tamarind is used for a number of purposes and this particular plant from CPRs is a source of living for good number of rural poor. The tender leaves of Tamarind (Chintha Chiguru) is marketed. The tamarind wood is used for fuel purposes, for domestic as well as for preparation of bricks.

Again wood is secured from CPRs for making different agriculture implements and for house building purpose. Also raw material for rope making (Hemp) is secured from CPRs. Different kinds of ropes are required in the agricultural operations and the raw material for rope making is secured from common property resources only.

In some villages, still there is practice of making pots, as a profession by some people (kummar). The raw material in the preparation of pots is clay, which is normally collected from the CPRs of the village community.

4. CPRs and Environmental Protection

Environmental concerns[4] encompass a wide range of issues such as public health, control of air, water and land pollution management of renewable natural resources (plants, animals, soil and water) erosion control, conservation of unique habitats etc. The CPRs contribute for environmental protection. Forests are a powerful ecological unit effecting environment and are among the most natural and renewable resources. They render the climate more equitable. The forests help to prevent soil erosion and facilitate rainfall. This emphasizes the importance of preserving forests as well as other common property resources. More than anything else, CPRs play a crucial role in environmental protection.

5. Declining Common Property Resources

Gradually the CPRs are dwindling for different reasons. Research studies reveal that in the early phase of the planned development, attempts were made to step up food-grains production by way of extensive cultivation through bringing more land under the plough. In this process, some land under common property was converted as a cropped area and that way CPRs was encroached. Population growth and consequent pressure on the cropped land also resulted in degradation of CPRs. In the absence of adequate employment opportunities in the non-agricultural sector, there is a growing pressure on land and it results in converting a part of CPRs land for agricultural purpose. In the process of development, for the purpose of formation of new roads or widening of existing roads, construction of major irrigation projects, industrialization and urbanization etc., there has been gradual decline of CPRs with all adverse effects on environment.

In some areas, in the absence of effective village administration, the CPRs are very much encroached upon. In some places irrigation tanks are converted as Bus depots, some manage to get ownership right of the common property like grazing land and use the land for cultivation purpose or for some commercial purpose. In recent years, it is reported that in certain areas there is over exploitation of forest resources. The rural people must necessarily consider safeguarding irrigation tanks and forest resources as a social responsibility and that way protect CPRs.

6. Steps to Safeguard CPRs

The local people must be alert enough to safeguard their own CPRs. It means the civil society must be alive to the problem and act with all seriousness in safeguarding the fast deteriorating CPRs. Non-governmental organizations must evince interest in the matter, educate, and create necessary awareness among the rural poor impressing upon them the impending danger if CPRs are destroyed. The government machinery must not shirk in its responsibility of safeguarding of CPRs. The schemes like Joint Forest Management should serve as a forum to involve local people in protecting the CPRs. The village panchayats must evince interest in safeguarding CPRs as a part of their legitimate function.

REFERENCES

1. *Report of National Commission on Agriculture,* Part IX, Forestry, p. 33.
2. K. Venkata Reddy, *Agriculture and Rural Development,* Himalaya Publishing House, Bombay, pp. 97 to 100.
3. G. Prakash Reddy, "Common Property Resources in the Rural Communities of Anantapur Districts" — An Empirical Study in *The State of Rayalaseema* (ed.) A. Ranga Reddy, Mittal Publications, New Delhi, pp. 197 to 208.
4. Warren C. Baum, Stokes M. Tolbert, *Investing in Development,* Oxford University Press, p. 523.

10 FOREST RESOURCES AND JOINT FOREST MANAGEMENT AND PARTICIPATORY RURAL APPRAISAL

1. Introduction

Forest resources constitute a major component of total natural resources. Forest cover is critical for desirable environment and sound economic growth. Forest, thickly covered with trees has been a congenial abode for meditation, for a good number of saints and sages.

Forest resources have been contributing for the growth of economy and welfare of man in different ways. Forests offer mainly two types of benefits – products and services. The products include timber, fuel wood, minor forest produce, plants useful for medicinal purpose, honey etc. Also certain forest products serve as food for consumption. The services that forests provide include soils, nutrients water and air as discussed below.[1]

2. Protection of Soils

Tree foliage, i.e., leaves of a tree or plant thickly spread in forest area, reduces the intensity of the raindrops striking the soil, thus reduces the run off soil particles. The root system of trees and bushes bind the soil firmly. Once the vegetation cover consisting of trees, bushes or grass is removed, soil erosion takes place. Trees protect the soil by providing shade which prevents the soil from becoming too dry and friable during summer. Forested mountain slopes conserve rainwater and allow it to flow into streams, rivers or into subsoil springs in a regulated manner.

3. Care of Nutrients

Forests improve the quality of the soil by increasing its porosity. Leaves and other organic matter falling from the trees gradually decay and mix with the soil to form humus. Soils containing humus are very fertile and absorb large quantities of water. Apart from supporting plant growth, soils provide an ideal micro habitat for a variety of microorganisms which play a very important role in nutrient cycling.

4. Conservation of Water

The soil under forest cover acts almost like a sponge absorbing and holding the rain water and letting it sink into the soil. This helps to maintain the perennial supply of subsoil water in wells and springs. Tree roots go deep into the soil and suck up subsoil moisture, which travels though the stem to the branches and finally to the leaves. The leaves have tiny holes through which the water evaporates into the atmosphere. Thus trees not only help conserve water in the soil but also bring the subsoil water back into the atmosphere.

5. Protection of Air

Also crucial service is rendered by forest. The leaves of trees absorb large quantities of carbon dioxide from the atmosphere. Scientists believe that the amount of carbon dioxide in the atmosphere keeps increasing, the high carbon dioxide levels results in the rise of atmospheric temperature all over the globe and cause the polar ice-caps to melt. Chlorophyll, the green substance in the leaves, uses this carbon dioxide to manufacture food for the tree in the presence of sunlight. In this process, which is called photosynthesis, large qualities of oxygen are released into the atmosphere. Trees thus, absorb potentially dangerous carbon dioxide from the atmosphere and release life giving oxygen into it.

6. Forest Cover for Sustainable Environment

To ensure sound environment, expert opinion is that one third of the geographical area must be under the forest cover. In 1947, the area under forest was around 30% of the total geographical area of the India. But during the last 60 years there has been gradual forest degradation. As per the Forest Survey of India, the total green cover, including green cover outside forests comes to 77.83 million hectares or 23.68% of the total geographical area. The net forest cover was only 67.83 million hectares or 20.64% of the total land area.[2]

The reasons for degradation of forests areas are many: population pressure increases the demand for forest products; forest lands were cleared for agriculture and irrigation developments; trees were felled for formation of new roads or widening of existing roads; the loss of community grazing lands led to overgrazing of forests. Meeting industrial needs was another reason for the rapid depletion of our forests. Bamboo was sold to paper industry and that way depletion of forest cover. The demand for wood and wood products has led to increased smuggling from the forests. Major irrigation projects undertaken in some States contribute for large scale degradation of forest cover. The poor in the rural areas continue to make their living by way of selling firewood, secured from the forests. Clandestine timber trading continues. Wood cutting for domestic purposes also causes heavy damage to forest cover. In some areas, it is reported that unlawfully goats graze in the forest area causing damage to the green cover.

In India, per capita forest cover is only 0.064 ha against the world average of 0.64 ha (FAO). While 78% of the forest area is subjected to heavy grazing and other unregulated uses, adversity affecting productivity and regeneration, nearly 10 MH of forests area is subject to shifting cultivation.[3] Plan investment in forestry and wildlife sector so far, including state and central plan, has been about 1% the total plan outlay. The National Forestry Commission (2006), has recommended an investment of 2.5% of the plan outlay in the forestry and wildlife sector.[4]

7. Forest Resources and Sustainable Development

The need of the hour is sustainable development. It requires proper use of land, water and bio-diversity. The goal of development must be one of guaranteeing food security. Food security can be guaranteed only by ecological security. Forest and soil are crucial in ecological security. It is rightly said that "managing forests for sustained yield is essentially eco system management."[5] In this context an integrated approach becomes necessary.

8. Integrated Approach and Protection of Forest Cover

Integrated approach addressing simultaneously forest protection, agriculture, energy and ecological security is found to be a right step. Forest and land use policy must integrate with agriculture, energy and rural development strategies. Agricultural development with emphasis on irrigation enhancement, fertilizer and pesticides inputs, relying so heavily on cash crop and energy development planning which essentially caters to the galloping urban and industrial demands have to be totally reoriented. Only then forests are saved from total destruction for timber or fuel for needy and greedy people.[6]

The Integrated approach and protection of forest cover is said to be feasible if only joint forest management is attempted. In this integrated approach, two important questions need to be addressed. (1) Ecological stability must necessarily be enhanced and (2) the goods and services that the village community requires must be taken care of. Maintaining ecological stability and meeting the basic needs of the people are to be viewed inseparable.

9. Joint Forest Management (JFM)

Joint Forest Management (JFM) is a scheme of associating village communities with the Forest Department in protection and development of forests. Under JFM, Vana Samrakshana Samithis (VSS) are formed, composed of local people to protect, develop, manage and enjoy certain benefits from the forest. In this arrangement VSS and forest department share usufruct yield crops on the basis of an agreed plan and both have responsibility of protecting the forest. Thus, local villager's access to the forest, earlier denied, is being restored by JFM. The local villagers if act with a sense of responsibility can prevent smugglers and clandestine timber trade.

10. Difference between JFM and Forest Department Management

Differences between JFM and the management exclusively by Forest Department can be noticed. Management by the Forest Department is a centralized management, while JFM is a decentralized one. The Forest Department often takes unilateral decisions, while in JFM there is participatory decision making. The working of Forest Department is guided by revenue orientation and production motives. The JFM is guided by resources orientation and sustainability. The Forest Department Management is through large working plan while under JFM management is through micro-planning. In the practice of management by forest department, focus is on single species and single product, while under JFM focus is shifted to multiple specialized products. For the forest department plantation is the first option, while under JFM focus is on input management and regeneration.

It is not that easy to ensure effective functioning of JFM. The forest officials have had their own way of managing and protecting the forest resources. The local people, particularly Tribal people, who have special links with forest may not so easily adopt themselves to the system of JFM. Naturally the local people, especially Tribals are suspicious of the intentions of Forest Department officials in the scheme of JFM. More than any other organization, the NGOs must play a key role in bringing together the forest officials and the local people and facilitate the effective functioning of JFM. There are variations in the relationship between people and forest in different regions. Also poverty-stricken people should not suffer under JFM. With JFM, there is a possibility that some people will be denied of their traditional access to the forest lands and products and there is possibility of strong opposition in certain pockets. It is therefore necessary that JFM is so planned that the local people are not denied of their basic needs. In rural areas, fuel wood and fodder are essential requirements. In villages with high cattle population fodder needs are most pressing. In JFM, it is important to pay attention to fuel wood and fodder and identify ways and means to secure them without over exploitation of forest resources. Without adversely affecting the interests and livelihood of the poor and Tribal people appropriate steps are to be taken in JFM. It is necessary to motivate the villagers to protect the common grazing lands. Also, barren lands around the villagers be used to raise fodder. In this way, JFM scheme can be made effective to deliver desired results.

11. Forest Degradation as Matter of Concern

Forest degradation is a matter of great concern in view of its adverse effect on environment. Integrated approach addressing simultaneously interrelated issues relating to forest, land and water is necessary and not at all isolated approach and isolated plan. Poor quality lands, barren and uncultivated wastelands must be brought under tree crops and that way reduce the pressure and destructive exploitation of forest resources. A clear cut policy of conservation of natural resources so as to ensure ecological balance avoiding conflict situation between the survival needs of the people, industrial requirements and conservation of nature for preservation of ecological balance is urgently needed. This policy must be so desired as to make integrated approach. Protection and enrichment of forest is the need of the hour.

12. Forest Resources and Eleventh Five Year Plan

It is encouraging to note that the Eleventh plan is committed to the task of protection of forest resource through participatory approach. It is stated in the plan "enhancing the green cover will be integrated with livelihood opportunities". It is also stated that the policy is one of strengthening participatory processes. National afforestation programme is categorical in the policy statement "Forest Department Agencies at the district level are to function as a link between the Ministry of Environment and of JFM committees for the scrutiny of projects, release of funds and implementation of the sanctioned programmes.[7]

13. Wildlife and Biodiversity

Preservation of wildlife and forest biodiversity should necessarily receive adequate attention. Effective forest cover, biodiversity and ecological security are linked and hence, an integrated approach is required to address these three areas simultaneously. The 1952 National Forest Policy provided for setting up of Sanctuaries and National Parks for preservation of wildlife. Enactment of the Wildlife (Protection) Act 1972; launch of Project Tiger in 1973; Project Elephant Act 1992; and Biological Diversity Act 2002 are all milestones in the preservation of wildlife and biodiversity. Biodiversity it is stated "provides building blocks for sustainable food, health and livelihood systems. It is the feedback for biotechnology industry and a climate-resilient farming system."[8] In view of its importance, a Convention on Biological Diversity (CBD) was adopted at the 1992 UN Conference on Environment and Development in Rio de Janeiro. The goals of the convention are conservation sustainable use and equitable sharing of benefits. It recognizes that a country's biodiversity is the sovereign property of its people. India is a signatory to the CBD and has enacted the National Biodiversity Act, which is in force since 2002. The UN has designated, 2010 the International Year of Biodiversity. To generate awareness about genetic resources conservation, each year May 22 is observed as the International Day for Biological Diversity. Inspite of all these, M.S. Swaminathan laments that "genetic erosion continues globally, 12% of birds, 21% mammals, 30% of amphibians, 27% of coral reefs and 35% of conifers cycado face extinction."[9]

There has been degradation of wildlife along with degradation of forest cover. Forest is the safe home for a number of wildlife. With destruction of forest there is no safety to the wildlife. The decay of wildlife need to be well understood. Just as there are vegetarian and non-vegetarians in the human being so also with regards to animals. Elephant, camels, deer, sheep, goat, giraffe, squirrels, pigeon, monkeys etc. are vegetarian animals. As against this, tiger, lion, fox, eagles, hyena, snakes, crane, crab and so on are non-vegetarian animals. With destruction of forest the vegetarian animals are denied of their normal minimum food in the forest area. Hence, these animals are forced to move to the nearby habitations causing damage to fields etc. Again, man who is not content with non-vegetarian stuff in their own habitation, is hunting vegetarian animals and thus, non-vegetarian animals in the forest are denied their food.

Analysis of this situation clearly indicates that man is responsible for degradation of forest and man is the destroyer of animals and plants in the forest. For some hunting is a prestigious hobby. Poor man is indulging in hunting to make a living. Clandestine timber trade is normally undertaken by the rich who have political strength. The poor, to make his living, is indulging in fuel wood cutting. In the villages, livestock wealth is owned by rich and poor. Through the unlawful entry of goats into forest for grazing, forest is destroyed. It is evident that both rich and poor are equally responsible for degradation of forest resources.

14. Artificial Regeneration or Plantation Method

This method is an intervention in the natural process. This involves several operations such as clearing the scrub growth, pitting planting and subsequent cultural operations like weeding, soil workings etc.

15. Micro Plan

The micro plan is an important basic document in JFM. It consists of the local participation; protection of forest and the distribution of usufruct rights between the members. It should clearly spell out the works to be carried out, physical targets to be achieved, the manner in which the VSS would protect the process, the role of the forest department in the management of VSS, the mechanism to meet the local needs for fuel wood, fodder, and minor forest produce and the method by which the usufruct rights would be shared.

As a part of JFM villagers need to be motivated to identity the activity and likely participants for each activity. In this stage discussions are to be initiated with the community on prioritization of the options identified. This leads to conducting of intensive topical appraisal exercises with the community. The focus is on the technical, management and financial aspects. Based on the exercises villagers are able to prioritize the different activities identified under JFM. To assess the likely benefits; share of the community and external investment in the proposed activity; to identity activities that can be completed with less cost and with local expertise and finally identify ways and means of operationalizing the plan and the responsibilities taken by the village community.

A micro plan requires, among others, (a) Brief description of the socio-economic structure of the village with special reference to land use all cropping pattern, grassing reaction fuel wood, fodder, biomass needs and rainfall situation and extent of irrigation facility. (b) Traditional systems of natural resource management. (c) Treatment plan along with budget for different works. (d) Plans for fodder/fuel. (e) Responsibility of the VSS member as planned by the villagers. The map of the locality chosen for JFM.

The micro plan is prepared by the VSS and sent for approval to the Divisional Forest Officer. The approval of the micro plan is essential; The micro plan is prepared through PRA exercise and the plan may be for a period of 7 to 10 years. The concerned offices and stake holders are involved in the preparation of the micro plan and in getting the information. Expert advice should necessary be obtained on matters like fodder, fuel wood, medicinal plants etc.

16. Financing the Micro Plan

The number of activities that can be taken up depends upon the availability of funds and financial support expected from different sources. For certain components of the JFM plan, financial support is expected from the government departments. For soil and water conservations works normally district administration will have to respond positively. Similarly for schemes like protecting forest area the district administration will have to respond positively. In Tribal areas, Tribal Development Agency is a source of finance. The Jawahar Rozgar Yojana is another source for VSS activities. For schemes like planting in forest areas, samplings could be obtained from the forest department nurseries. Despite all these sources, JFM still suffers from want of sufficient funds. The NGOs may come forward to assist the scheme in a big way.

PART - B
PARTICIPATORY RURAL APPRAISAL (PRA)

The participatory rural appraisal is a technique used to collect information from the village for preparation of a micro plan. PRA is basically a set of innovative tools, to evaluate the resources available and plan village development activities. It facilitates the easy flow of information from the village and helps to ensure their participation. In order to appreciate PRA we must understand different issues associated with existing planning methods. Planning involves the optimization of resources like land, water, soils trees, grasses, seed, animals, rainfall, money and people. Understandably special skills are required for optimum use of different resources. Normally different experts sit together and pool the knowledge to draft village plans. Survey Method with questionnaire is followed to get required information in identifying appropriate plan for utilization of different resources. PRA stresses people's participation to design, implement and review village development plans. PRA facilitates the process of gathering information and involves different sections in the community, PRA can be used to plan any development work including watershed development programme, village sanghams, thrift and credit schemes etc.[10]

Difference between Survey method and PRA method need to be well understood. In a Survey method, the investigator build up a report with the respondents, ask questions and write down the information and also seek his or her views on how best to improve the farm. In PRA method, the focus is centered round the objective of gathering information on the resource and simultaneously looking for ways to facilitate participation and make them feel that the plan for management of the resource is their own.

The PRA method emphasizes the following: (i) Group discussion by the people, (ii) diagramming and visual sharing by the people, (iii) attitude and interactions within the community, (iv) involving their participation in the work plan. (v) understanding potential conflicts between various groups in the villages in implementing these priorities, (vi) a prioritized plan and (vii) a frame of plan of action for implementation, management, monitoring and evaluation of the programme.

In PRA, information is gathered through different methods. Each method has to be adopted to obtain particular type of information. Some of the methods or techniques useful in the preparation of the micro plan are as follows.[10]

1. Social Resource Mapping

The objective of this method is to understand the geography and the physical condition in the village. It becomes necessary that PRA team is equipped with required materials. The PRA staff must select suitable place and time where all communities especially the poor can assemble and participate by freely giving their views. Women participation must necessarily be ensured in the preparation of a map, consisting of all details of resources. The social map delineates the interior of the village with all particulars. When social map is complete, a picture emerges of the social, economic and physical structure of the village. The people conducting the exercise should be patient and not at all be in a hurry. They simply sit and watch giving villagers to take the initiative.

A study by the Centre for Environmental Concern, Hyderabad observed that when a such a map being drawn in a tribal hamlet, the villagers showed the route taken by wood smugglers. This information can be useful in planning forest cover protection. When told of the route, one PRA team chided the villagers for being silent witnesses to the degradation of their forest. The villagers then as a unit fell silent and chose not to participate in the matter. It is said that a lot of persuasion becomes necessary to get them involved in further PRA exercises.[11]

In Resource Mapping, necessary information and opinions of the people are obtained on the following matters: (i) location and the condition of the forest lands, (ii) pastures which exist, (iii) common pr:operty land, (iv) wastelands, (v) soil in different areas, (vi) water resources and (vii) irrigated and dry land.

The staff must identify those people who actively participated in the discussion and gave required information. Such "informants" must necessarily be associated in rest of the PRA exercises. In all matters, people conducting the exercise should act with tact and precede slowly involving villagers. The best method is that the villagers take the initiative.

2. Transect Method

This method involves: (a) observation of physical characteristics like erosion, water logging, soil depth, soil type and people's perception of these issues. (b) questioning and eliciting required information. (c) intensive discussion with people to elicit their views and suggest solutions. (d) local people are involved to identity natural resources, problems in development of these resources, local solutions tried and options identified.

3. Root Stock Survey

The objective of this type of survey is to estimate the density of viable route material of different plants species as presently available in the area selected for JFM. The root stock survey helps to estimate the potential of availing the natural regeneration stock among the concerned. The potential for rapid regeneration of the green cover on the land taken for JFM depends on the availability and the condition of the root stocks. Species selection needs to be planned carefully and not let by commercial consideration. In this matter women need to be consulted and their participation becomes necessary.

4. Matrix Scoring

Matrix ranking and scoring is used to find out local attitudes and perceptions. These relate to soil and water conservation measures, varieties of cereals, types of fertilizers, trees, vegetables, fruits, wood and income earning opportunities and so on. In this method finding out one or more informants who are knowledgable and willing to participate and discuss the problem under consideration. In the course of discussion with such people, the concerned officials beside giving the rank or score of different items like fodder, trees, fuel wood, fruits etc. the informants are asked to rank or score each item to spell out first rank and second rank and so on. The highest score indicate the most preferred one.

5. Historical Profiles/Time Trends

The objective here is to understand key changes over generations in land use, rainfall, erosion, population, tree cover, income, opportunity time common property resource and so on. Required information may be obtained by asking older people in the community to describe changes over a period of time since their younger days.

6. Seasonal Calendars

Seasonal calendars are used for understanding of local livelihood systems. They show the patterns month by month of rainfall, crop sequences, water use, fodder availability, income, debt, migration, labour demand, labour supply, health and diseases, prices of different products and so on. This type of local calendar is helpful to take JFM plan to ensure at least some produce is available all thorough the year. Here we go by local calendar and not regular calendar months. Women are associated in collecting information in the preparation of seasonal calendar.

7. Venn Diagrams

Venn diagrams are drawn to help understand the current formal and informal institutions in the village and their roles in community development. They highlight gaps between institutions, opportunities for better communication and cooperation to resolve conflicts and some times the need for a new role for outside agencies in the villages. While drawing Venn diagrams we use circles of different sizes drawn on paper or cutout, each representing different institutions. We then ask informants to choose large circles for the most preferred institutions, small circle for less preferred institutions. The circles are to be placed and analyzed in the context of the village. In that exercise the villagers are asked to identity the role of different institutions and their interaction with the people.

8. Conducting Participatory Rural Appraisal (PRA)[12]

The team conducting PRA may consist of 4 to 5 members from different disciplines. It is necessary that there is representation of women as well as experts from government departments like soil conservation forest managements, animal husbandry. The team members must have clear understanding of PRA and the micro plan and identify methods and procedures to be followed: Necessary information about the village chosen for study and the forest area need to be obtained from the local Mandal office and from Forest department. Informal consultation with all concerned is necessary before the visit of the team to conduct PRA. On reaching the concerned village, the team members need to be introduced to the villagers, explain the purpose of the visit and seek their cooperation. The work of PRA begins with village and social map. All the PRA maps, drawing and information obtained, should be noted with explanatory notes after obtaining required information, Microplan is prepared involving villages at all stages. Team members should identify themselves as one with the villagers and they should not begin sermonizing or lecture. On the other hand let villagers lead the discussions. The participants must necessarily be representing cross section of the village. The villagers must be given an opportunity to come forward with practical suggestions, plan for the development of their own village.

Conclusion

The scheme of joint forest management and wild life care needs to the examined from different angles. The policy of forest protection and management should not in any way adversely affect the livelihood prospects of the poor. The objective of enhancing green cover must necessarily be integrated with livelihood opportunities.

JFM is the right forum which can effectively supervise and well manage the forest cover with integrated approach taking care of the interest of livelihood opportunities of the poor. The concerned village Panchayat need to be associated with JFM as well as in the PRA Programme. The concerned state Governments must be associated with JFM as well as PRA. The JFM need to be well strengthened to enable it to cover meaningfully the areas of wild life management, animal welfare, keeping in view the livelihood of the poor and without degradation of forest and wildlife. The

Biodiversity Act and JFM need not clash, if only both are committed to the common goal of ecological security, food security and livelihood security. The biodiversity act envisages action at 3 levels: (1) At the Panchayat level, biodiversity committee is responsible for conservation and for operationalizing the programme, (2) The state level biodiversity Boards and (3) The National level biodiversity authority. These three units should ensure that all development programmes are subjected to a biodiversity impact analysis linking economic advance with biodiversity loss. The biodiversity act and its provision should not ignore JFM, since both have common aim of ecological security, food security and livelihood security. The PRA is equally interested in the overall ecological security, food security and livelihood security. The biodiversity act, JFM and PRA schemes should function with proper understanding with the common goal of ecological security, food security and livelihood security.

REFERENCES

1. Centre for Environment Concerns, Hyderabad on Joint Forest Management in Andhra Pradesh.
2. *Eleventh Five Year Plan Report,* p. 193.
3. Centre for Environmental Concerns, Hyderabad on Joint Forest Management in Andhra Pradesh, p. 12 & 13.
4. *Eleventh Five Year Plan Report.*
5. *Eleventh Five Year Plan,* Chapter IX, "Environment and Climate Change", p. 193.
6. Eleventh Five Year Plan, *Ibid.,* p. 65.
7. *Ibid.*
8. Sathis Chandra Nair and N.D. Jayal, Forest Policy, *The INTACH Environmental Series,* p. 4.
9. Swaminathan, M.S., Biodiversity, Development, Livelihoods", *The Hindu,* Saturday, May 22, 2010.
10. Centre for Environmental Concerns, *op. cit.,* pp. 10-16.
11. *Ibid.,* p. 21.
12. *Ibid.*

11 ENVIRONMENT AND CLIMATE CHANGE

1. Introduction

Until recently developing countries generally regarded environmental quality as a luxury that could be afforded only after they had attained considerably higher level of economic development. Views have changed dramatically for good. Many development planners now recognize that sound management of the environment is a necessary component of economic development, not an obstacle to it. Today, almost all countries have Environment Ministries, well committed to safeguard the environment.

2. Meaning of Environment and Environmental Concerns

Environment in its broad sense refers to all those physical and social conditions that affect the behaviour and development. Environment encompasses the natural world covering people, animals, micro organisms, and plants, etc. Basic resources like land, atmosphere, water, assume critical importance in Environment. Environment concerns encompass a broad range of issues including, control of air, water and land pollution, sound management of renewable natural resources, efficient use of natural resources, recycling and erosion control; Public health and occupational safety and conservation of unique habitats; especially rare and endangered species. The environmental problems faced in the developed countries, are also found in the developing countries. Their relative importance varies depending on local circumstances, the country's state of development and the resource base. The problems often differ more in degree than in kind.[1] The industrial pollution is more pervasive in well developed industrial countries. Similarly industrial pollution and other environmental issues are emerging as a problem in selected cities of emerging countries like India.

3. Deteriorating Trend of Environment

Research studies reveal that environmental quality is deteriorating both in developed and developing countries and that adversely affects the human welfare posing threat to sustainable development. Among others, most serious environmental problems, particularly in developing counties are those relating to land, pressure on the cropped area and overgrazing by livestock; cultivation of unsuitable soils; large scale deforestation; air and water pollution; development-related pressures on the stock of wildlife; increased poverty; industrialization and urbanization; fast declining ethical values; poor awareness of the need for environmental safety.

The environmental problem centered round land, are severe in developing countries, where most people make their living directly from the land. Environmental degradation adversely affects all sections of society, more particularly the poor. Poverty, unemployment, and population pressures often compel the poor to resort to unsound farming, grazing or fishing methods or to settle on ecologically fragile, marginal lands. It is rightly observed that "a vicious cycle of increased poverty and further ecological degradation, often follows, in which potentially renewable resources including top soil and fuel wood, are depleted for current use."[2]

Problems arise when tribal people or vulnerable ethnic minorities are disturbed from their traditional pattern of life. They are in a way custodians of the forest resources and source of information concerning the practical uses of little known plant and animal species.

Now the major issue is how to avoid environmental damage or reduce it to an acceptable minimum without showing the pace of development. The environment understandably, constitutes "natural capital"; its benefits, water flow, soil protection, breakdown of pollutants – support and enhance economic development. When poorly planned it may result in the depreciation of country's national capital even when carefully planned, the process of economic development inevitably causes some damage to the natural ecological system and generate wastes and pollutants. The objective of environmental management should be to achieve a balance between human demands on the natural resources base from the present as well as future generations, and the environment's ability to meet these demands.[3]

In almost all the countries, the objective of planning is sustainable development. This demands considerable change in attitude towards the environment. The policy measure and people at large, must be well committed to the environmental protection and achieve ecologically sustained development.

4. Environmental Protection

In environmental protection strategy, permeation is more important and less costly than remedial action, which sometimes may not be feasible at all. Remote Satellite technology need to be used to assess the environmental potential. Like data on agricultural production, foreign trade, tax revenue and national income, data on the environment should become a regular input into development planning. That way appropriate strategy for environmental protection can be ensured.

5. Climate Change

Climate change is possibly the greatest global environmental threat the world is facing. The nature and impact of the climate change needs to the analysed and appropriate remedial measures have to be initiated immediately.

Climate change refers to a change in the state of climate than can be identified by changes in the mean and or the variability, of its properties, and that persists for an extended period, typically decades or longer. Climate change may be due to natural internal processes or external forces or to persistent anthropogenic changes in the composition of the atmosphere or in land use (IPCC). The impact of a changing climate can be noticed in the form of more droughts, more floods, more strong storms and more heat waves – drawing resources away from development. World Development Report[4] highlights the concept of Sustainable Development. Sustainable Development is viewed "as Development that meets the needs of the present without compromising the ability of future generation to meet their own needs". World Development Report predicts that by the present century end, it could lead to warming of 5°C more compared with pre industrial times. This predicted climate change under various conditions is likely to have implications on the food production, water supply, settlements, health, energy security, etc. (World Bank, 2000).

6. Impact of Climate Change in India

India is a large developing country with nearly 700 million rural population directly depending on climate sensitive sectors, (agriculture forests and fisheries) and natural resources (such as water, biodiversity mangroves coastal zone, grasslands) for their subsistence and livelihoods. Their adaptive capacities to the negative impact of climate change are said to be very low. Recognizing this, Indian National Active Place for Climate (INAPC) 2008, identifies development and poverty eradication as the best forms of adaptation to climate change. In developing countries, adaptation responses are closely linked to developmental activities. Adaptation to the climate change involves deliberate adjustments in natural or human systems and behaviours to reduce these risks to people's lives and livelihoods. Mitigation of climate change involves actions to reduce greenhouse gas emissions in the short-term and development choices that will lead to low emissions in the long-term.

7. Climate Change and Ecosystems

Climate change is likely to impact all the natural ecosystems as well as socio-economic systems as shown by the National Commission Report of India of the United Nations Framework Convention on Climate Changes (UNFCC) as indicated below.

(i) An annual mean surface temperature may rise by the end of century ranging from 3 to 5 degrees celsius.

(ii) A 20% rise in All-India summer monsoon rainfall and further rise in the rainfall is projected over all States except Punjab, Rajasthan and Tamil Nadu, which show a slight decrease.

(iii) Extremes in maximum and minimum temperatures are also expected to increase particularly over the west coast of India and west central India.

Floods, droughts and cyclones are the mean extreme climatic events in India. The total flood-prone area is said to be 12.16% of total area, and 12% of the population is affected by drought annually. Western parts of Rajasthan and the Kutch region of Gujarat are chronically drought affected. Drought conditions have also been reported in Karnataka, Andhra Pradesh, Orissa and Bihar States (Sharma and Smakhtin, 2004).

8. Some Projections on Climate Change Impact

1. More recent studies suggest a 2 to 5 per cent decrease in yield potential of wheat and maize for a temperature rise of 0.5 to 1.5°C in India. (Agarawal, 2003).
2. Agriculture will be adversely affected not only by an increase or decrease in the overall amounts of rainfall, but also by shifts in the timing of the rainfall.
3. Agriculture will be worst affected in the coastal regions of Gujarat and Maharashtra, where agriculturally fertile areas are vulnerable to inundation and salinization.
4. In Rajasthan, 2°C rise in temperature was estimated to reduce production of pearl millet by 10-15% (Ramakrishna *et. al.,* 2002).
5. Changes in the soil, pests and weeds brought by climate change, will also affect agriculture in India (Teri, 2002).
6. All along the western Indian Coast line, tropical eco systems and species such as mangroves coral reeds are threatened by changes in temperature, rising sea level and increased concentration of carbon dioxide in the atmosphere.

9. Adaptation to Climate Change[5]

Some specific reasons are listed justifying adaptation to climate changes as follows:

(i) Climate change cannot be totally avoided.

(ii) Anticipatory and precautionary adaptation is more effective and less costly than forced, last minute emergency steps.

(iii) Climate change may be more rapid and more pronounced than current estimates suggest. Unexpected events are possible.

(iv) Immediate benefits can be gained by removing maladaptive polices and practices.

(v) Climate change brings opportunities as well as threats.

The World Development Report very rightly indicates the responsibilities of both high income countries and of developing countries. The high income countries the Report says, can and must reduce their carbon foot prints. They cannot continue to fill up an unfair and unsustainable share of the atmosphere pollution. The developing countries in the process of development must be equally cautious. What is important for all the countries is one of pursuing growth and prosperity without causing dangerous "climate change".

10. Climate Change and Sustainable Development

Development that is socially, economically and environmentally, sustainable is a challenge. Economic growth alone is not enough if it does not reduce poverty and increase the equality of opportunity and failing to safeguard the environment eventually threatens the economic and social achievements.

India's post-1980 deceleration of rice productivity is said to be attributable not only to falling rice prices, but also to adverse climate phenomena from the local pollution and global warming.[6] Today's global warning was caused overwhelmingly by emission from the rich countries.[7] It is moral and ethical responsibility of the developed. Countries to take a lead in ensuring that the level of warming is kept within 2°C of the pre-industrial level, by willingly committing themselves to deep, drastic and legally binding cuts in their greenhouse gas emissions. Similarly it is the moral and ethical responsibility of developing countries to ensure that their right to development is exercised within the framework of sustainable development. The right to development is not a license for deforestation or unabated use of fuel, energy and technologies that exacerbate climate change. Among the priority measures are: the shift to sustainable agriculture, sustainable forestry, and sustainable transport systems with the caveat that they should achieve required food production without damage of forest lands.[8]

REFERENCES

1. Warren C. Baum and Stocks M. Tolbert, *Investing in Development,* Oxford University Press, 1985, p. 523.
2. *Ibid.,* p. 525.
3. Inter-Governmental Panel on Climate Change (IPCC), Climate Resilient Development – Synthesis Report, Juriss Agency for Development and Cooperation, p. 7.
4. *World Development Report,* The World Bank, p. 1.
5. Climate Resident Development – Synthesis Report, *op. cit.,* p. 9.
6. *World Development Report,* p. 4.
7. *Ibid.,* p. 55.
8. Towards Climate Justice, Social Equality and a Justifiable Future, *Climate Action Network Southeast Asia (CANSEA),* p. 3.

12 ETHICS, ENVIRONMENT AND SUSTAINABLE DEVELOPMENT

1. Introduction

The need of the hour is not just development, but eco-friendly development, development without destruction of the environment. Now all countries all over the world are interested in sustainable development. The concept of sustainable development is viewed from different angles and concept is defined in different ways. Of all the definitions of the concept of 'Sustainable Development', the most acceptable one is that given by World Commission on Environment.

2. Definition of Sustainable Development

The World Commission on Environment and Development, 1987 (Bruntland Commission) viewed, "the sustainable development is development that meets the needs of the present, without compromising the ability of future generation to meet their own needs." This definition looks at sustainability in a dynamic framework and it is now widely accepted definition. It is both a call and caveat that every generation has the responsibility of using life supporting resources like land, water, forest resources etc., rationally in meeting their present requirements and then passing on the same to future generations without damaging the carrying capacity of the resources.

3. Ethics and Sustainable Development

Sustainable Development is ethical in nature, because using the basic resources rationally now and passing them on to future generation safely is certainly an act of good behaviour on the part of an individual. In other words, it is an ethical behaviour. Protection of environment certainly is an ethical behaviour of great appreciation because sustainable development is an act of responsibility in between two generations. The implication is that we can conceive practices for managing the resources well, taking care of the present as well as the future generations and this is certainly a moral step or ethical in nature of high degree. According to the Indian Tradition, right from the Vedic times, the environment and life of people formed an inseparable part of the living system. The flora and fauna of the country formed an integral part of the spiritual tradition in India and they were looked upon as a part of Nature which they respected, in fact, worshipped as a custom and sacred duty.

Gradually, so manly factors had set in and relationship of man with nature has taken a different turn away from the principle of Ethics. With Industrial Revolution from the later part of 18th century in England and with foreign rule in India spread over 200 years, the traditional ways changed. Industrial revolution of the west had its impact on India too.

4. Degradation of Environment

With industrialization, began gradual degradation of natural resources and environment. The concentration of industrial centres in selected cities, has attracted rural labour and there has been gradual exodus of rural population to urban centres, creating environmental problems. Apart from this, the problem of population explosion coupled with poverty of high magnitude contributed for environmental degradation.

Environment covers flora and fauna, forests, bio-diversity, wildlife, greenery around, pollution of different kinds. There seems to be gradual deterioration in the ethical standards at different levels. The economic growth is found to be associated with certain environmental problems. Apart from environmental problem arising from the greenhouse gases and global warming, associated with industrialization and urbanization, the society at large faces environmental problem arising from damage to the basic life supporting resources like land, water, forest, bio-diversity, oceans and atmosphere. Research studies reveal that the human population is growing at a rate often exceeding the population supporting capacity of the ecosystem. It is also noticed that the rich-poor divide is widening leading to the extensive co-existence of unsustainable lifestyles and unacceptable poverty.[1] In the name and process of development, there has been over exploitation of forest resources; the poor regularly exploit forest for firewood for domestic use as well as for making a living, clandestine timber trade continues, damage to bio-diversity and wildlife is on a large scale. Major irrigation projects cause heavy damage to forest and that way create environmental problem. Irrigation tanks are encroached upon, felling of trees and destruction of greenery in the name of development; excessive exploitation of ground water through tube wells or by way of deepening of existing wells has become the order of the day all leading to destruction of forest resources and environmental degradation.

Introduction of green revolution technologies during mid 1960s, necessitated the use of high cost external inputs like machinery, high yielding variety of seed, hybrid seed, chemical fertilizer and pesticides and resorting to monocropping in the place of multiple cropping. Livestock-linked farming almost disappeared resulting in the use of only external inputs. The cultivation has become very costly and often farmers are forced to incur losses both ecologically and economically. With excessive application of chemical fertilizer, the soil is degraded. Practices like application of tank silt, organic manure have almost disappeared leading to loss of soil fertility with all adverse consequences of declining yields, apart from great damage to environment.

Many right thinking persons, who were very much concerned with the fast deteriorating environment, viewed environmental protection as a moral responsibility. Ethics now is linked with environment and development.

5. Gandhiji's Concern for Environment

Gandhiji all through his life, in word and deed, had been guided by the ethical values. In his writings he makes mention of seven sins as stated in earlier chapters. Gandhiji's concern for environmental preservation and ecological balance is evident from his "Sarvodaya Principle" – Welfare for all, not only all humans but also to the entire creation, *"Sarvejana Sukhinobhavanthu, lokasamastha sukhinobhavanthu"* is the vedic aphorism. Environmental preservation received special attention of Gandhiji. He desired planting of shade giving trees all around the village, preserving land for grazing and protecting the forest, should be taken as a community responsibility. He wanted people to live in harmony with the nature and cautioned of impending danger if forests are destroyed. To ensure social harmony, he insisted on revival of traditional cultural activities and regular meetings at the community halls to deliberate upon common problems. His emphasis was on simple living. The goal of any development, according to Gandhiji should be welfare of the poorest and weakest guided by the principle of "Antyodaya", putting the last "first". At the same time, he felt that the development is only to meet the needs of the people and not their greed. Gandhiji proposed Trusteeship Ideology.

Those who have surplus resources and wealth more than the minimum needs, it is suggested, to handover the same to the Trust voluntarily, and the earnings from such Trust, should be utilized for the welfare of the poor and the needy. He warned that if inequalities of income are wide, it would lead to social unrest. Gandhiji's stress on "production by masses" and not "mass production", making use of the best of the modern knowledge and native experience is compatible with the loss of ecology as rightly pointed out by Schumacher in *"Small is Beautiful"*. Gandhiji's stress on minor irrigation and cottage industries clearly indicates his concern for environmental preservation.

6. Ethical Basis of Permaculture

Mallison[2] in his monumental work *Permaculture* speaks of the ethical basis of Permaculture in three major principles.

(a) Care of the Earth: Provision for all life systems to continue and multiply.

(b) Care of people: Provision for people to access those resources necessary to their existence.

(c) Setting limits to population and consumption: By governing our own needs, we can set resources aside to further the above principles. This ethic is a very simple statement of guidance relating to relationship of man with nature. Mallison says Permaculture is a philosophy of "working with rather than against nature". Mallison[3] indicates some rules of use of Nature. (1) One rule of guidance is the "Rule of necessitous use" — It simply means we leave any natural system alone, until we are, of strict necessity, forced to use it; (2) Another rule is Rule of conservative use. It means, having found it necessary to use a natural resource, we may insist on every attempt to: (i) thoroughly replace lost minerals or any other natural resource and (ii) make an assessment of the long-term, negative, bio-social effects on society and act to buffer or eliminate these Mallison insists that we release much of the landscape for the sole use of the wildlife and for re-occupation by endemic flora. "Respect for all life forms is a basic and infact essential ethic for all people."[4]

7. Environment and Government

The very fact that a separate Ministry of Environment and Forests is formed at the Centre indicates importance attached to the Environment by the Government. The objectives of the Ministry are[5]:

(i) Conservation and survey of flora, fauna, forests and wildlife,

(ii) Prevention and control of pollution

(iii) Afforestation and regeneration of degraded areas

(iv) Protection of environment and

(v) Ensuring the welfare of animals.

These objectives are well supported by a set of legislature and regulatory measures aimed at the preservation, conservation and protection of environment. National Forest Policy of 1988, Policy Statement on Abatement of Pollution 1992, and a National Environment Policy, 2006, clearly indicate the concern of the government and steps in this direction.

Forest resources survey is undertaken regularly and steps are taken to ensure the protection of forest area. As per the state of forests report 2005, forest cover in India is only 20.60 per cent of the total geographical area of the country. This falls short of the optimum level of one-third of geographical area under the forest cover. Attempts are initiated for joint forest management associating the participation of local people and community in protecting the forest cover.

8. Wildlife Conservation and Environment

Wildlife conservation is attempted as a part of environmental protection. India is a member of World Heritage Convention responsible for listing of world heritage sites, which include both cultural and natural sites. Wildlife wing of the Ministry of Environment is associated with conservation of the Natural World Heritage sites.[6]

The concern for environmental quality has become the utmost issue in the context of increasing urbanization, industrial and vehicular pollution as well as pollution of water courses due to discharge of effluents without conforming environmental norms and standards.

Hazardous Substances Management Division (HSMD) is engaged in activities such as framing necessary rules relating to environmentally sound management of hazardous wastes/chemicals, plastics and municipal wastes.

National River Conservation Directorate (NRCD) under the Ministry of Environment is functioning with the objectives of: (1) improving the water quality of the rivers and (2) construction of low cost sanitation toilets, construction of electric crematoria, the water quality of river Ganga is being monitored at 27 locations, the quality monitoring rivers like Yamuna, western Yamuna Canal, Gomati, Sutlej (Punjab), Cauvery (Tamil Nadu), Tungabhadra in Karnataka and waterways of Chennai.

The National Wetlands conservation programme initiated in 1987, National Afforestation and Eco-development Board set up in 1992, G.B. Pant Institute of Himalayan Environment and Development established in 1988, Wildlife Research Programme, Wildlife Education and Training Programme, Fellowship Scheme of Environment and Biodiversity, Forestry Education, Training and Extension programme etc., are all actively involved in environmental protection.

With environmental degradation, soil and land are adversely affected. In this connection, watershed development programme was initiated. The programme began functioning effectively with the participation of stakeholders and the community at large. Hence, participatory watershed programme was designed and is now in operation.

9. Participatory Watershed Programme for Environment Protection

Soil erosion and land degradation coupled with declining per capita availability of land and water are posing serious threat of environment. Hence careful and concerted efforts are needed for efficient and effective management of natural resources for increased productivity of the soils. Watershed development projects have been launched by several government and non-government agencies. The objectives of the scheme are: soil conservation, improving the land productivity and promoting appropriate technologies for efficient and sustainable use of natural resources. However many watershed projects around the world have not performed well because of the poor community participation.[7] It is rightly pointed out that for achieving the desired participation of people, the roles of community organization, groups and other stakeholders are crucial. Local people must play an active role, starting from project design, moving to implementation and the project maintenance. In this context, a participatory watershed management approach is considered as the ideal for achieving food security and sustainability.[8]

India began watershed development programme in the 1970s, addressing problem of controlling land degradation and increasing productivity of soils. In 1994, the Ministry of Rural Development issued guidelines for implementing watershed programme in the country. Now the guidelines aimed to tackle the concerns related to the realization of the full benefits of watershed work. The revised policy was essentially people-centered aiming awareness raising with NGOs and community participation. However, research study reveals that "there was a lot of mismatch between the needs of the stakeholders and the activities for implementation of watershed development. Such watershed projects often failed to achieve the intended targets in the absence of people's "participation."[9] A case study[10] identifies the impact of watershed programme in terms of five capital assets as follows. The impact of

- **(a) Physical Capital:** There was a positive impact on physical capital in terms of enhanced availability of irrigation in the study area.
- **(b) Natural Capital:** Impact on this capital was positive. The value of land assets has gone up; there was improvement in the fodder availability; increased availability drinking water and the proportion of area under irrigation has increased marginally.
- **(c) Human Capital:** Impact on this capital asset was little though the expenditure on education has increased significantly.
- **(d) Social Capital:** The study attempted to examine impact on social capital in terms of migration and gender. It was reported that migration of people was less during the implementation period of the programme. The other positive impacts were strengthening the self-help groups (SHG) through better employment and wages for woman.
- **(e) Financial Capital:** Impact on this capital asset was positive in terms of increased land productivity, increased employment and increase in household income.

The Participatory Watershed programme, wherever it was implemented with the participation of local people and stakeholders, had yielded good results in terms of diversification of rural economy, increasing cropping intensity, improved fodder production, rising water table increased availability of drinking water ensuring sustainable rural livelihoods. There is a strong case for effectively implementing this programme, particularly in areas of rainfed cultivation.

10. Eleventh Five Year Plan and Environmental Protection

Protection of the environment has become a central part of sustainable development strategy in the Eleventh Plan. The Eleventh Plan proposes integrating environmental considerations into policy making in all sectors of the economy. The plan recognizes the need for setting up on independent statutory body on sustainable development with the specific responsibility of guiding government policies and programmes for making them socially and environmentally sustainable. It is indicated that this body should comprise eminent environmental experts and citizens well versed in environmental research. It is also proposed that the State Pollution Control Boards should be restructured into statutory Environment Protection Authorities with the mandate of framing regulations for environmental protection. At the district level, the scheme of Paryavaran Vahinis, or committees of concerned citizens, should be revived to serve as environmental watchdogs, monitoring the environmental situation in the districts. It is also proposed to associate local self governments in different programmes associated with environmental protection. The Eleventh plan gives importance to the programmes of afforestation and Joint Forest Management. The plan fixes the target of 33% of forest cover. Towards this end, under NREGP plantation in the waste lands of the reserved forests is undertaken. There are also proposals for prevention and control of air pollution. Vehicular pollution is a major source of air pollution in the major cities of the country and therefore, the plan proposes to give priority in planning for a clean urban environment.

It is proposed to take appropriate steps in the direction of water quality. The different standards of water quality, required for drinking, bathing and irrigation will be identified and steps will be taken to ensure the quality of water.

Untreated sewage dumped into our rivers is a major cause of river pollution. The total sewage generation in the country is about 33,000 MLD (million liters per day) Against this, the total average treatment capacity is only 6,190 MLD and 40% of that capacity is in Delhi. This underlines the urgent need to expand the sewage treatment capacity in the country.[12]

11. National Lake Conservation Plan (NLCP) and Environmental Protection

The NLCP came into operation during the Ninth Plan as 100% centrally funded scheme. The objective was restoring the polluted and degraded lakes of the country. By 2007, the scheme covered 10 lakes and 10 more are likely to be completed by very soon. The activities covered under NLCP include interception and treatment of pollution loads entering the lake, lake cleaning such as desilting, deweeding, catchments area treatment, lake front eco-development like bunding, fencing and creation of facilities for public recreation and entertainment.

12. Disaster Management and Environment

The Tenth Five Year Plan (2002-07) recognized disaster management as a development issue for the first time and accordingly initiated steps both for prevention and mitigate the adverse effects of different types of disasters. The National Disaster Management Authority (NDMA) under the Chairmanship of the Prime Minister has been set up as an apex body responsible for laying down of policies, plans and guidelines on disaster management so as to ensure timely and effective response to disasters. It is evident that from time to time, different schemes/programmes have been initiated with financial allocation for protection of environment viewing the problem as a development issue.

Table 12.1 presents selected Schemes/Programmes in the Tenth Five Year Plan with details of outlay and actual expenditure.

Table 12.1

Financial Allocation and Actual Expenditure in the Tenth Plan on Schemes/ Programmes of Environmental Protection (2002-07)

(In Crores)

S.No.	Schemes/Programmes	Outlay	Actual Expenditure
1	2	3	4
1	Central Pollution Control Board (CPCB)	100.00	140.06
2	Industrial Pollution Abetment through Preventive Strategies	5.00	1.48
3	Common Effluent Treatment Plants (CETP)	25.00	20.10
4	Environmental Management in Heritage Pilgrimage and Tourism Centers including Taj Protection	170.00	25.00
5	Environmental Health	10.00	1.03
6	Environmental Impact Assessment (EIA)	13.00	12.35
7	Industrial Pollution Prevention Project (IPPP)	10.00	13.55
8	Hazardous Substances Management	70.00	30.71
9	G.B. Pant Institute of Himalayan Environment and Development	35.00	37.00
10	Conservation and Management of Mangroves, Coral Reeds and Wetlands	54.00	57.11
11	Biodiversity Conservation	12.00	15.51
12	Research and Development	24.00	20.78
13	Environment Education and Awareness	125.00	118.16

14	National Natural Resource Management System (NNRMS)	7.00	22.74
15	Environment Management Capacity Building Projects (EMCBP)	48.98	46.14
16	Integrated Forest Protection Scheme	445.00	208.01
17	Development of National Parks and Sanctuaries	350.00	234.65
18	Wildlife Institute of India (WII)	50.00	46.07
19	Biodiversity Conservation and Rural Livelihood Improvement Project	N.A.	2.03
20	Protection of Wildlife outside Protected Areas	60.00	0.00
21	Total Forests & Wildlife	1600.00	1287.53
22	National Afforestation Eco-development Board (NAEB)	80.00	71.60
23	National Afforestation Project (NAP)	1115.00	1179.07
24	National Action Programme to Combat Desertification	30.00	0.00
25	Eco-development Forces	75.00	42.96
	Grand Total	**5945.00**	**5119.14**

Source: *The Eleventh Five Year Plan Report,* pp. 215-217.

From Table 12.1 it is evident that the total expenditure on different schemes/programmes falls short of the outlay to the extent of ₹ 826 crore. The actual expenditure is less than allocation of difficult schemes/programmes in the case of Industrial pollutions, CETP, Environmental Management, Environmental Health, Hazardous substances, Management, capacity building project, Integrated Forest Protection Scheme, Development of National Parks and Sanctuaries, WII, Forests and wild life, NAEB, and Eco-development Forces.

In the case of schemes like protection of wildlife and programme to combat desertification, disappropriately no amount is spent. However, it is encouraging that in the case of certain schemes/programmes, the actual expenditure exceeds the outlay on the particular item.

Conclusion

The problem of environment is a burning problem, problem not only of this generation, but also of the future generations. Sustainable development requires proper care and protection of all the basic resources like, land, water, forest resources, atmosphere and climate change. No doubt, the government has the responsibility of protecting environment, paving the way for sustainable development. At the same time, every citizen has his or her own responsibility of taking care of the earth and contribute for desirable environment. Any problem associated with environment, is obviously a problem created by man and not an outsider. Jiddu Krishnamurthy (J.K) rightly states that it is the individual who creates the problems and the solution lies in the creator of the problems; J.K. was categorical in his observation, "You and I are the problem and not the world, because the world is the projection of ourselves and to understand the world, we must understand ourselves. The world is not something separate from you and me. To transform the world, we must begin with ourselves". We have to understand for ourselves, our relationship with Nature and our responsibility for the Nature, or environment of what it is. If each individual acts with a sense of responsibility, keeping self or personal interest aside, certainly we can take care of the environment. Gandhiji said long back, that we have enough resources to meet our needs, but not greed. As Mallison pointed out we have to regulate for ourselves our own consumption and that way take care of the earth and environment. This is the way, right approach and right direction for sustainable development. Conducting ourselves on the right lines and in the right direction is a simple ethical practice one can easily follow. Ethics, Environment and Sustainable Development should be viewed as an integrated strategy and everyone should act with a sense of responsibility and this is the right step consistent with Indian culture and heritage.

REFERENCES

1. Swaminathan M.S., *A Century of Hope,* Harmony with Nature and Freedom from Hunger, East-West Books Pvt. Ltd., Chennai, 1999.
2. Malison, *Permaculture,* The Deccan Development Society and Permaculture – India, October, 1990, p. 2.
3. *Ibid.*, p. 3.
4. *Ibid.*
5. *India 2009* on Environment, Publication Division, Ministry of Information and Broadcasting, Government of India, pp. 282-320.
6. *Ibid.*, p. 295.
7. Johnson N., *et. al.,* "Addressing Water Needs of the Poor in Watershed Management and Research", Internet site.
8. Budumuru Yoganand and Tesfa G. Gebramadhin, "Participatory Watershed Management for Sustainable Rural Livelihoods in India", Research Paper presented at Southern Agricultural Economics Association Annual Meeting, Florida, Feb. 5-8, 2006.
9. *Ibid.*
10. Reddy, V.R. *et. al.,* "Participatory Watershed Development in India: Can it Sustain Rural Livelihoods?", Development and Change, 2004.
11. *Eleventh Five Year Plan Report,* Section 9, p. 191.
12. *Ibid.*, p. 197.

PARTICIPATORY WATERSHED MANAGEMENT FOR SUSTAINABLE RURAL LIVELIHOOD

1. Introduction

Proper Management of Natural Resources aiming at protection of environment and food security is the need of the hour. From time to time, different schemes have been grounded in the direction watershed management is one such scheme addressing specifically water management. Watershed Management approach aims at well streamlining the water sources without causing environmental degradation. Rain water runs-off in different directions, often unutilized or under utilized. Watershed management scheme attempts to locate right place, channelise the run-off water into one stream, preserve at one well defined place. In so doing, water is well preserved and groundwater potential in the neighbouring is improved. The water so conserved becomes an assured source of irrigation for farming, ensuring livelihood.

2. Definition of Watershed

Watershed is defined as "Natural hydrologic entity that covers a specific area of land surface for which the rail fall run off flows to a defined drain channel stream or river at any particular point". India began experimenting with this scheme in 1970s, for controlling land degradation and increasing the productivity of soils. With experience gained, changes have been introduced from time to time. Initially, watershed projects concentrated on the soil and water conservation issues. A decade later, it was realized that physical works alone would not lead to the desired objectives of watershed development and it must also take into account the social, financial and institutional aspects of rural development. India is a vast country with heterogeneity of agro-climatic characteristics across different regions.

A uniform type of watershed may not function well. Depending upon the size of the watersheds, they are broadly divided into Micro and Macro watersheds. Watershed with areas upto 1250 hectares were classified micro watersheds and those with still more extent of area were classified under macro watersheds.

The experience with the functioning of watersheds in the earlier stages revealed that the scheme should not be simply of a government. On the other hand it should be of participatory nature. In 1994, the Ministry of Rural Development, Government of India issued a set of guidelines for implementing the watershed programmes. The direction from the Government was that the scheme must be essentially a people-centered, participatory involving all stakeholders and NGOs.

3. Paradigm Shift in Watershed Management

In the beginning, resources were allocated by the central and state governments for watershed development. This top down approach was found to be not conducive for including the stakeholder's participation in designing the programme. There was a lot of mismatch between the needs of the stake holders and the activities for implementation of watershed programme. Such watersheds failed to achieve the intended targets in the absence of people's participation. Realizing this gap, participatory watershed management has emerged as a new paradigm for watershed development in India. This new approach of decentralized nature with increased participation of all stakeholders and community at large, is expected to rise to the occasion and face the challenges effectively by strengthening the capacity of local people.

4. Participatory Watershed Management: Defined

The concept of 'participatory watershed management' emphasizes an inter-disciplinary, inter-sectoral and multi-institutional mechanism (Rhoades All Elliot) Participatory watershed management has been defined as a process which aims to create a self supporting system which is essential for sustainability (Wani *et.al,* 2005). The need for roles of community organizations and stake holders is said to be essential. The participatory watershed management approach is considered as the ideal for achieving food security and sustainability[1].

Participatory watershed management provides opportunities to the stakeholders to jointly negotiate their interests, set priorities, evaluate, opportunities, implement and monitor the outcomes (Yogananda and Gefremedhin, 2006).

In India, participatory watershed management has roots in the non-government sector. The scheme of participatory watershed management took a shape in a small village called Ralegan Siddhi in Maharashtra state of India with the initiative by Anna Hazare. He was responsible for bringing many social changes in the village particularly soil and water conservation measures; family planning; a ban on alcohol, and felling of trees; protection of grazing land and voluntary labour for community welfare and other measures which helped in restoring natural resources base of the village (Kere *et.al.,* 2002) This ultimately led to people participation in watershed management and the evolution of participatory management looking beyond bio-physical aspects to social and institutional aspects following a bottom up approach. It is now widely accepted that the communities must participate to enhance the productivity of natural resources in a sustainable fashion (Turton *et al.,* 1998).

Management of watershed development had undergone changes from time to time. From the main concentration on biophysical Criteria during 1970s, the Scheme turned towards improving the providing of natural resources. With clear guidelines from the Ministry of Rural Development in 1994-95, the scheme begin looking beyond soil and water conservation, to include environmental protection, optimum utilization of watershed's natural resources, employment generation and betterment of social conditions of the rural poor. The guidelines clearly stated that the District Rural Development Agency (DRDA) or concerned Zilla Parishad was responsible for programme implementation. Partnership based community participation is central in the programme. The guidelines also encourage the involvement of Users Groups (UGs) and Self-help Groups (SHGs).

5. Impact of Watershed Projects

A case study identifies the impact of water shed programme in terms of five capital assets as follows: The impact of

(a) Physical Capital: There was a positive impact on physical capital in terms of enhanced availability of irrigated in the study area.

(b) Natural Capital: Impact on this capital was positive. The value of land assets her gone up; there was improvement in the fodder availability; increased availability of drinking water and the proportion of area under irrigation has increased marginally.

(c) **Human Capital:** Impact on this capital was little through the expenditure on education had increased significantly.

(d) **Social Capital:** The study attempted to examine impact on social capital in term of migration and gender. It was reported that migration of people was less during the implementation period of the programme. The other position impact was strengthening the Self-help Groups (SHGs) through better employment and wages for women.

(e) **Financial Capital:** Impact on this capital asserts was position in terms of increased land productivity, increased employment and increase in household income.

The participatory watershed programme, wherever it was implemented with the participation of local people and stakeholders had yielded good results in terms of diversification of rural economy, increasing cropping intensity, improved fodder production, rising water table, increased availability of drinking water ensuring sustainable rural livelihood.

Many studies on the performance of the scheme reveal that watershed projects had a positive impact on crop productivity. With increase in the irrigated area under the scheme assured, crop production became a reality. Encouragingly enough productivity gains were reported to be higher even in the case of rainfed crops. Average yields of rainfed crops like soybeans and legumes increased by more than 200% (Renfro, 2005). The watershed programme have helped in improving soil moisture content leading to diversification of farming. It is also noticed that under this scheme cropping intensity has increased by 13%-25% (Renfro, 2005). Another important impact of watershed development was on controlling soil erosion as revealed by many studies (Kerr *et. al.*, 2002) Apart from this, Watershed development projects helped significantly employment opportunities to rural people under the conditions of labour-intensive vegetable crop cultivation and other horticultural crops (Reddy *et. al.*, 2001) However, employment generation from the scheme varies across regions depending on cropping intensity and the labour-intensive crops. Understandably, employment generation in the rural sector minimized migration of landless labour and other poor category of labour to urban centres.

Conclusion

In view of positive and lasting effect of the participatory watershed programme, there is a strong case for extending the programme to the uncovered areas, particularly drought-prone and rainfed cultivation areas. The non-governmental organizations have a crucial role to play in associating people with watershed programme and make it a participatory programme.

REFERENCES

1. Budumuru Yoganand and Tesfa G. Gebremedhin, "Participatory Watershed Management for Sustainable Rural Livelihoods in India", Research Paper presented at Southern Agricultural Economics Association, Annual Meeting, 2006.

2. Reddy, V.R. *et.al.*, "Participatory Watershed Development in India, Can it sustain Rural Livelihood"?, Development and Change, 2004.

WOMEN EMPOWERMENT AND MICRO FINANCE

1. Introduction

Women constitute nearly 50 per cent of the total population of India. Many studies reveal that nearly 70 per cent of different farming activities are carried by women only. Despite significant role of women in the maintenance of family, besides bearing the sweat of the farm work, their role is considered just an extension of household domain and remain mostly non-monetized. Even though, a vast majority of workers employed in agriculture continue to be women only, they do not get equal wages along with their male counterparts. Women work relatively for lower wages. Their employment is seasonal and subject to market fluctuations and hence, most of them remain poor. The increasing incidence of rural women's poverty may be attributed to their unequal access to economic opportunities in the labour market and also due to prevalence of gender discrimination and inequality in the society. In the planned development it was expected that women will equally benefit along with men by different schemes initiated by the government. This has been belied by the ground realities. The Ninth Five Year Plan document recognizes that inspite of series of development measures and constitutional legal guarantees, women far lagged behind men in almost all sectors. It is in this context, the concept of empowerment of women assumes critical importance.

2. Concept of Women Empowerment Defined

The concept of "Empowerment of Women" has gained currency in recent years, particularly in emerging countries like India. Empowerment of women may be viewed from different angles and described in multifarious ways. The term 'empowerment' indicates the process of "giving power to" developing conditions for "generating power within". Conceptually, therefore, the term empowerment has multidimensional focus and can be described as a process wherein a group or individuals are enabled to enhance their status in the society, on the one hand, and over all participation in growth and development, on the other. The process envisages greater access to knowledge and resources, greater autonomy, in the decision making process at home as well as in the matters concerning society.[1] The goal of women empowerment is simply providing necessary strength to them. This strength may be giving political or economic authority or provision of health and nutrition care or social empowerment. Empowerment of the poor is considered as an essential element in the poverty eradication. Empowerment of women particularly rural women has become an important issue in the strategies of balanced development with social justice. Research studies reveal that economic empowerment results in women's ability to influence or make right decision, increased self-confidence, better status and role in household etc.

3. Concept of Micro Finance

'Micro finance' is emerging as a powerful instrument for poverty alleviation in the economy. The Task Force on Micro Finance defined micro finance as "provision of thrift, credit and other financial services to the poor in rural, semi-urban areas, for enabling them to raise their income levels and improve living standards. Micro finance is said to be a powerful tool which can be used effectively to address poverty and empower the socially marginalized poor and strengthen the social fabric. And when it is directed towards women, the benefits accruing out of the micro financing activities are expected to multiply manifold.[2] In India, Micro finance scheme is dominated by Self-help Groups (SHGs) — Bank Linkage Programme as a cost-effective mechanism for providing financial services to the "unreached poor," which has been successful not only in meeting financial needs of the rural poor women, but also strengthen collective self help capacities of the poor leading to their empowerment. Rapid progress in SHG formation has now turned into an empowerment movement on among women across the country.[3] Micro finance is found to be necessary to overcome exploitation, create confidence for economic self-reliance of the rural poor, particularly among rural women who are mostly invisible in the social structure.[4]

The tem micro finance is of recent origin and is now widely used in addressing issues related to poverty alleviation financial supports to micro entrepreneurs, gender development etc., The term 'micro' literally means "small". But the Task Force has not defined any amount as "micro". However, as per Micro Credit Special Cell of the Reserve Bank of India, the amount that can be borrowed is upto the limit of ₹ 25,000, this amount could be considered as Micro Credit and this amount could be gradually increased upto ₹ 40,000, over a period of time. If the amount borrowed exceeds ₹ 25,000, then it is known as Macro finance.

The term micro finance, some times is used interchangeably with the term micro credit. However, the difference between the scope of the two terms need to be identified. The micro credit refers to provision of loans in small quantities, the term micro finance, on the other hand, has a broader meaning covering in its ambit, other financial functions and services like thrift/saving, credit, insurance etc. Micro finance is basically banking through groups. The essential features of the approach are to provide financial services .through the groups of individuals formed either in joint liability or co-obligation mode. In micro finance, savings/thrift precedes credit as credit facility is limited to the amount of savings/thrift. The groups has a role in credit appraisal, monitoring and recovery. The micro finance model was first initiated by Bangladesh Grameen Bank and is being used now widely with reference to women empowerment.

4. Woman Empowerment in India

As per the census of 2001, female population constituted 48.2% of the population of India. Article 14 of the Indian Constitution provides equal rights and opportunities to both men and women in all fields, i.e., political, social and economic. Article 15 prohibits discrimination against any citizen on account of age, sex, race, caste etc., Article 16 provides for equality of opportunity in the matter of public appointments for all citizens. These different provisions of the constitution, are meant to strength the hands of women and increase their capacity to utilize effectively what is due to them as natural rights and justice. The central and state governments have been implementing various programmes for the empowerment of women.

5. Programmes for Empowerment of Women

As early as 1953, during the First Five year plan, the government has established the Central Social Welfare Board (CSWB) for implementing the welfare programmes for betterment of women. Organizing women with Mahila Mandals was one such programme. In the Second Five Year Plan, the government has initiated steps towards the policy of equal pay for equal work and created provision for training of women to enable them to claim equal wages. Fully convinced of the role of education as an effective instrument for empowerment, the Third Five Year Plan, concentrated mostly on the female education. During the Fourth Five Year Plan, period too emphasis was on the female education. The Fifth Five year plan designed schemes to train women in selected skills so that they could withstand in the labour market and get higher wages. During the Sixth Five year plan, the government has appointed the working groups on employment for women in agriculture and rural development and adopted a multi disciplinary approach with a thrust on the three core sectors, i.e., health, education and employment. The Seventh Five Year Plan, initiated the Beneficiary Oriented Scheme (BOS) for empowering women towards their rights and equality. During the Eight Five Year Plan period, the government implemented various developmental schemes, with human force covering women in the development process. The Ninth Five Year Plan, aimed at empowering women through series of schemes, such as

(i) Adoption of Women's Component Plan (WCP) to ensure that benefits from different developmental schemes reach women also. It was stipulated that not less than 30 per cent of benefits and funds flow to women from all the women related schemes.

(ii) Setting up of self-help groups (SHGs) and linking them with Banks.

(iii) Sthri Sakti Puraskar was initiated for the first time to honour five distinguished women annually for their outstanding contribution in the direction of empowerment of women.

(iv) Setting up of Task Force on women under the Chairmanship of Deputy chairman of the Planning Commission to review the existing women specific and women-related legislations and suggest amendments if necessary. The Task force also suggested that the year "2001 be celebrated as Women's Empowerment Year." With celebration of 2001, as women's empowerment year, it was hoped that necessary awareness be generated among women so that they could take their rightful place in the mainstream of the nations social, political and economic life.

(v) Recasting of Indira Mahila Yojana as "Swayasidha', an integral programme for empowerment of women through a major strategy of converging the services available in all the women related programmes.

(vi) The Ninth Plan gave importance to the formation of Self-help Groups (SHGs) and through this scheme promoted the habit of saving. By linking this programme with Banks ground was well prepared for economic empowerment of women.

6. Tenth Five Year Plan and Empowerment of Women

The Tenth Five Year Plan, addressed the women Empowerment through three fold strategy with specific prescriptions of the National Policy. The threefold strategy include (a) Social Empowerment, (b) Economic Empowerment and (c) Gender Justice.

Social Empowerment:

(i) Provision of easy and equal access to basic minimum services of primary health care and family welfare with special focus on the much neglected poor and under privileged segments of population and universalizing Reproductive and Child Health (RCH) services.

(ii) Moving in the direction of reducing Infant Mortality Rate (INR) to 30 per thousand and Maternal Mortality Rate (MMR) to 100 per lakh live births by 2010.

(iii) Supplementing health care and nutrition services through the Pradhan Mantri Gramodhaya Yojana (PMGY) to fill the critical gaps in the existing primary health care infrastructure and nutrition services.

(iv) Consolidating the progress made under the female education and carrying it forward for achieving the goal of "Education for Women's Equality as advocated by the National Policy on Education, 1986.

(v) Providing easy and equal access to free vocational education for women and girls at all levels in the field of technical and vocational education and training in upcoming and job oriented trades.

(vi) Increasing enrolment and retention rates and free education for women and at all levels.

(vii) Extending the existing network of regional vocational training centres to all the states and industrial training institutes for women with residential facilities in all districts.

(viii) Encouraging the media to well project the problems of women. An attempt was also made to change the mind-set of people towards the girl child birth and her education and portray positive thinking.

(ix) Gender sensitizing ensuring that the rights and interest of women are taken care of, besides involving them in planning and implementation of the different projects.

7. Economic Empowerment of Women and Ongoing Programmes

(i) Provision of training and employment and income-generation activities with both 'forward' and 'backward' linkages with the objective of making all women economically independent and self-reliant.

(ii) Organizing women into self-help groups under various poverty alleviation programmers, *viz.*, Swarnajayanthi Grama Swarvodaya Yojana (SGSY), Indira Awas Yojana (IAY), National Social Assistance Programme (NSAP), Integrated Rural Development Programme (IRDP), Development of women and children in Rural Areas (DWACRA), Support of Training Employment Programmes for Women (STEP), Mahila Samakhya Programme (MSP) etc.,

(iii) Ensuring that women in the informed sector are given special attention with regard to working conditions, particularly equal wages to women along with men.

(iv) Issue of joint pattas for husband and wife under the Social Forestry and Joint Forest Management Programmes Also steps were initiated to ensure the benefits of training and extension in agriculture and allied activities, animal husbandry, poultry, fisheries etc., reach women in proportion to their number.

(v) Ensuring that the employers fulfill their legal obligations towards women workers in extending childcare facilities maternity benefits, special leave, protection of women from occupational hazards; freedom to women workers for formation of labour unions to safeguard their own interests

(vi) Improving the skills of women workers displaced from traditional sectors due to advancement of technology so that they could secure jobs in the new and expanding areas of employment.

(vii) To ensure 30 per cent was reserved for women in the Public sector services.

(viii) Increasing access to credit for women through Self-help Groups linked with financial institutions.

8. Gender Justice to Women

To eliminate all forms of gender discrimination and ensure justice to women through:

(i) Complete eradication of female infanticide through effective enforcement of the Indian Penal Code, 1860, as well the Pre-natal Diagnostics Technique (Regulation and Prevention of Misuse) Act 1994, with most stringent measures of punishment.

(ii) Safeguarding reproductive rights of women to enable them to exercise their reproductive choices

(iii) Working out strategies in the direction of employment opportunities and thus remove inequalities in employment both in work and accessibility.

(iv) Defining the Women Component Plan (WCP) clearly and identifying different schemes under each Ministry ensuring flow of funds and benefits.

(v) Initiating action for enacting new women specific legislations amending the existing women-related legislation, if necessary.

(vi) Expediting action with regard to reservation of not less than 1/3 seats for women in the Parliament and in the State Legislative Assembly so that, their voices are heard and interest of women are well taken care of.

(vii) Arresting the ever increasing violence against women and the girl child with the support of the law in force or through concerned law enforcing authorities.

(viii) Preparing Gender Development Index collecting necessary data and information, well compiling the same for the purpose of analyzing the status of women at regular intervals.

Thus, the government has been implementing various programmes during the planning era for the empowerment of women such as IRDP; Support of Training and Employment Programs and women (STEP), Development of Women and Children in Rural Area (DWACRA), Mahila Samakhya Programme (MSP), Prime Minister Rojgar Yojana (PMRY), Women's Credit Fund (WCF), Women's Development Corporations (WDCs) and Self-help Groups (SHGs), etc. effectively contribute for empowerment of women.

Many studies reveal that meaningful empowerment of women can be achieved through making women economically independent and self-reliant. The Self-help Group Programme, well supported by the institutional credit is found to be an effective programme in the direction of women empowerment.

Conclusion

Indian Society is still man-dominated and women degraded society. There is a strong case for empowerment of women. Still some 70 per cent of rural women remain illiterate. Poverty, illiteracy and ill-health and civic inertia are all centered round poor rural women. They are not aware of the different schemes in operation intended for their betterment. SHGs-Bank linked scheme, no doubt, facilitates easy flow of credit and to that extent micro finance can contribute to solving problems of inadequate housing and urban services as an integral part of poverty alleviation programmes. Credit no doubt is important for development, but cannot by itself, enable every poor woman to over come all poverty-based problems. Making credit available to women does not automatically mean they have control over its use in a situation of man-dominated family.

In the direction of empowering poor women, basic issues like provision of literacy, health and motivation for development become crucial. In rural areas, planned effort to achieve 100 per cent literacy particularly of rural women is a basic requirement in the direction of empowerment of women. No doubt, political empowerment in terms of providing 1/3 seats in all the law making bodies is necessary to enable the voice of women is heard through their representatives. But that alone is not sufficient. Economic empowerment is necessary. SHGs-Bank Link Scheme facilities economic empowerment. Equal level of wages as in the case of NREGP to both men and women labour is another scheme contributing for economic empowerment. That apart, women must be well educated and qualify themselves to be earning members and then only women can enjoy self confidence and respect in the family and society at large. Economic empowerment by way of providing legal right to women for equal share along, with male children from the ancestral property of the family is necessary. Such step becomes an effective step in the direction of economic empowerment of women. With economic empowerment ground is well prepared for political empowerment and social empowerment and then now women can emerge to the stage of full-fledged empowerment.

REFERENCES

1. S.P. Jain, "Empowerment of Rural Women through Local Self Government in Afro-Asian Countries" in *Dynamics of Sustainable Development*, G. Raghava Reddy and S. Subramanyam (Eds.), Serials Publication, New Delhi, 2003.
2. Anil K. Khandelwal, "A Micro Finance Development Strategy for India" in *Economic and Political Weekly*, March 31st 2007, p. 1127.
3. Tiyas Biswas "Women Empowerment through Micro Finance, A Book for Development", *National Institute of Technology*, Durgapur.
4 *Ibid.*

AGRICULTURAL LABOUR

1. Introduction

Agricultural labourers constitute an important segment of Rural Labour Force. Agricultural labourers are drawn from socially and economically backward sections of society, like Scheduled Castes and Scheduled Tribes. They are unorganized and they remain poor. Agricultural labour in India is widely scattered over six lakh villages and their number is fairly big. As such, it is not easy to have an effective organization. Further, the agricultural operations being seasonal in nature, the demand for agricultural labour is also seasonal. During the sowing and harvesting seasons, almost all the labourers, including child labourers, are fully employed. In many areas where cultivation is primarily dependent on rains, employment of landless agricultural labour is confined only to a short period and the rest of the period most of them remain unemployed. Their position is better in irrigated tracts as compared to that in the dry areas. In view of these different factors, the bargaining power of the agricultural labourer is very weak and hence the wage level remains low.

In the agricultural operations, normally, the landlord directly supervises the work and therefore, there is direct contact between the agricultural labourer and the landlord. In such situation, the landlord extracts much work and there is no possibility of escaping from the work by the worker. It is particularly so, in the case of small-size landowner. Agricultural labourers are sometimes asked to do, apart from farm work, domestic work like processing of food stuffs, cooking, cleaning of vessels, carrying of food to the farm for workers on head load, etc.

In certain areas with abundant supply of labour and without gainful employment throughout the year, agricultural labour is found to be migratory, moving in search of gainful employment. Problems of this nature are not normally found in the case of industrial labour.

2. Agricultural Labour Defined

The definition of industrial labour cannot be applied to agricultural labour. It is not possible to make such distinction among agricultural labourers as unskilled, semi-skilled and skilled workers. In the agricultural sector, one has to do a variety of jobs. A farm worker is expected to have certain skills in carrying out different types of jobs. The type of relation found between employers and employees in the manufacturing industries cannot be found in agriculture, because of the prevalence of subsistence farming. In view of the special nature of agricultural operations, it is very difficult to define and determine the size of agricultural labour.

The different Labour Enquiry Committees and different census have adopted different criteria to identify the agricultural labour. Unlike industrial labour, agricultural labour cannot be defined accurately. However, attempts were made to define agricultural labour by different experts and committees appointed by the government. The first Agricultural Labour Enquiry Committee of 1950-51, regarded those people as agricultural workers who were engaged in agricultural operations on payment of wages. An attempt is also made to define agricultural household. In case the major occupation of a large number, i.e., half or more numbers of a household, are working on payment of wages in agriculture, then that household would be treated as agricultural labour household.

The second Agricultural Labour Enquiry Committee of 1956-57, took a broad view of the agricultural activities and included those workers engaged in allied activities like animal husbandry, dairy, poultry, piggery etc. The committee stated that if 50 per cent or more of its income is derived as wages for work rendered in agriculture, only then, it would be classified as agricultural labour household. The change over from 'work' to 'income' seems more scientific. The Census of India 1961, included in the category of agricultural workers, those who worked on the farms of others and received payment either in cash or kind or both.

3. Different Types of Agricultural Labour

There are broadly three types of agricultural workers in India: (a) Small and marginal farmers who have very small holdings whose income from their own land is quite inadequate and hence they are forced to work on the farms of others to make both the ends meet; (b) Tenants who work on the land taken on lease but this is not their main source of income. The major source of income is by way of wages for working on the lands of others and; (c) Share croppers who, besides sharing the produce of land cultivated by them, also work as labourers.

It is evident that agricultural labour covers all those landless and assetless poor, small and marginal farmers and rural artisans who work as wage earners on the farm of others.

In 1951, the percentage of cultivators to total workers was 69.9 and it has come down to 54.4 per cent in 2001. Regarding agricultural workers, in 1951 there were only 27.3 million or 28.1 per cent to the total workers. Gradually, the agricultural workers increased. In 2001, there were as many as 106.8 million or 45.6 per cent to the total workers. It can be inferred that small and marginal workers now converted themselves into agricultural labourers.

4. Mounting of Agricultural Labour

The size of agricultural labor has been on increase for different reasons:

(i) High rate of population growth;

(ii) Decline of cottage industries and village handicrafts;

(iii) Growth of indebtedness due to low income leading to transfer of land from the small and marginal farmers to creditors;

(iv) Disintegration of the joint families,

(v) Displacement of some people from subsidiary occupations;

(vi) Eviction of the small farmers and tenant cultivators and

(vii) Capitalist agriculture due to the technological development, abolition of intermediaries and the expansion of marketing facilities.

5. Agricultural Wages

The 25th Round of the National Sample Survey has clearly revealed a gloomy picture of the agricultural wages. Rural works programme of 1970-71, was started with the aim of generating adequate employment with reasonable wages. Food for Work Programme was launched in 1977, to provide opportunities of work for the rural poor, particularly during the slack seasons. The workers were paid partly in cash and partly in foodgrains from the stocks in warehouses. The Rural Landless

Employment Guarantee Programme (RLEGP) was launched in 1983, with a view to alleviating poverty, among the rural landless workers ensuring reasonable level of wages.

6. Minimum Wages

The Minimum Wages Act was passed as long back as in 1948, and since then, there has been demand for payment of minimum wages to the agricultural labourers. At present, excepting Jammu & Kashmir, Nagaland and Sikkim, legislations have been enacted in all other states fixing the minimum wages to agricultural labour. However, it appears that the Act has not been effectively enforced for different reasons mainly because of, excessive supply of rural labour and unorganized situation of rural labour. The Mahatma Gandhi National Rural Employment Guarantee Programme (NREGP) launched in 2006, aims at giving legal guarantee to work for 100 days in a year. The minimum wage fixed is ₹ 100 for both men and women. This is the first act of fixing minimum wages for agricultural labour with equal wages for men and women. It is hoped that this legislation will benefit the rural poor.

7. Bonded Labour

A special problem affecting rural worker is "bonded labour". There are agricultural workers bound to work for a landlord throughout their lifetime for petty wages. Bondage begins with some amount of money or foodgrains borrowed from the landlord under pressing conditions. The small amount borrowed multiplies itself with compound interest and the workers continue to work for several years to payback their debts which never happens and thus they become bonded labour.

The system of bonded labour in India has been in existence for hundreds of years and the system is still in vogue in some form or other. The system of bonded labour in rural areas is the peculiar product of the agricultural system. There are a large number of landless agricultural workers that belong mostly to Schedule Castes and Scheduled Tribes who have to live on meager wages. They are in pressing need of money and do not have any assets to offer as security for the purpose of securing a loan. They pledge themselves as security for small amounts borrowed which carries an abnormally high rate of interest and the accumulated credit over several years becomes so burdensome that the borrower never succeeds in repaying the debt during his lifetime. The landlords fully exploit the illiteracy and ignorance of the borrower, in working out the loan amounts due. The borrowers and even their descendants are doomed to bonded labour and perpetual slavery. There are instances in Andhra Pradesh where a petty landowner, who borrowed a small amount of ₹ 100 and who never succeeded in repaying it, lost his tiny holding, became labourer and then a bonded labourer. Not only the person who actually borrowed became a bonded labourer, but his sons also were reduced to the status of bonded labour. Instances of this type are recorded in other parts of rural India.

The bonded labourers are exploited by the moneylender-cum-landlord in several ways. They have to work for petty wages only on the farms of the landlord. They have to do any work entrusted to them, including domestic work. They are made to work for longer hours till late in the night. It is reported that bonded labour still exists in certain States.

There are difficulties in identifying bonded labour and arriving at the actual number of bonded labour in different States. There are no written records of the time and amount borrowed, accrued interest, amount repaid towards principal and interest, outstanding loan amount, the type of attachment of the worker to the concerned landlord, etc. Understandably, the landlord is not interested in furnishing the required information. The bonded labourer is afraid to disclose the conditions of his bonded service because of the predominant grip of the landlord over him. The existence of bonded labour even now in certain areas is a reflection of State Government's inability to tackle the problem seriously.

After Independence, attempts have been made to abolish the bonded labour, because it is exploitative, inhuman and violative of all norms of social justice. A legislation was enacted known as the Bonded Labour System (Abolition) Act, 1976. Covering the entire country. The bonded labour system was abolished and every bonded labour was freed and discharged from any obligation. The National Sample Survey (32nd round) identifies as many as 3.5 lakh bounded labour. It is hoped that the bonded labour will no longer continue with growing awareness.

The planning commission had launched a series of programmes aiming at the betterment of rural labour such as Rural Works Programme, Cash Scheme for Rural Employment, Pilot Intensive Rural Employment Programme, Small Farmers Development Agency, Marginal Farmer and Agricultural Labour Development Agency, Minimum Needs Programme, National Rural Employment programmes etc., aimed at creating employment opportunities for rural labour. In addition to these programmes, some state governments have launched certain schemes for employment generation and betterment of the conditions of labour. The Employment Guarantee Scheme of Maharashtra of 1974, is worth mentioning.

8. Employment Guarantee Scheme (EGS) of Maharashtra

The EGS of Maharashtra represents a bold acknowledgement of a citizen's right to work assuring employment to the poor throughout the year. The EGS of Maharashtra aims at providing gainful and productive employment – gainful to the individual and productive to the economy to all unskilled persons in the rural areas and in "C-class" Municipal areas who need work on their own". The main features of this scheme are:

(i) Any adult person residing in the rural areas and willing to do any unskilled manual work under the EGS has to get his name registered with the registering authority. On receipt of this application, the concerned officer would allot the work;

(ii) The beneficiaries of the scheme will have no choice as to the type of work and where they have to be employed;

(iii) The employment is normally provided at the Panchayat Samithi or the Block level.

(iv) Works are undertaken departmentally and not through contracts;

(v) At least 60 per cent of the expenditure of the Programme must be spent on wages to the labourers and

(vi) Wages are determined keeping in view, the working hours and minimum needs of the poor.

The EGS programme had taken up different irrigation works, land development and soil conservation, flood control, forest works, road works etc., The EGS was described by Dr. Kumuduni Dandekar as "a kind of income guarantee to the poorest of the poor in rural areas and hence, seems to alleviate the biggest problem of reducing poverty".[1]

The different schemes under Social Security, Food Security, Poverty Alleviation Programme, if implemented effectively would certainly take care of the interests of the rural in general and agricultural labour in particular.

9. The Mahatma Gandhi NREGP

The NREGP with equal level of wages guaranteed to men and women must also be implemented effectively. However, NREGP should not be detrimental to the interests of local farm producers.

Conclusion

In spite of different schemes, the lot of agricultural labour has not improved. In certain pockets, still there is bonded labour and agricultural labour is very much exploited. There are no organized agricultural labour unions to fight for fair wages. Except in NREGP, women do not get equal wages along with male workers. Women labour continue to be much exploited segment of rural society.

REFERENCE

1. Dandekar Kumuduni, "Prospects of Employment Guarantee Scheme" in *Lokrajya,* October 6, 1983, p. 55.

16 SMALL-SCALE AND MICRO ENTERPRISES

1. Introduction

Since times immemorial small-scale and cottage industries have been playing an important role in the rural economy of India. In the past, village artisans and craftsmen contributed richly, in terms of goods and support services for achieving a self-sufficient village economy.

In the past there was a system of specialized artisans concentrated in selected villages, and such villages were like workshops for handlooms and handicrafts of a high quality. There was a saying that "India produced cobweb like cotton fabrics, which even the modern organized mills cannot duplicate, which inspired some people to call Indian textiles as "the work of fabrics and insects" rather than of human beings. Workers both men and women occupied a pride of place in this production set up. Gradually, the native indigenous technology disappeared. However, rural artisans with special skills remain even today, without much of support from the state and the society of large and they continue to survive with meagre incomes.

2. Changed Nomenclature

With enactment of Micro, Small and Medium Enterprises Development (MSMED) Act, 2006, Small-scale and Cottage Industries are given a different name. This Act of 2006, introduces the concept of enterprise, in the place of industry. Village and small industries are now known as Micro and Small Enterprises (MSE); Small and Medium enterprises (SME) and Micro, Small and Medium Enterprises (MSME). In the Act, investment limits have been prescribed for manufacturing and service enterprises.[1]

3. Definition of Micro Enterprises

The New definition is as follows:

A. Manufacturing Enterprises:

(i) A Micro enterprise, where the investment in plant and machinery does not exceed ₹ 25 lakhs;

(ii) Small enterprises, where the investment in plant and machinery is more than ₹ 25 lakhs and less than ₹ 5 crore.

(iii) Medium enterprises, where the investment in plant and machinery is more ₹ 5 crore and less than ₹ 10 crore.

B. Service Sector

(i) A micro enterprise, where the investment in equipment does not exceed ₹ 10 lakhs

(ii) A small enterprise, where the investment in equipment is more than ₹ 10 lakhs, but does not exceed ₹ 2 crore

(iii) A Medium enterprise, where the investment in equipment is more ₹ 2 crore, but does not exceed ₹ 5 crore.

In India, Micro and Small Enterprises[2] (MSES) (till recently known as village and small enterprises) account for almost 40% of the total industrial production, 95% of the industrial units (along with medium enterprises) and 34% of the exports. They manufacture over 6000 products ranging from handloom sarees, carpets and soaps to pickles, papads and machine parts for large industries. In 2003-04, the contribution of Small-scale Industries (SSI) Sector alone to the GDP was 6.71%. Apart from contributing to the GDP, as per Eleventh Plan Report, they are instruments of inclusive growth which touch upon the lives of the most vulnerable, the most marginalized — women, Muslims, SCs and STs and the most skilled. Being the largest source of employment after agriculture, the MSE sector in India enables 650 lakh men, women and children living in Urban slums, upcoming towns, remote villages and isolated hamlets to use indigenous knowledge and entrepreneurial skills for their livelihoods. The Eleventh Plan recognizes the MSE sector as an important component of the industry that needs infrastructure, credit and policy support.[3]

4. National Importance of Micro and Small Enterprises (MSEs)

The micro and small enterprises have certain inherent advantages for the overall development and they have an important part to play in the national economy;[4]

(i) **Utilization of Local Resources:** MSEs facilitate tapping of the local resources like raw material, labour etc for productive purposes. They can mobilize the rural savings which may otherwise be spent on unproductive purposes;

(ii) **Limited Capital:** MSEs can be started with a limited capital. In a country like India, where investment capacity is confined to selected rich in urban centres, MSEs can be taken up by rural people even with limited capital. The rural people can easily mobilize small investment and start MSEs.

(iii) **Less Risk:** There is normally not much of risk involved in MSEs. The output from these industries is limited and it is generally intended for local market only. Often, production is undertaken with a clear knowledge of the market condition. Hence, risk element is negligible.

(iv) **Short Gestation Period:** In the use of MSEs, the time lag between investment and returns is very short and therefore, these industries would give yields within a short period.

(v) **Eco-friendly:** MSEs will not create slums, housing problems and problems of sanitation. Environment as such is not adversely affected by these industries and they are eco-friendly enterprises.

(vi) **Goods of Consumer's Choice:** In MSEs goods of individual consumer's choice and taste can be produced. For example, jewellery, or specialised any other type of handlooms, products etc., can be produced meeting the tastes of different women or other category of consumers.

(vii) **Earning of Foreign Exchange:** Micro enterprises facilitate earning of the foreign exchange through export of products from these categories of enterprises. India earns considerable foreign exchange through, for example, exports of Kondapalle toys and Dharmavaram sarees (Andhra Pradesh). In 2001-02, through the export of coir products alone, foreign exchange to the extent of ₹ 320 crore could be obtained.

(viii) **Employment Generation:** MSEs are generally labour intensive and they create more employment opportunities to the rural people. In 2001-02, from coir industry alone, 5.50 lakh persons were employed.[5] The MSE sector in India has grown significantly since 1960. In 1960 there were only 12376 MSEs, providing employment to 10 lakh people. By 2007, there were 128 lakh MSEs, employing 295 lakh people. If the units in the Khadi industries, village industries and coir industries are taken into account, the employment is estimated to be over 332 lakh. With the inclusion of handlooms, handicrafts, wool and sericulture, the total job in the MSE sector in India goes upto 650 lakh.[6]

(ix) **Prevention of Exodus of Rural Poor:** In 2004-05, as per planning commission, 28.30 percent of rural population were below poverty line, while the urban population below poverty line was 25.70%. The agricultural sector cannot support any more number. The landless rural poor are forced to move to the urban centres. Under the circumstances, MSEs are a potential source of gainful employment with assured income. With the development of MSEs, employment opportunities can be generated for the rural poor and the exodus of rural poor to the urban centres can be prevented, which will finally facilitate the development of rural areas.

(x) **Reduction in Regional Disparities:** The concentration of major industries in selected urban centres has resulted in regional imbalances and disparity between the urban and rural in the level of development. Through MSEs, the concentration of industries can be checked and industrial dispersal can take place leading to even development of different regions in the country. Further, the growth of MSEs, will prevent the migration of people from the rural to urban centres.

(xi) **Entrepreneurial Development:** The MSEs are in a position to encourage young and promising entrepreneurs in rural areas to set up micro units. If only necessary incentives, in terms of provision of adequate credit, raw material, marketing facilities etc. are provided, certainly a cadre of entrepreneurs can emerge in the rural areas which will finally facilitate the development of rural areas.

The importance of MSEs has been well recognized in the context of development of India. The Reserve Bank of India has advised the banks to ensure that at least, 40 per cent of their total advances are extended to priority sectors and small scale sector is one such priority sector.[8] The growth of MSEs is not uniform across states in India. Tamil Nadu (14.5%) makes the maximum contribution to employment followed by Maharashtra (9.7%), Uttar Pradesh (9.5%) and West Bengal (8.5%).

The MSE sector in India is heterogeneous, dispersed and mostly unorganized. Due to the unorganized nature of the sector, entrepreneur and artisan/workers face difficulties in accessing the government schemes. Consequently, the MSE sector faces problems as listed below:

5. Problems of Micro and Small Enterprises (MSEs)

Major problems, among others, are as under:

(i) **Non-availability of Quality Raw Materials:** The large-scale industries compete with the village and cottage industries for raw materials etc. Non-availability of quality raw materials such as dyes and yarn (especially for handloom and power looms) is a problem. Though the National Small Industries Corporation (NSIC) and State Small Scale Industries Development Corporations are providing some raw materials, their efforts are not in consonance with the requirement.[9] In the absence of proper organization, for channelising the flow of the quality raw material from the large-scale sector to the small-scale sector, those engaged in rural industries are exploited by the middlemen and they have to incur heavy cost on the raw material.

(ii) **Low Level of Technology and Insufficient Market Research:** The development of village industries is hampered by their present low level of technology and skills. Technology used now need to be improved upon. They do not have facilities of reason and training, nor they have ability to invest on such schemes. Most MSEs do not have money to invest in market research and are unable to carry out design and technical improvements to keep up with market demands. Unlike big business, they cannot invest in advertising and packages. This limits their ability to tap the markets and attract consumers.[10]

(iii) **Lack of Adequate Credit:** In recent years, the percentage share of the MSE, in the institutional credit began declining from 15.1% in 1990-91 to 8.0% in March 2007. Though, the Nayak Committee set up by RBI in 1991-92 had recommended working capital to the small scale enterprises at 20 percent of their annual turn over, SSI and KVI units together received only 13.3% of their production value from Scheduled Commercial Banks in 2005-06. Lending to micro enterprises has fallen from 51.2% in 2002-03 to 45.1% at the end of 2005-06. The difficulty in arranging collateral security continues to be a problem. MSEs are forced to depend on money lenders for finance at a high rate of interest with all its adverse effect on the business.

(iv) **Marketing:** Marketing is a major problem for micro enterprise. They cannot afford to invest big amounts on market research, advertisement etc., in order to promote sales. The organized large scale sector produces products which are close substitutes for the products of the village artisans. In the absence of technology upgradation and achieving economies of scale, the village industries are not able to compete with the large scale sector and sell at a competitive price. Recent surveys have indicated that the demand for artistic goods produced by micro units is declining as the upper strata of the society prefer to use prestigious imported goods like electrical appliances, etc., in preference to artistic goods of the village industries.

(v) **Lack of Transport Facilities:** Village industries face the problem of transport, both to get the raw material and to market the final products. Some villages still don't have link roads connecting the nearly towns or urban centres. In such a situation, small size industrial units face series of problems in running the enterprises.

6. Karve Committee Recommendations

In June 1955, the planning commission constituted the village and small industries committee under the chairmanship of Karve, with the aim of examining the problems of small-scale and cottage industries. The committee made the following recommendations:

(i) Modernization and development of small and cottage industries without causing unemployment;

(ii) Imposition of ceiling on the production of consumer goods by the large scale industry;

(iii) Reserving the production of certain common consumption items like woolen goods, vegetable oils, sugar and khandsari, leather footwear and matches, etc., to small scale sector;

(iv) Liberal financial aid to the rural industries through cooperatives, commercial banks, state government and Reserve Bank of India,

(v) The more drastic recommendation is one of levying a cess on the mill cloth for the purpose of financing small industries.

Some of the recommendations have been accepted and implemented. However, the recommendation of the committee relating to imposing a ceiling on mill production of consumer goods and levying a cess to mobilize resources for small industries was subject to criticism.

7. Industrial Policy Resolution of 1956

The Industrial Policy Resolution of 1956 emphasized on the need for dispersal of industries aiming at reduction of disparities in level of development. In order to achieve this goal, among other steps, it was emphasized that cottage and small industries should be supported, ensuring more equitable distribution of income and facilitating an effective mobilization of local resources. The Industrial Policy declared that industries of smaller size must be dispersed all over the country.

8. Khadi and Village Industries (KVIs)

The KVI sector continues to provide part-time for some and full-time employment for some others utilizing the local skills and resources. The production of Khadi cloth increased from 1.5 million square metres (m.sqm) in 1995-96 to 85 m.sqm in 2001-02. The production turned out by the village industries in monetary terms, increased from ₹ 3504 crore in 1995-96 to 7896 crore in 2001-02. The employment in KVI sector increased from 57 lakh in 1995-96 to 66 lakh persons in 2001-02.

9. Coir Industry

The coir industry provides employment to more than five lakh persons and manufactures finished products like threads, mattings, mats, carpets, rugs etc. The government has initiated a scheme of Mahila Coir Yojana, under which financial assistance is made available to women artisans to the extent of 75 per cent of the cost, debt, subject to a maximum of ₹ 7500. The coir industry provided employment to 5.50 lakh persons in 2001-02. In the same year earnings from exports went up to the extent of₹ 320 crore.

10. Handlooms

A major thrust in recent years has been on expanding marketing opportunities for the handlooms. Towards this goal, schemes pertaining to training, design development, exhibition and publicity etc., are being undertaken. Training for craftsmen is now entrusted to NGOs and cooperative societies, besides departmental training programmes. In 2001-2002, Handloom industry provided employment to the extent of 72 lakh persons. The earnings from exports of handloom items was to the tune of ₹ 2150 crore.

11. Powerlooms

In order to boost up the export of powerloom fabrics, the government has set up a separate export promotion council called 'the Powerloom Development and Export Promotion Council (PDEXCIL). A Group Insurance Scheme for powerloom workers in association with LIC of India was introduced during 1992-93. Employment was provided to 75 lakhs persons in 2001-2002. The earnings by way of exports was estimated to be ₹ 11,000 crore in 2001-2002.

12. Handicrafts

A major thrust in recent years, has been on expanding marketing opportunities for handicrafts. Towards this goal, schemes pertaining to training, design development, exhibitions and publicity are being undertaken. Training for craftsmen now entrusted to NGOs and cooperative societies, besides departmental training programme. In 2001-02, employment to the extent of 72 lakh persons was created. In the same period, earnings by way of export of handicrafts was to the tune of ₹ 10,600 crore.

13. Wool Development

The woolen industry in India is small in size and widely scattered. The Central wool Development Board (CWDB) set up in 1987, with headquarters at Jodhpur in Rajasthan looks after the development of this sector. During the Eighth Plan, CWDB had set up three more industrial service centres in U.P., Punjab and Himachal Pradesh. Employment generated in 2001-2002 (in unorganized sector) was to the extent of 5 lakh persons. The export earnings from this sector was to the extent of ₹ 1,036.8 crore in 1996-97.

14. Food Processing Industry

India, as the largest producer of milk, second largest producer of fruits and vegetables, a major producer of spices, groundnut and rapeseed and the fourth largest producer of wheat has a tremendous potential for development of food processing industry. The food processing sector includes paddy processing, maize processing, pulse processing and making of cereal based products. During the planning era, food processing industry received necessary encouragement from the government as a component of agricultural development.

15. Sericulture

The Sericulture industry has employment potential. The employment in sericulture was estimated at 67 lakh in 2001-02. In the same period, it was estimated that the earnings by way of exports would be 1650 crore.

16. Fisheries

Fisheries production has an important place. Besides, a rich source of protein, it provides income and employment to millions of fishermen and farmers, particularly in the coastal regions. If modern technologies are used, there is a vast untapped potential which can be exploited. The fish exports from India are on the increase. The export earnings which were ₹ 5,117 crore in 1999-2000 increased to ₹ 7,620 crore in 2007-08. Fish production increased from 6.8 million tones in 2006-2007 to 7.3 million tones in 2007-2008. Fishing, aquaculture and allied activities are reported to have provided livelihood to over 14 million persons in 2006-2007.[11]

17. Horticulture

India is the second largest producer of fruits and vegetables in the world, next to China. The production of horticulture crops was 207.01 million tones during 2007-08.[12] A major programme, namely the "National Horticulture Mission (NHM)" was launched during the Tenth Five Year Plan, with effect from 2005-06, for holistic development of horticulture sector.

18. Animal Husbandry and Dairying

The livestock sector is an important component of rural industrial sector. The livestock sector contributed over 5.26 per cent to the total GDP during 2006-07. In 2007-08, this sector contributed 104.8 million tones of milk, 53.5 billion eggs, 44 million kg. wool and 2.6 million tones of meat.[13] The 17th livestock census of 2003 has placed the total livestock population at 485 million and total of poultry birds at 489 million.

The government implemented Livestock Insurance Scheme as a centrally sponsored scheme during 2005-08, in hundred selected districts in the country. In the Eleventh plan, there is provision of ₹ 153.43 crore for the livestocks insurance scheme.

India produces more than 47.3 billion eggs per year, with per capita availability of 42 eggs per annum. The value of exports was around ₹ 441 crore during 2007-2008.[14]

The Eleventh Five year Plan looks at MSE sector as an engine for sustained and inclusive economic growth and employment. The strategy is two-pronged; focusing on livelihood and social security. In the Eleventh Five Year Plan, focus is on women empowerment. In the handloom sector alone,[15] 60.6 per cent weavers are women, 10.76 percent belong to SCs, 25.5 per cent to STs and 42.65 per cent to OBCs. In the direction of credit support to MSE sector, it is decided to provide bank loans upto ₹ 5 lakh to MSEs without collateral security at interest rate of 8 per cent. It is also proposed in the Eleventh Five Year Plan that neglected areas such as occupational health insurance, and so on will be taken up as an integral part of the MSE policy. Special emphasis would be laid on skill development and upgradation of the units across all sectors. Also, in the Plan Report, it is made clear that the government will take steps to develop adequate infrastructure in terms of roads, transport, water and other schemes.[16] An outlay of ₹ 11500 crore has been allocated to Ministry of MSME, out of which ₹ 7,500 crore for agro and rural industries and the rest of amount of ₹ 4000 crore exclusively for MSME programme. It is also made clear in the plan document that different schemes already grounded in the earlier plan will be continued in the Eleventh Plan period also.

Conclusion

In spite of several measures in the direction of promotion of micro units during the planning era, some of the basic problems remain untackled effectively. The different schemes are likely to benefits, certain category of micro enterprises like modern small industries, including power looms, and not adequately the traditional industries like Khadi, village industries, handlooms, sericulture, handicrafts, coir industry etc., benefiting the rural poor. The weaker sections engaged in the traditional industries have to be identified, their problems in running the small units have to be carefully analyzed and appropriate steps should be taken immediately. Generally, traditional industries run by rural poor find it difficult to secure required credit from the institutional agencies, mainly because of inability to offer adequate security. The rural entrepreneurs of micro projects are not well equipped with necessary skills. The District Industries Centres, Small Industries Service Institutes (SISI), National Institute of Small Industry Extension and Training (NISIET), Central Institute of Tools Design (CITD) should be actively involved in providing extension services ensuring necessary knowledge and skills needed for the entrepreneurs of small enterprises. The Special Economic Zones or any other scheme under globalization policy should not be detrimental to the interests of MSEs and the rural poor.

REFERENCES

1. *Eleventh Five Year Plan Report,* Vol. III on Industry, p. 197.
2. *Ibid.,* p. 195.
3. *Ibid.*
4. K. Venkata Reddy, *Agriculture and Rural Development,* Himalaya Publishing House, Bombay, 2001, pp. 236-237.
5. Planning Commission, *Annual Plan,* 2001-02.
6. *Eleventh Five Year Plan,* p. 197.

7. *Agricultural Statistics* 2008, pp. 54-55.
8. Government of India, Ministry of Information and Broadcasting, *India 1985*, p. 40.
9. Eleventh Five Year Plan, *op. cit.*, p. 201.
10. *Ibid.*, p. 202.
11. *Economic Survey,* 2008-09 p. 108.
12. *Ibid.*, p. 183.
13. *Ibid.*, p. 186.
14. *Ibid.*, p. 187.
15. Eleventh Five Year Plan, *op. cit.*, p. 203.
16. *Ibid.*, p. 205.

17 RURAL BANKING AND CREDIT

1. Introduction

Finance is an essential requirement for almost every activity. Both producers and consumers in rural as well as urban areas need finance for the day-to-day requirement and for all productive activities. If required finance is not available out of one's own income, then one needs credit. In rural areas of India, credit requirements are more pressing rural population is composed of agricultural producers, tenant cultivations, village artisans, landless labour etc. All these categories, except rich landlords are in need of credit. Among the agricultural producers, the earnings of small and marginal farmers and tenant cultivators, are not sufficient enough to meet even the minimum requirements of life. In their case, the prospects of 'plough back' are very weak. Therefore, they need credit not only for productive purposes, but also for day-to-day minimum consumption purposes.

2. Different Types of Agricultural Credit

More than any other sector, agricultural sector needs credit support as two-thirds of the population, is dependant on this sector and this sector accounts for 19% of GDP. Agriculturists need credit of the following types: (i) *Credit for Consumption Purposes:* Some of the agriculturists are not able to produce the required quantity of different food items. They may not have sufficient finance of their own to purchase food and other important consumption items. In such a situation, they have to necessarily borrow from some source or other for meeting the consumption requirements also. (ii) *Production Credit:* Credit is needed for day-to-day production activities, apart from credit for investment on permanent assets. The productive credit may be classified into three categories: (a) Short-term credit, (b) Medium-term credit and (c) Long-term credit. Credit required for purchase of seeds, manures for payment of wages etc., may be called short-term credit. Such type of credit is repaid immediately after harvest of a crop. The agriculturist is also in need of credit to purchase agricultural tools and implements, live stock, to dig or deepen a well etc., and credit for these items is termed as medium-term credit. This type of loan is repaid normally within a year. The credit required for purchase of costly agricultural machinery like a tractor or purchase of additional land etc., may be called long-term credit. Such type of credit is repaid normally in easy installments spread over 10 to 15 years. Government has recognized credit as an essential requirement for revitalizing agriculture and there have been positive developments in this area. The total credit to agriculture increased from ₹ 62,045 crore during 2001-2002, the terminal year of the Ninth Five Year Plan, to ₹ 2,00,000 crore during 2006-07, the final year of the Tenth Five Year Plan. The share of commercial banks in total agricultural credit increased from 54% in 2001-02, to around 69% in 2005-06. The share of investment credit increased from 35% in 2001-02, to around 41% in 2004-05, despite the negative growth achieved by the long-term cooperative credit structure.[1]

3. Sources of Agricultural Credit

Agricultural credit is available from different sources. The different sources of credit may be classified into two categories. They are: (a) Non-institutional sources or private sources: private sources include professional money lenders, agricultural money lenders, commission agents, merchants, friends and relatives, (b) Institutional sources: including government, cooperatives, commercial banks, including Regional Rural Banks.

Table 17.1 presents the extent of credit available to the cultivators from different sources.

Table 17.1

Extent of Credit Available to the Cultivator Households from Different Sources

(Per Cent)

Sources of Credit	1951	1961	1971	1981	1991	2002
1	2	3	4	5	6	7
Non-institutional Sources	92.7	81.3	68.3	36.8	30.6	38.9
Moneylenders	69.7	49.2	36.1	16.1	17.5	26.8
Institutional Sources	7.3	18.7	31.7	63.2	66.3	61.1
Cooperative Societies/Banks	3.3	2.6	22.0	29.8	23.6	30.2
Commercial Banks	0.9	0.6	2.4	28.8	35.2	26.3
Unspecified					3.1	
Total	**100.0**	**100.0**	**100.0**	**100.0**	**100.0**	**100.0**

Source: All India Debt and Investment and NSSO

It is evident from the Table 17.1 that the share of Non-institutional sources particularly, moneylenders has been higher till 1970s. From that time onwards, it began declining. While it was 92.7% in 1951, it has come down to 38.9% in 2002. The role of moneylenders too has been casually declining. But yet, in 2002, moneylenders provided 26.8% of the agricultural credit. Mostly, small and marginal farmers borrow from moneylenders and they are very much exploited by the money lender.

Regarding institutional sources, cooperative share was only 3.3% in 1951. It declined to 2.6% in 1961, but from this period onwards, the share of cooperatives in the agricultural credit has been increasing and in 2002, it was 30.2%. Commercial Banks share in the total agriculturists' credit, has been gradually increasing. While it was only 0.9% in 1951, it jumped to 26.3% in 2002. The overall institutional credit share in 2002, was 61.1%. Farmers face the problem of indebtedness and it requires a detailed analysis.

The maters of concern are inadequacy of institutional credit to the agricultural sector and dominant role of moneylenders in agricultural credit. The small and marginal farmers who normally suffer for want of adequate credit, borrow from moneylenders and indebtedness is of higher magnitude in the case of small and marginal farmers.[2]

4. Evolution of Institutional Credit

The evolution of institutional credit to agriculture could be broadly classified into four distinct phases.

1. 1904-1969 — Predominance of cooperatives and setting up of RBI
2. 1969-1975 — Nationalization of Commercial Banks and setting up of Regional Rural Banks (RRBs)
3. 1975–1990 — Setting up of NABARD
4. 1991 onwards — Financial sector reforms as a part of New Economic Policy of Liberalization and Globalization.

Moneylenders for a long time, dominated the rural credit scene. The need for institutional credit was realized soon in view of seriousness of rural indebtedness.

5. Rural Indebtedness

A considerable number of people in rural areas do not have gainful employment throughout the year. Therefore, they do not have sufficient income to secure food and other necessities of life. They are compelled to borrow not only for productive purposes, but also for consumption purposes. The amounts borrowed are not utilized for asset-building or to increase the yield from the land or generate incremental income and often they fail to repay the loans. Often, rural people search for new sources of borrowing in order to repay the out standing loans. They borrow mostly from private sources, under certain terms and conditions, which are quite unfavourable to the borrowers. From time to time, the debt burden goes on increasing and this is what is known as "Rural Indebtedness".

According to the All India "Rural Credit Survey" (1951-52), of the Reserve Bank of India, about 63 per cent of Rural families were in debt and the average amount of debt for family was ₹ 283. As per the National Sample Survey 59th round[2] (Jan-Dec 2003) in All India, 48 per cent of farmer household were indebted. At state level, there were variations. In states like Andhra Pradesh, Gujarat, Haryana, Karnataka, Kerala, Madhya Pradesh, Maharashtra, Punjab, Rajasthan, Tamil Nadu, West Bengal and in Union Territories, percentage of farmer households indebted exceeded 50 per cent. The average outstanding loan amount in All India stood at ₹ 12,585. The NSS 59th round[3] January-December 2003, identified that all categories of farmers with different size of holdings of less than 0.01 hectare and more than 10 hectares of holding were in debts. The average indebtedness of all size of holdings in All India stood at 48.6 per cent, even farmer households with more than 10 hectares, the indebted farmer households stood at 65.1 per cent. In the case of households with landholdings between 4 and 10 hectares indebted households stood at 66.4 per cent. In case of households with landholdings between 4 and 10 hectares, indebted households stood at 65.1%. In the case of other extreme, with those less than one hectare, the percentage of indebted households was 45.3%. Regarding those between one and two hectares, the indebted household percentage stood at 51. It is thus evident that in all categories with different size of holdings, indebted farm households were recorded. From this angle, the seriousness of the problem can be well understood.

6. Causes for Rural Indebtedness[4]

There are innumerable causes of socio-economic-political nature leading to heavy rural indebtedness from time to time. The causes for rural indebtedness are many and important causes are as follows:

(i) **Ancestral Debt:** The most important cause of the existing indebtedness is the ancestral debt, which is handed over from father to son, generation to generation. Many agriculturists start their life with a heavy burden of ancestral debt and drag on the loan for the whole of their lifetime. For their part, they may add to the debt burden by way of fresh loans they raise for religious and social obligations. The Royal Commission on Agriculture remarked that "The Indian peasant is born in debt, lives in debt, dies in debt and bequeaths debt." This is really true in many cases of agricultural families in rural areas. In fact, rural people are so accustomed to the practice of taking over debt from their father and to pass it over to their sons, that they accept indebtedness as a settled fact and a natural state of life.

(ii) Subdivision and Fragmentation of Holdings: For various reasons, land is divided and subdivided from time to time. When the holdings are small, cultivation becomes uneconomical. Even in the best of years, the yield from land becomes insufficient for the maintenance of the farmer and his family. The holdings are so small and the margin of safety is so narrow that any misfortune may throw the peasant into the debt trap from which he can never release himself. A series of bad years, the death of cattle or mere carelessness may lead to additional debt. In the best of years, the surplus produce may be too small to provide the required amount to repay the loan amount.

(iii) Vagaries of the Climatic Conditions and Other Calamities: Agricultural production suffers from frequent failure of rains and resultant famines. Indian agriculture is described as a "Gamble in the Monsoon". This makes agricultural production quite uncertain. The failure of crops either due to drought, floods or any other unforeseen events, lands the farmer in miserable condition. He is forced to borrow money from the moneylenders who exploit him in several ways.

(iv) Ignorance and Illiteracy of Cultivators: Ignorance, illiteracy and superstition of the rural folk act as a hindrance to the progressive agriculture. Finally, it results in debt burden. Illiteracy forms one of the principle obstacles to a farmer's progress and thus he us easily lured into the clutches of the shrewd and unscrupulous moneylenders. It has been said that "The moneylender tempts him to borrow, the lawyer to quarrel and the trader to waste".

(v) Failure to Provide for Deficiency: Agriculture is subject to the law of diminishing returns and in the absence of application of modern inputs, the yields remain low. As most of the cultivators are in poverty and as they get low income from the tiny plots, which is hardly sufficient to meet their necessities, they are not in a position to keep aside funds for depreciation of agricultural tools and equipment, cattle, etc. Uncertain weather conditions, loss of cattle due to diseases, fluctuations in farm prices, etc., force the farm producers to borrow and the loss is not so easily recouped.

(vi) Low Income of Cultivator: Poverty and lack of capital are the twin evils from which the entire rural economy suffers. It is a well-known fact that the average income of the rural community is very low and insufficient to cover even domestic minimum consumption expenditure. Because of low and insufficient income in many cases rural people are underfed and suffer from malnutrition. Physical deficiency, resulting from such conditions, makes them an easy prey to diseases and this situation also forces them to borrow.

(vii) High Rates of Interest: The interest rates charged by the village moneylenders are often very high and the compounding interest tends to perpetuate the indebtedness of the farmer. The interest rates vary from about 10 to 30 per cent in different states and in different situations. High rate of interest would land the borrower in heavy indebtedness.

(viii) Extravagant and Improvident Borrowing: Although the Indian farmer is said to be leading a simple and frugal life, sometimes he indulges in extravagant expenditure. He spends money wastefully on social ceremonies like marriages, funeral rites etc. The long series of festivals observed, particularly by the Hindus, land the rural poor in indebtedness.

(ix) **Litigation:** Calvert estimated that nearly 2.5 million persons attended court every year, either as parties or as witnesses, and that way ₹ 3 to 4 crores were wasted in the process. Litigation involves not only more expenditure, but it drags on for several years. One has to spend huge amounts by way of payments to lawyers, court fees, expenses to secure required document, sometimes to hire witnesses, etc. It is not uncommon to hear of the suits dealing with petty land disputes over a fraction of an acre running over several years in lower courts and then going for an appeal in the High Court. Prolonged litigation involves huge unproductive expenditure and this also leads to indebtedness.

(x) **Distress Sales:** Sometimes, out of pressing cash requirements, agriculturists sell away their produce even at a lower price without keeping for domestic requirements like food, seed, etc. The same persons once again buy, often by borrowing the foodgrains, seed etc., paying higher prices. This type of distress sales also accounts for indebtedness.

(xi) **High Medical Expense:** Towards medical expenses, the rural poor are forced to borrow heavy amounts and this also leads to indebtedness. The Primary Health Centers are not of much use as are not well equipped with doctors and drugs and often the rural poor are forced to go to corporate hospitals driving them into further debt burden. The local money lender may come to the rescue immediately and the terms and conditions are quite unfavourable to the borrower. This situation ultimately may drive to indebtedness.

In a situation of heavy debt burden and unable to repay the heavy debt incurred, out of distress, some resort to suicides. According to NSS 59th round. In the ten year period between 1997 and 2006, as many as 1,66,304 farmers committed suicide in India. It is also estimated the 27% of farmers did not like farming because it was not profitable. In all 40% felt that given a choice they would take up some other career. The problem of rural indebtedness demands effective steps by the government.

7. Evils of Indebtedness[5]

The evils resulting from indebtedness may be enumerated as follows:

(i) The chronic state of indebtedness affects the cultivators in many undesirable ways. The very low standard of living with miserable life of some rural people is the outcome of indebtedness.

(ii) Debts also prevent the orderly and profitable marketing of agricultural produce.

(iii) A large part of the income of the peasant is spent on repayment of outstanding debts and he is left with a very meagre amount quite insufficient for his subsistence.

(iv) Indebtedness causes loss of property and transfer of land from cultivators to non-cultivators. This lead to grave economic and social consequences in the rural areas.

(v) The worst social and moral effect of indebtedness is that it results in the servitude of the debtor, sometimes resulting in bonded labour. The rural households who have lost their lands to moneylenders due to heavy indebtedness feel alienated from society and feel frustrated. The different consequences of this situation are of serious nature.

8. Remedial Measures[6]

It should be recognized that indebtedness is a hindrance to rural development and therefore, debt relief measures should become an integral part of rural development programmes. Debt relief must necessarily be in two ways, one by way of reducing the burden of existing debt and another is preventing of the emergence of new debt.

Reducing the Burden of Existing Debt:

A large part of outstanding debt in many cases is inherited one. A small amount might have been borrowed decades ago at a high rate of interest. Since even a part of interest amount has not been cleared, the compounded interest together with principal amount, accumulated over a period of time results in heavy debt. The debt is dragged on for several decades and some times the amount paid by way of interest would be much higher than the original principal amount borrowed. The moneylenders sometimes resort to malpractices and even if the part payment is made, it is not recorded in the promissory bond. The moneylenders exploit the borrowers by way of forcing them to render services without payment. Appropriate steps towards the cancellation of certain types of inherited debts become necessary to help the rural poor. If the moneylenders had received payment by way of interest much more than the original principal amount, there is justification for cancellation of such debts. Further, if the inherited debt amount is in excess of the value of the property inherited, scaling down of such debt is justified. Also, scaling down of debts of long standing and where it is almost impossible for the debtor to repay is justified.

Preventing the Emergence of New Debts:

Apart from reducing the burden of existing debt, it is equally important to take necessary steps to prevent the emergence of new debt. This is possible in three ways.

(a) Curbing the unproductive expenditure through educating the rural poor with relevant knowledge of the dangers of incurring debts for various unproductive social and religious ceremonies and avoiding going to courts over petty disputes;

(b) Raising of the earning capacity of rural work force through the provision of appropriate skills and technology; and provision of supplementary occupation/enterprise to every farm household;

(c) Making arrangements for timely and adequate institutional credit to all categories of rural people for meeting both production as well as minimum subsistence requirements of the rural poor without giving room for the role of moneylenders.

9. Role of State and Rural Credit

The problem of mounting debt burden and farmers resorting to suicides must be addressed by the state with all seriousness. Write off the debt is not a solution to this recurring problem. Perhaps this step encourages the farmers to borrow huge amounts from the banks and look for further write off. Even after the write off of bank loans, it was reported that farmers continued to commit suicide. It is because the borrowers who borrowed from the moneylenders are not benefited by the step of write off of the bank loans. In tackling the problem of indebtedness the following steps become necessary.

(i) Provision of adequate and timely institutional credit to all categories of farmers with adequate supervision on the end use of credit.

(ii) The role of moneylenders and local merchants in meeting the credit of form producers must be eliminated.

(iii) Facilities must be created for every rural household to have a supplementary occupation like dairying, depending upon local resources

(iv) Livestock-linked farming should be revived and encouraged. Insurance cover both for crops and livestock is necessary.

(v) Revival of indigenous farm technology coupled with modern bio-technology, accomplied by gradual reduction in the use of chemical fertilizers and pesticides is necessary.

(vi) Encouragement of SHGs, particularly of rural poor is necessary to mobilize resources for asset-building.

(vii) Investment in watershed development and water saving technologies in the rain fed areas would certainly help in improving the incomes of the rural households

(viii) Ways and means must be explored to reduce input costs and government takes the responsibility of supplying qualitative inputs, particularly quality seed.

(ix) Strengthening of the primary health facilities in rural areas as a part of development strategy.

(x) The Minimum Support Price (MSP) mechanism should be strengthened and scope widened covering all crops across the country. MSP of different crops need to be guided by rising input costs.

(xi) An efficient marketing system.

(xii) A comprehensive social security and employment security ensuring likelihood security is essential.

10. Multi Agency Approach in Rural Credit

10.1. Cooperatives and Rural Credit

Cooperation is nothing but a voluntary association of individuals for the achievement of a common good. It stands for the principle of "all for each and each for all". The basic principles of cooperation are: (a) Voluntary Association, (b) Democratic organization, (c) Self-help through mutual cooperation, (d) Common welfare and (e) Selfless service with honesty and integrity.

The need for cooperative approach, particularly with regards to rural credit was highlighted in the report of Frederick Nichelson in 1895. The Famine Commission of 1901, also stressed the need for credit society in India. Some individuals also made some efforts in popularizing the cooperative credit scheme.

The history of cooperative movement in India began with enactment of the Cooperative Society Act, 1904, by the Government of India. The main provision of the Act were:

(i) Any ten persons in a village or town may join together for the formation of a society and register the society for the encouragement of thrift and self help among the members;

(ii) The main objectives of a society are:

(a) To raise funds by way of deposits from members and loans from non-members, government and other cooperative societies and

(b) To disburse the money so obtained as loan to members or to other cooperative credit societies with the special permission of the Registrar of cooperative societies.

(c) To extend credit facilities in adequate measure at a cheap rate of interest.

The Cooperation Act of 1912, recognized, along with cooperative credit societies, cooperatives for marketing, farming, housing, etc. The Mechegan Committee of 1914, concluded that "In majority of cases, primary societies in India fall short of the cooperative ideal". However, the Godgil Committee of 1945, the Saraiya Committee of 1946 and the Manilal Nanavati Committee of 1947, have categorically stated that cooperatives are the most suitable agencies alternative to private moneylenders at the village level for providing credit for different purposes on reasonable terms and conditions.

The report of the All India Rural Credit Survey, 1954, made a number of recommendations of great signification for rural and agricultural development. The Mehta Commission of 1959, appointed by the government of India also made several important recommendations Accordingly, amendments have been made from time to time to the Cooperative Act so as to enable the organizations to function effectively. The Cooperative Credit structure of India consists of two parts. One provides short-term and medium term loans. The other provides long term loans. This part of cooperative credit organization has a three-tier structure. At the lowest or village level, there are primary agricultural cooperative credit societies, at the village level, central cooperative banks at the district level and state cooperative banks as the Apex. Table 17.2 presents the trend in cooperative credit to agriculture.

Table 17.2

Cooperative Credit to Agriculture

Year	Cooperative Banks	Share (% age)
1	2	3
1985-86	3874	55
1990-91	3408	39
1995-96	10479	48
2000-01	20801	39
2005-06	39404	22

Source: Economic Survey and NABARD various issues.

From Table 17.2 it is clear that the cooperative credit to agriculture stood of ₹ 3,874 crore in 1985-86 or 55%. It has come down to ₹ 3,408 crore in 1990-91 (39%). However, the amount increased to ₹ 10,479 in 1995-96 (48%) and from that time it began declining. In 2000-01, share of cooperatives in total credit was 39% and in 2005-06, it has come down to ₹ 39,404 crore or 22%.

Instead of depending upon cooperatives alone for short, medium and long terms agricultural finance, it was felt that other agencies also, besides cooperatives play their role. Therefore, along with cooperatives, the State Bank of India and its subsidiaries, commercial banks, and regional banks are associated in meeting the credit requirements to agriculture.

10.2. Involvement of Commercial Banks

The involvement of commercial banks in the agricultural finance began with the nationalization of Imperial Bank of India in 1955, which began functioning as the State Bank of India (SBI). The State Bank of India has set up a separate wing namely Agricultural Development Branch, with the aim of promoting agricultural growth. In this direction, "The village adoption scheme" has been introduced. Under the scheme, the credit requirements of the cultivators and the rural artisans in the adopted villages are met by the Bank.

Soon, the State Bank of India has set up Agricultural Development branches spread over the entire country. The State Bank of India provides advances against the warehouse receipts. This facility helps the agriculturists to wait for better price for his produce in the market. The State Bank of India group has been financing small scale industries, right from the stage of purchase of raw materials upto the stage of marketing, including exports.

The State Bank of India launched in 1977, an integrated rural development programme on a pilot basis with the aim of meeting not only the requirements of agriculture and allied activities, but also rural industries, rural housing, rural health and such other needs of the villagers in selected areas.

The Differential Interest Rate (DIR) scheme introduced by the State Bank of India covers mainly the physically handicapped persons, scheduled castes and scheduled tribes, orphans, etc., also, under this scheme, financial assistance is provided for productive purposes to the weaker sections of the community like barbers, cobblers, carpenters, rickshaw-pullers, basket-makers, blacksmiths, potters etc.

The State Bank of India also provides financial assistance for research schemes of rural development undertaken by individual researchers of institutions.

11. Social Control of Commercial Banks

Soon it was realized that all commercial banks in the country must necessarily be involved in financing the agricultural sector also consistent with the set objectives of planning and overall goal of establishing a Socialistic Pattern of Society, launched by Pandit Nehru, the first Prime Minister of India. To began with, social control of commercial banks was introduced by effecting changes in the composition of the Board of Directors of every Bank.

12. Nationalization of Commercial Banks

Social control over commercial banks has not resulted in active involvement of commercial banks in financing priority sectors including agriculture, small scale industry etc. In July, 1969, the 14 major commercial banks were nationalized with a view to giving purposeful direction to the resources owned by them. In April, 1980, six more commercial banks were nationalized and now it can be said that banking industry in India is almost in the public sector.

Now the nationalized banks also like State Bank of India took up village adoption scheme, As desired by the Reserve Bank of India, a scheme of financing agriculture by the commercial banks through Primary Credit Societies was taken up.

An expert committee constituted by the Reserve Bank of India under the chairmanship of Gunvant Desai made the following recommendations for the effective functioning of the commercial banks in the rural credit market:

(i) The agricultural credit schemes should be thoroughly examined from the point of view of economic viability, repayment capacity, infrastructure support to and functional linkage of the purpose of loan with other activities and development programmes, staff required to implement schemes, etc.

(ii) Land records should be updated and revenue passbooks should be issued to the farmers containing all relevant information of the landownership, source of irrigation, crops raised, taccavi loans taken, if any.

(iii) Producers in processing of loan applications should be simplified and standardized

(iv) The commercial banks should provide credit on the basis of the revenue passbooks.

(v) As far as possible, the branch managers should not be transferred within three years so that they can have adequate time to know fully the local conditions.

(vi) A branch manager should be given powers to sanction loans up to ₹ 5,000 for approved credit schemes without insisting on encumbrance certificate, etc.

(vii) Sureties should not be insisted in the case of applications from weaker sections.

(viii) Disbursement of loans in kind should be preferred to that in cash to prevent misuse of credit.

(ix) There should be proper supervision over the use of credit by the trained staff.

(x) There should be effective steps in the recovery of loans and overdues through tie-up arrangements with the marketing agencies, timely reminders to borrowers and stern action against willful defaulters.

(xi) The expert committee recommended that all state governments should enact legislation, incorporating all the provisions of the Model Bill, recommended by the Talwar Committee, and take steps to effectively implement them in order to avoid time-consuming procedures in processing applications and sanctioning the loan amounts.

(xii) It is also recommended that a simplified loan application form as evolved for Regional Rural Banks should be adopted by all commercial banks.

(xiii) The expert committee also recommended that commercial banks should establish a monitoring and evaluation cell "not only to keep a watch on the progress in lending to achieve the targets, but also to gather information on the difficulties experienced in implementing the schemes. Such information should be utilized both for follow-up action at various levels as well as to draw lessons in modifying the existing schemes and formulating new schemes".

(xiv) The objective of commercial banks should not be just making profits, but follow the principle of "growth with profits". The commercial banks should also enjoy the confidence of the depositors by utilizing the funds with necessary care and caution.

It is observed that since 1975, published reports on profits of public sector banks have been showing a declining trend both in relation to income and deposits. It is further estimated that profitability is likely to decline still further in relation to both income and deposits in future.[7] However, since commercial banks are expected to play a crucial in rural development, their efficiency should not be judged on the basis of level of profits alone at any point of time. The commercial banks need to be actively and extensively involved in financing of rural sectors, both agricultural and non-agricultural. The State Governments must assist the commercial banks in a number of ways, particularly by way of appointment of paid secretaries to the societies, right of inspection of borrowing societies, settlement of disputes, through the Registrar of cooperative societies, and providing necessary powers to the banks to proceed against the defaulting members of a cooperative society. In charging a differential rate of interest, care should be taken to ensure that the weaker sections are really benefited by such step. Further, the commercial banks are in a better position to provide both short and long-term loans meeting production and investment requirements. Therefore, commercial banks must be actively involved in financing of rural development schemes through a "single window credit system." Also, the commercial banks are in a better position to support the required staff both for extension service and supervision of credit.

13. Commercial Banks and Weaker Sections in the Reform Era

Commercial banks must necessarily have a social concern addressing, particularly the needs of weaker sections. Weaker sections may be defined as those who are socially and economically backward and who are wanting strength or vigour to sustain themselves. The inputs required to strengthen such people may be different in different cases such as financial assistance, tertiary, health, motivation for development, self-confidence, and more importantly ethics.

No doubt, as a result of planned development of nearly six decades, there has been significant improvement in different sectors of economy. However, the benefits of development have not trickled down significantly so as to reach the needs of the poor and downtrodden. Still some 40% remain illiterate; some 70% has no proper sanitation and 20% lack access to drinking water. The conditions of rural poor are still more miserable with higher degree of illiteracy, poor medical facilities, etc. Along with growth, the rich-poor divide is widening leading to "co-existence of unsustainable lifestyles and unacceptable poverty". It is noticed that economic growth is taking place without generating employment. In the era of economic reform, over 63% of the work-force engaged in farming has to share only 19% of GDP while a little over one-fifth of the work force engaged in service sector obtains a little over half of the GDP. The poverty ratio in India in 2004-2005 was 27.50 per cent. Rural poverty is estimated at 28.30 per cent (Planning Commission). At disaggregated level as per the study of National Bank of Agriculture and Rural Development, the poverty ratio in All India in 1993-94 was concentrated in agricultural labour with 54.4 per cent, followed by 42.2 per cent in non-agricultural labour, 32.8 per cent in artisans, 29.7 per cent in self-employed in agriculture and 21.2 per cent "others". With all probability this trend continues. The rural poor have been by passed both in terms of growth and distribution of gains of growth. MacNamara, former President of World Bank rightly observed, "Growth is not equitably reaching the poor and the poor are not significantly contributing to growth". Research study reveals that the benefits of development so far have been pocketed mostly by urban rich and rural poor have failed to be participants in the development process and get benefits. As a consequence, we notice a clear cut distinction between urban and rural centres of growth and the well to-do sections and weaker sections of the community.

During the Economic Reform Phase, between 1993 and 2007, in 15 years period, some 4750 rural bank branches were closed down in the name of 'consolidation' (*The Hindu*, 28th March, 2008). The credit lending by the commercial banks to the priority sectors declined from 39 to 32 per cent as against the Reserve Bank stipulation of 40 per cent. The share of rural credit as a percentage of total credit by all scheduled commercial banks has been declining and this declined to 7.93 per cent in March 2007. As per NSS 59th Round of 2003, the share of institutions in total rural credit accounted for 61 per cent and that of private sector accounted for 39 per cent. It is evident that still more than one third of rural credit is met by the private sources only. The money lenders still dominate the rural credit scene, despite cooperatives and Regional Rural Banks. The National Bank for Agriculture and Rural Development (NABARD) set up in 1982, gives special attention to different priority sectors like agriculture, small-scale and cottage industries. The NABARD provides refinance assistance for the development of tiny, small-scale and cottage industries and rural handicrafts. There is a provision for rescheduling of loans and advances to rural artisans, cottage industries, handicrafts upto seven years if NABARD is convinced of the necessity of rescheduling.

Self-help Groups need to be extended covering all weaker sections as it facilitates mobilization of small savings, promotes thrift in the rural poor, improves recovery of loans and promotes asset-building by the weaker sections. Though, there are costs involved in the promotion of SHGs, definitely, the social benefits of SHGs outweigh the private and social costs. Hence, the case for further expansion of SHGs.

Loans for Infrastructure Building

Weaker sections are very much handicapped in marketing and getting remunerative prices due to lack of infrastructural facilities. The commercial banks must advance loans to the village panchayats for creation of infrastructural facilities such as village godowns, cold storage facilities for grading, packaging etc. This arrangement goes a long way ensuring remunerative prices to producers and easy recovery of loans by the banks.

14. Commercial Banks and Rural Artisans

The Bank credit for rural artisans has been on the decline during the Reform era. In 1996, the percentage share of Bank credit for rural artisans out of total artisans in India was 52.15. In 2006, the percentage share declined to 27.57. Generally traditional industries run by rural poor find it difficult to have access to various institutional agencies providing credit, mainly because of their inability to provide adequate security. Even the group credit guarantee scheme, introduced by the Reserve Bank of India, is not able to protect adequately the interests of rural artisans and other weaker sections. There is a wide communication gap. It appears that the promising rural entrepreneurs are not properly posted with adequate up-to-date information nor are they assisted adequately by the commercial banks. It is necessary that the young technical officers associated with the banks should have the requisite practical experience so as to guide the rural entrepreneurs properly. The District Industries Centres need to be strengthened so that they can deliver the goods effectively in terms of timely and adequate credit along with technical services. Small Industries Service Institutions (SISI), National Institute of Small Industry and Training (NISIET), Central Institute of Tools Design (CITD) should be actively involved in providing extension services to different categories of village industries. Further preferential treatment is to be given to the traditional industries by the introduction of appropriate technology, grant of tax concessions, provision of timely and adequate institutional credit at lower rate of interest, regular supply of raw material, and provision of organized market channels with a thrust on weaker sections.

15. Regional Rural Banks

Inspite of rapid expansion of branches of commercial banks and increase in the membership of primary agricultural credit societies, it was realized that institutional credit was not available in adequate measure to the weaker sections in the rural areas. Basing on the recommendations of M. Narasimhan Committee, regional rural banks were set up in 1976. The role of the regional rural banks would be "to supplement and not supplant the other institutional agencies in the field.[8] The regional rural banks are found to be well suited for the purpose of meeting the credit requirements of the rural poor. It is hoped that the RRBs would help the rural poor in securing adequate institutional credit and thereby reduce the dependence of the rural poor on the village money lenders. The RRBs are able to identify the weaker sections of the rural community, especially the scheduled castes and render them financial assistance. The RRBs are providing loans with a policy of differential rate of interest. Again, the physically handicapped are provided cheap finance for the purchase of artificial limbs, hearing aids, etc., subject to a maximum of ₹ 2,500.

The Danthurala Committee, after reviewing the performance of RRBs has concluded that "within a short span of 2 years, the RRBs have demonstrated their ability to serve the purpose for which they were established.[9]

16. The Review Committee (Danthwala Committee) Recommendations

The Danthwala Committee appointed by the Reserve Bank of India in 1977 made the following recommendations.

(i) The RRBs should become an integral part of the rural credit structure.

(ii) The jurisdiction of a RRB should be over one district only with 50 to 60 branches, each branch covering a population of 20,000.

(iii) There should be an appropriate organization set up in the Reserve Bank of India to look after the performance of RRBs.

(iv) The Reserve Bank of India, the sponsoring bank, the State Government and the local participants should contribute by way of shares.

(v) The RRBs should earmark at least 6 per cent of the loans for the benefit of small farmers, rural artisans and other category of rural poor.

(vi) It is recommended that specialized training should be imparted to the personnel of the RRBs by the sponsoring banks.

In a multi-agency approach RRBs also have been playing a crucial role. As per Reserve Bank of India Occasional Paper, 2007, the share of Commercial Banks including RRBs in 2005-06, reached to 68 per cent. The RRB branches should be set up where the role of cooperatives is inadequate to meet the credit needs of the rural poor. It is necessary that the RRBs and cooperative institutions function side by side without clash of interest.

All commercial banks, including sponsoring banks, should handover, in a phased manner, to the RRBs, such business which they can effectively handle and serve the rural poor. However, demarcation of business for each institutional agency with flexibility becomes necessary to avoid unhealthy competition in the rural credit system. All the credit giving agencies, cooperatives, commercial banks, RRBs, NABARD should function without clash of interest aiming at meeting the credit needs of the rural poor. In the process of expanding branches of different institutions priority must be given to uncovered areas. In the exercise of amalgamation of institutions, the rural poor should not suffer in any part of the country in India.

17. National Bank for Agriculture and Rural Development (NABARD)

The Reserve Bank of India, in consultation with the Government of India, set up a Committee to Review Arrangement for Institutional Credit for Agriculture and Rural Development (CRAFICARD) in 1979, under the chairmanship of B. Shivaraman, The CRAFICARD recommended that the NABARD be set up. The objective was that such a National Bank would be able to give undivided attention to the problems of providing production and investment credit to different rural sectors like agriculture, small-scale and cottage industries, etc. The NABARD began functioning from July 1982.

18. Functions of the NABARD

(i) The NABARD will provide refinance assistance to the State Cooperative Banks, the Regional Rural Banks and any other financial institution approved by the Reserve Bank of India. The NABARD also provides refinance assistance to carry on agricultural operations including marketing of agricultural produce. Further, refinance assistance is provided to carry on any activity for the rural development, including agricultural development.

(ii) The NABARD will provide medium and long-term loans. The medium term credit is provided for periods not less than 18 months and not exceeding 7 years to the State Cooperative Banks and Regional Rural Banks. The long-term credit is provided not exceeding 25 years to the state cooperative banks and regional rural banks.

(iii) The NABARD will provide loans and advances to state governments not exceeding 20 years to enable them to subscribe directly or indirectly to the share capital of cooperative credit societies.

(iv) The NABARD is empowered to undertake inspection of cooperative societies and RRBs without prejudice to the powers of the Reserve Bank of India in this respect.

(v) The NABARD will maintain research and development fund out of the profits made. The fund is utilized in undertaking research in different aspects of programmes of agriculture development to suit the requirements of different regions.

So far, the NABARD financed different schemes such as Minor Irrigation, Land Development, Farm Mechanization, Plantation and Horticulture, Poultry, Sheep, Goat, Piggery, Fisheries, Dairy Development, Forestry, Storage and Marketing and IRDP programmes.

There is a feeling that the different credit institutions validating Regional Rural Banks continue to benefits mostly the rural rich. The NABARD must necessarily give attention to this aspect of the problem so as to ensure that the weaker sections are fully and adequately covered by institutional credit. There are variation in the availability of institutional credit per hectare of gross cropped area in different states. It was as high or ₹ 9403 is Tamil Nadu, ₹ 7,666 in Kerala, ₹ 5,352 in Punjab and ₹ 4,604 in Andhra Pradesh while it was as low as ₹ 311 in Assam, ₹ 667 in Rajasthan and ₹ 698 in Madhya Pradesh during 2001-02.[10]

During the Tenth Five year plan, the total credit flow to agriculture and allied activities was projected at ₹ 7,36,570 crore. Accordingly the ground level credit flow to agriculture has grown to reach over ₹ 2,60,540 crores (36 per cent of the projected level) during the first three year period (2002-03 to 2004-05) of Tenth Plan, indicating a wide gap in supply of credit, requiring a large increase in credit, particularly in investment credit to achieve the desires growth level.[11]

19. Recent Policy Initiatives

The Finance Minister in his Union Budget 1995-96, speech stated that "Inadequacy of public investment in agricultural is totally a matter of general concern. This is an area which is the responsibility of states. But many states, have neglected investment in infrastructure for agriculture. There are many rural infrastructure projects which have been started, but are laying incomplete for want of resources. They represent a loss of potential income and employment to rural population.

Rural Infrastructure Development Fund (RIDF) was set up in NABARD. Commercial banks make contributions towards the fund, the scope of RIDF has been widened to enable utilization of loan by Panchayat Raj Installations (PRIs), Self-help Groups (SHGs), Non-government organization (NGOs) etc., since 1999-2000.

Two innovations, *viz.,* Micro finance which was formally heralded in 1992, through self-help groups and Kisan Credit Cards Scheme (KCCS) of 1998-99. Self-help Group is a group comprised of small/marginal farmers, landless agricultural labourers, rural artisans, women folk and other micro-enterprises, organized themselves as a voluntary action or by the initiative of NGOs. Banks many adopt such SHGs and supplement its resource base by providing them loans. These SHGs in turn will extend finance to their members for taking up suitable economic activity as well as meeting their pressing consumption needs. The success of SHG depends primarily on the motivation and dedication level of each group members and the extent of repayment of loan taken regularly.

The KCCS was introduced in 1998-99, with a view to facilitating the flow of timely and adequate short-term credit to the farmers. The KCCS has emerged as the effective mode of credit delivery to agriculture in providing timely and adequate credit with minimum transaction costs and documentation. The cooperative banks, RRBs and commercial banks evinced interest in encouraging both the schemes to the betterment rural poor.

So far as SHG-Bank Linkage Programme was concerned by 2006-2007, some 29.24 lakh SHGs were formed and bank loans amounted to ₹ 18,041 crore and cumulative refinance was to the extent of ₹ 5,459 crore.[12] Agency-work KCCS indicates by 2005-2006, cooperative banks share was 30.41 millions (51.5%) RRBs 6.88 millions KCCS, Commercial Banks 21.80 million, the total being 59.09 millions. In terms of percentage distribution cooperatives claim 51.5%, RRBs 11.6% and Commercial Banks 36.9%.[13]

Notwithstanding the rapid spread of micro finance programme there has not been even distribution of SHGs across the states. More than 50 per cent of the total SHG credit linkages in the country are concentrated in the southern states. In the states which have a large share of the poor, the coverage in comparatively low.

20. Mounting Overdues

Through the multi-agency approach, massive network of rural branches of commercial banks, cooperative banks and regional rural banks has been established for improving the access to credit. However, this expansion also created a major problem of mounting overdues. During 2001-02, RRBs overdues stood at 30%, overdues of commercial bankers stood at 17% and overdues of cooperative bankers stood at 16%.

Several factors account for overdues with RRBs, CBs and cooperatives. These include: (i) lack of rescheduling/rephasement of the loan during the period of natural calamities, (ii) inadequacy of loan and improper timing of loan disbursement, (iii) Repayment period is short, (iv) absence of moratorium, (v) not giving sufficient time to repay, particularly for the small and marginal farmers, (vi) failure to provide working capital loan in adequate amounts and (vii) the borrowers of all categories are not effectively supervised with reference to end use of credit.

21. Supervised Credit System

The need of the hour is provision of adequate and timely credit to all weaker sections without the role of private moneylenders. However, the credit should not become a "hangman's rope" for the borrower, nor should it jeopardize the interests of banking institutions. Towards this end, there is need for Supervised Credit System. In the words of H. Belshow, "supervised credit is a technical term to describe a particular system of credit aiming at improving both production and family living, which coordinates credit provision with extension and aims at effective supervision, but usually goes further in making improved provision for marketing and other ancillary services". It simply means the provision of credit should be accompanied by effective supervision of credit use, coordinated with extension and marketing services. The extension service must be in the direction of guiding the farm producers to adopt livestock-based farming technology to ensure the level of chemical residues in agricultural food stuffs remained within the tolerable limits harmonized with international standards and specifications as required under the liberalized policy. No doubt, the commercial banks have on their rolls, technical persons, named as Agricultural officers. They have to devote time for technical service and effective supervision over the end use of credit. The supervisors should also help the borrowers in getting remunerative price for their produce so as to ensure regular repayment of the loan amount. A well-thought-out credit system with effective supervision along with extension and marketing services meeting credit needs of all weaker sections is immediately called for.

22. Reserve Bank of India and Rural Credit

Reserve Bank of India's Occasional Paper observes, "an assessment of agricultural credit situation brings out the fact that the credit delivery to the agriculture sector continues to be inadequate. It appears that the banking system is still hesitant on various grounds to provide credit to small and marginal farmers. The situation calls the concerted efforts to augment the flow of credit to agriculture, along side exploring new innovation in product design and methods of delivery through better use of technology and related process."[14] The cooperative credit structure needs revamping to improve the efficiency of the credit delivery system in rural areas. The commercial banks and RRBs must focus attention on covering unreached population. The Occasional Paper[15] rightly observes, that the experience of micro finance proved that the "poor are bankable" and they can and do save in a variety of ways and the creative harnessing of such savings is a key success factor. There is also need to explore the possibility how SHGs can be induced to graduate into matured levels of enterprise. The SHG-Bank linkage programme needs to introspect whether it is sufficient for SHGs to only meet the financial needs of their members or whether there is a further obligation on their part to meet the non-financial requirements necessary for setting up business and enterprises.

Conclusion

Provision of adequate and timely institutional credit covering mainly the interests of small and marginal farmers so as to prevent them from falling into the clutches of money lenders is an important requirement. To have an adequate impact on agricultural production, the provision of credit should be accompanied by, and coordinated with, extension services and supervision. Cooperatives are better suited to exercise supervision, and effective supervision should be invariably treated as a legitimate function of the apex banks and central banks. As financiers of societies, they should take necessary steps to ensure sound working of the primary credit societies and primary land development banks. Through proper supervision, the mounting cooperative overdues can be minimized. The government should help the cooperatives, to introduce suitably designed supervised credit programme. To start with, such schemes may be introduced in places where cooperatives are functioning well and gradually extend the scheme to other places. The commercial banks may be insisted upon to have adequate technical staff on their rolls. It is paradoxical that agricultural graduates remain unemployed in some states, in a situation of inadequate technical staff, including extension personnel.

The Regional Rural Banks sponsored by the commercial banks may be utilized for introducing the supervised credit scheme. For effective implementation of the programme, well-trained persons should be employed as local supervisors. Young persons with high school education preferably from agricultural families with practical farm experience may be chosen for training to serve as credit supervisors. One year specialized training in agricultural economics with emphasis on farm credit, cooperative organization, farm management, techniques of supervised credit etc., is necessary. The technical aspects of farm operations such as crop choice, use of fertilizers, control of pests and plant diseases, should be included in training courses. The Reserve Bank of India, has to necessarily play a crucial role in helping cooperatives and commercial banks to introduce supervised credit scheme.

The credit giving institution, while providing short-term credit to weaker sections, should also take into consideration the normal credit requirements for consumption purposes. However, a greater proportion of the credit should be in kind. Further, it should examine carefully the technical feasibility and productive capacity of the proposed farming operation with the help of credit. Technical advice, if required, should be provided to the loan applicant at this stage to evolve a suitable plan. Follow-up action must begin with the release of the first installment of the loan. The credit supervisor must work closely with the borrowers. He has to supervise and control the use of the loan amount throughout the crop season. The subsequent installment of the loan amount should be released after the bank is fully satisfied with the utilization of the loan amount. Tenant cultivators, agricultural labourers, rural artisans and other weaker sections should be made creditworthy by helping them to take to certain gainful occupations through supervised credit programme. Attention should also be given to marketing and it should be made obligatory on the part of the borrower to sell his produce with the knowledge of the concerned bank. The credit supervisors have to assist the borrowers in marketing products. Even an element of effective supervision over credit use will go a long way in using the credit for productive purpose and reap the benefits. A well thought out supervised credit system with organizational efficiency aiming at the benefit of the weaker sections is immediately called for.

REFERENCES

1. *Eleventh Five Year Plan*, p. 20.
2. *Agricultural Statistics, 2008*, p. 313.
3. *NSS – 59th round*, January-December 2003.
4. K. Venkat Reddy, *Agriculture and Rural Development*, pp. 194-195.
5. *Ibid.*, p. 196.
6. *Ibid.*
7. Desai, S.S.M., *Rural Banking in India*, p. 287.
8. Reserve Bank of India, RRBs, p. 74.
9. *Ibid.*, p. 5.
10. Ramesh Golait, Reserve Bank of India, *Occasional Papers*, 2007.
11. *Ibid.*
12. *Report on Trend and Progress of Banking in India.*
13. *Ibid.*
14. Reserve Bank of India, *Occasional Papers*, p. 97.
15. *Ibid.*

18 RURAL MARKETING

1. Introduction

Marketing occupies an important place in the process of development. The word "Market" is used in many senses and therefore, it is not easy to define it. To an agriculturist, "Market" means sale of his produce and purchase of goods of his need. To the wholesale businessman, 'market' means finding different techniques to get more buyers for his goods. For the industrialists, it is one of producing new type of goods and discovery of new markets for goods manufactured.

The common usage of the term 'market' however, is a place where buyers and sellers meet and exchange goods. For example, we speak of vegetable market, fruit market, flower market, etc. The function of a market is said to be one of bringing buyers and sellers together for exchange of goods and services. In modern times, with advancement in transport, and communication technology, exchange of goods takes place without the actual meeting of buyers and sellers at any specific place. Therefore, the term 'market' is now used in a wider sense to signify any area, may be a country or the whole world, where buyers and sellers are in close contact with one another, price making forces operate and buying and selling of goods takes place without actual meeting of buyers and sellers.

2. Definition and Scope of Rural Marketing

Rural marketing refers to all those activities comprising the exchange of rural products and services with urban goods and services, through which the consumption and production requirements of the rural people are fulfilled and urban manufacturers get market outlet for their products and services with profit. Rural marketing covers the problems of both the sale of rural products and services and purchase of urban products and services to the mutual advantage of all those concerned. The different steps and techniques in marketing in general will equally hold good in rural marketing.

From time to time, the definition of "rural marketing" has been undergoing change and the scope of it is widened. At one stage, rural marketing has been defined in a narrower sense to include only marketing of agricultural crop production and of animals and of the produce of animal husbandry and livestock and in return purchase domestic consumption requirements, and farm inputs. The advent of commercial farming, necessitated the use of modern inputs like high-yielding variety of seed, chemical fertilizers, pesticides, tractors etc. Hence, such of those organized industries manufacturing and marketing these inputs have entered the rural market. Now the scope of rural marketing is widened. Later, it has been realized that the sale of products of rural artisans and rural industries and purchase of requirements to this sector, should also constitute an important component of rural marketing and thus the scope of rural marketing is further widened. In recent years, we notice Rural India in transition. There is a visible transformation of traditional and subsistence agriculture into modern and commercial farming. Now almost all categories of farmers are interested in raising some cash crops, and sell a part of their produce. Further, there is a gradual shift from agriculture to non-agriculture occupations within the rural sector. These developments, resulted in appreciable increase in disposal rural incomes. There is a marked increase in the income, particularly of middle and upper classes, during the post-green revolution period. Green Revolution contributed for marketable surplus and in the process of marketing, rural people come closer to the urban life and due to ''demonstration effect'', the lifestyle of rural people is fast changing. All these developments brought about a significant change in marketing process.

Now, there is flow-out of more goods and services from rural to urban areas and flow-in of goods and services from urban to rural. Also, we notice within the rural settlements, there are transactions of significant nature. Rural marketing now has three way process:

(i) Flow out of goods and services of different kinds and nature from rural to urban;

(ii) Flow out of goods and services from urban to rural areas and

(iii) Flow of goods and services within rural sector from one social group to another. All these transactions ultimately fulfill the needs and preferences of customers.

In the process of marketing, consumers' choice or preference becomes more critical. Under a system of free enterprise, it is the consumers who decide what goods and services shall be produced and in what quantities, what shall be the form and quality of a product and what type of packaging, etc. It is for this reason, the concept of "consumers sovereignty gains currency. In a situation of keen competition among many producers of similar products, with an eye on market, Consumers' tastes or preferences or choice need to be carefully analyzed. Every time a consumer makes a purchase, he is in fact voting for the continued production of that commodity. An increases in consumers' demand for a commodity will raise its price and so encourage producers to increase their output of it. If consumers' demand declines, prices will fall and producers will reduce output. Since consumers' demand determines the quantity of a commodity that can be sold at a price, even a monopolist is subject to sovereignty of the consumer. The monopolist, though can either fix the price of his product or decide how much he will produce, he cannot do both of these things at the same time. It is because consumers' choice or customers' preference is ultimately the deciding factor.

Even in farm sector, under the commercial farming, there is almost monopolistic competition type of situation. Though similar goods are produced, attempts are made at marketing stage, to differentiate products through several devices such as cold storage, supply in such size and such type of packaging of consumers' preference, well suited advertisement in the rural marketing can be seen in the retail trade. Each retail trader at village fair or in a town, attempts to attract customers through pleasing manners, better packaging. Home delivery of products, etc. In view of customers' preference being a deciding factor, the concept of marketing may be defined as the process of defining customers' tastes and preferences and accordingly organize resources to turn/out certain products or services that facilitate easy sale and assured returns.

Therefore, proper understanding of consumer behaviour becomes vital in achieving the marketing goal. The psychology, income levels, preferences, and spending pattern of the different category of producers and consumers in rural areas are different from those of the urban manufacturers and buyers. Full understanding of these matters and related issues with the background of rural socio-economic conditions become more critical in the rural marketing process.

In a broader sense, rural marketing may be defined as a process of marketing rural production — both farm and non-farm, including products of rural industries and artisans and also, marketing of urban products of consumables and consumer durables, meeting the consumption and production requirements in the rural settlement to the satisfaction of customers and sellers concerned. The scope of the subject of rural marketing is very wide. It is concerned with the problem of marketing surplus and prices. It is interested in identifying market segments and suitable techniques of sales promotion in a situation of large scale illiterate customers. The problem of price parity between the prices of farm products and non-farm products is also examined in the study of rural marketing. The problems of transport, grading, storage, and warehousing assume greater importance in rural marketing. Rural marketing covers marketing finance, too, For effective marketing, information about prices, etc., is essential. Therefore, market information is an integral part of the study of rural marketing. The study of rural marketing embraces the problem of marketing efficiency and the methods of increasing efficiency in marketing.

In India, the state is now playing an important role in the marketing field. Rural marketing covers all those steps taken by the state in improving marketing conditions in rural areas. Agricultural price policy, setting up of Regulated Markets, the purchase of products directly by the government through the food Corporation of India, etc., come under the subject matter of rural marketing. Thus, the scope of rural marketing is very wide.

3. Importance of Agricultural Marketing

An efficient system of marketing[1] is essential for the economic development of a country. Marketing becomes an important instrument in improving the income of the individual producers of all categories, as well as rendering different services as follows:

(i) Development of marketing leads to the integration of different sectors of the economy such as agriculture, industry, transport, etc.

(ii) Marketing helps in the creation of different utilities like form utility, time utility, place utility, etc.

(iii) Well-organized markets will encourage even the small farmers to turn out marketable surplus, Further, marketing facilities encourage the small farmer to raise cash crops and earn money Through a well-organized marketing system, it becomes possible to make available foodstuffs, etc., at different centers and secure non-farm products.

(iv) With the advent of the Green Revolution, there is possibility of a bigger marketable surplus. In fact, the marketing problem is described as a ''second generation problem'' in the spread of Green Revolution. The first generation problems being supply of high-yield variety seeds, chemical fertilizers, pesticides, etc. The outcome of the application of these modern inputs is a substantial increase in marketable surplus. Therefore, the second generation problem is one of creating marketing facilities so as to enable the farmers to sell their produce at remunerative prices. It is also said that marketing helps in introducing new crops and improved methods of cultivation.

(v) Through marketing, rural incomes increase thus leading to higher rural savings. This, is turn, will help modernization of agriculture and capital formation for industrial development.

(vi) Marketing helps to maintain stability in the prices and also economic stability in the country. In the marketing process, goods are purchased and stocked. They are purchased at a time when there is more production. By so doing, it is possible to check the rise in the price level. In a situation where there is low production, the price level is bound to increase. At this stage, the goods stocked may be released to the market, and this will help to check the rise in the price level. In this way, through marketing, it is possible to control fluctuations in the price level and maintain price stability.

(vii) Marketing helps to achieve and maintain a higher standard of living. To achieve a particular standard of living, the supply of different goods and services is necessary and there must be a continuous supply of these different goods and services. An efficient marketing system will help to provide the different goods and services, whenever they are required and wherever they are required. In the words of Richard Buskirk, "Marketing delivers the standard of living to the people by satisfying the multitude of need and desires of the consumer."

(viii) Marketing helps to increase employment opportunities to the members of the society both in production and marketing activities.

Needless to say, the development of agricultural marketing is essential in modernizing agriculture. Through well-developed rural marketing, it is possible to develop the rural economy. Therefore, development of marketing in rural areas assumes paramount importance in the context of rural development.

4. Marketing Functions

In the process of marketing, goods pass through several hands before they reach the consumers. There are different middlemen between producers and consumers who perform various functions. The important marketing functions are as follows:

Assembling

In the marketing process, the first step is to purchase the produce from different producers and assemble them at selected places. In the rural areas, agricultural production is the main economic activity. There are a large number of agricultural producers scattered over a wide area. From each producer only a small quantity of marketable surplus may be available. Purchasing from different producers and assembling at selected places is a marketing function performed by several middlemen. In some cases, the producer may directly sell to the consumer. For example, a farmer may supply sugarcane directly to the sugar factory. In this case, there is no role of middlemen and the assembling problem does not arise.

Transportation

Individual producers are not in a position to make their own transport arrangements and take the produce directly to the wholesale trader. Therefore, a large number of intermediaries who purchase the produce from different production centres make transport arrangements and take it to the wholesale market.

Storage and Warehousing

The agricultural produce is seasonal. Therefore, the produce has to be purchased in the seasons when it is available and it has to be stored. The stored or stocked produce is released to the market, wherever there is demand. The storage and warehousing facilities become important in rural marketing. The middlemen or wholesale traders or retailers may create storage or warehousing facilities. The state can also create warehousing facilities. On the recommendations of the All-India Rural Credit Survey Committee of 1951-52, the Government of India has constituted the central Warehousing Corporation. This corporation has been taking steps to construct warehouses or godowns in different parts of the country. These warehouses are useful to the individual farmers also to stock their produce. However, all farm producers have no access to these godowns.

Classification and Grading

The function of classification and grading helps in sorting out the commodities according to size, quality, colour, weight, etc. This step helps to fix the prices according to the quality or grade of the commodity. Classification and grading help the producer to get a fair price for his produce and help the consumer to get good quality produce.

Equalization

The process of equalization refers to matching the flow of supply with the rate of demand. Agricultural Commodities are seasonally produced, while they are consumed in all seasons. Hence, the middlemen purchase the agricultural produce at a time when it is produced and stock the produce. They release the stock whenever and whenever there is demand. In doing so, they are helping in equalizing seasonal supplies with regular demand. This type of equalization process is an important function in marketing.

In the process of equalization, exchange functions are performed. The exchange functions include buying and selling. In marketing, buying and selling are fundamental functions, without which there is no marketing at all. The buyer must know the sources of supply from which he can purchase the required quantity at a particular price. In the words of Richard, "Buying function is largely one of seeking of the sources of supply, assembling of products and activities which are associated with the purchase of goods, raw materials, etc." The selling, on the other hand, is a step in assisting or persuading a prospective buyer to buy a commodity. Selling is defined as "the process which stimulates demand or desire, finds the buyer, advises the buyer and negotiates with him to bring about a transfer of title." Without buying, there is no need for surplus production. Without selling, there is no possibility for the consumers to get the required commodities for the satisfaction of their desires. In the absence of buying and selling, there is no economic activity and consumers cannot satisfy their ever-growing desires. Therefore, buying and selling become fundamental marketing functions.

Processing

Agricultural products are not acceptable to the consumers in the same form as they are available from the field. They have to be processed, like converting paddy into rice, wheat into wheat powder, etc. This processing function is equally important.

Financing of Marketing

There is a time-lag between assembling of commodities and their sale to the consumers. During this period, somebody's money is tied up in these stocks. This creates the problem of financing of marketing, which is an important marketing function.

Market Information

There is a wide gap between the place of production and place of consumption. Hence, the function of market information assumes importance. This function involves all the activities of collecting appropriate information regarding the prices of different commodities in different market centres and take this information to the different agencies in different places who are engaged in the sale of the products. The market information is useful both to the producer and the consumer. It also helps the government in formulating price policies.

Dispersion or Distribution

Produce purchased and stocked must be taken to the consumers. The consumers also are many in number and they are distributed over a wide area. Therefore, here again, some persons, like wholesale traders or retailers, take necessary steps to take the produce to the consumers. This part of marketing process is known as dispersion.

Risk-taking

'Risk' here is the possibility of incurring loss by some unforeseen happenings. Such possibilities are always involved in any business transaction. Goods may get destroyed due to fire, floods, cyclone, or there may be a fall in the price, etc. These risks cost heavily sometimes and must be borne by someone. Some physical risks like loss of produce due to floods, etc., may be covered under the Crop Insurance Scheme. But the risks arising from price fluctuations will often result in loss. Sometimes, due to change in the fashion, certain goods may not have adequate sales. Further, sometimes it may not be possible to collect bills from customers to whom credit sales are made. Thus, marketing is exposed to risks which have to be borne by someone in the process of marketing.

In the rural areas, it is a well-known fact that there are no adequate storage facilities to individual farmers and they are not in a position to classify the produce and store it safely. Further, transport continues to pose a serious problem in the rural areas. Some of the farm producers do not have correct information about the prices of different products. In view of this, the rural producers often find it difficult to sell at remunerative prices.

5. Marketing Costs

The study of marketing costs, particularly in the rural areas, assumes importance. The marketing costs are an index of marketing efficiency. If the marketing costs are low, the producer will be able to get a major part of the price paid by the consumer. The marketing costs help in assessing the share of different intermediaries in the total marketing costs. Further, marketing costs will help the government to formulate an appropriate price policy, assuring legitimate share to the producer from the money paid by the consumer and also safeguarding the consumer against unduly high prices, Lastly, the knowledge of marketing costs helps to evaluate the incidence and the effects of different taxes at different stages of marketing.

The marketing costs include all those expenses incurred by several agencies in the process of marketing, from the time the produce levels the producer till it reaches the consumer. These expenses are mainly handling charges at assembling, storing, grading, packing, transporting, etc. The marketing costs are incurred by several middlemen who undertake different marketing activities with a profit motive. Marketing costs, therefore, include also profit margins of the intermediaries. It should be remembered that the marketing of agricultural produce is costlier than that of industrial products.

In the case of agricultural products, it so happens that there is much difference between the price received by the producers per unit of a product and the price paid by the consumers for that unit of product. Various reasons account for it:

(i) In agriculture, there are many production centres and the middlemen have to spend considerable amount in transporting the goods from the production centres to the marketing centres.

(ii) Most of the agricultural products have to be processed and given a shape acceptable to the consumer. This again results in increase in the marketing costs.

(iii) It is not possible to store some of the agricultural products for a long time. Even if storage facility is created the storage cost will be high.

(iv) In India, there are many small farmers and if their retained costs are also taken, into consideration, the cost of production is bound to be higher.

6. Problems in Marketing of Agricultural Products

Marketing of agricultural produce is more complex than that of industrial products for different reasons:

(i) Agricultural products are generally perishable in nature Therefore, it is not easy to stock them safely or postpone their sale or consumption for a long period. In view of this, the farm producer is not in a position to secure always a remunerative price for all his products.

(ii) There are a large number of farm producers scattered over a wide area. From each production centre, only a small quantity is a available for sale. There is no proper organization of the Indian farmers. In view of this, middlemen assume greater importance in agricultural marketing. The middlemen purchase the small quantities from different producers, stock and process them and finally sell them to the consumers. Since the middlemen perform these different functions, the marketing costs go up considerably. Hence, the price paid by the consumer is much higher than the price received by the farm producer. In the case of marketing of industrial products, the role of middlemen is quite insignificant. Hence, there is not much of a difference between the price at the production level and the price at the consumption level.

(iii) Most of the agricultural commodities, except commodities like fruits, vegetables, etc., are not consumed in the form they are produced. For example, paddy, tobacco, jute, cotton, etc., cannot be used as they are. They have to be processed and given a form or shape acceptable to the consumers. In other words, in the process of marketing, form utility is created. This function is performed by middlemen. This type of problem is peculiar to agricultural products.

(iv) Some of the agricultural commodities are bulky in relation to their value. Therefore, their storage and transportation become difficult and costly. This nature of agricultural products poses a problem in marketing them.

(v) The agricultural producer cannot adjust production to the changing prices or demand as in the case of industrial products. Once a crop is raised, the farmer has to allow the crop to grow and harvest it, irrespective of changes in price levels. Even if there is a fall in prices, the farm producer cannot think of stopping the growth of the crop in the middle. Such situation would also result in loss. Since the farm producer is not able to adjust production to the changing demand, he has no control over prices. Therefore, very often, the farm producer is not able to get a remunerative price for his products. Further, it is very difficult to have a common understanding among a large number of farm producers in controlling the level of production and in that way, to control supply and prices.

(vi) Grading and standardization is a major problem. These facilities are not available to many producers in the rural area and therefore, the farmer is not able to get a better price for better quality products.

(vii) Many rural people do not have storage facilities. Large losses in agricultural products are incurred each year by dampness and damages by ants, pests and rodents. This loss is mainly due to insufficient and defective storage facilities.

(viii) Lack of transport facilities is another major problem in marketing of rural products. Many villages do not have approach roads to the nearby towns and therefore, transport of products between rural and urban areas becomes difficult and costly. The individual farmer is not able to make his own arrangements to transport the products to a nearby market centre and get a remunerative price. He is forced to sell to a local merchant and accept the price paid.

(ix) An important requisite of a good marketing system is the availability of accurate market information. The market information now made available through the Press and All India Radio is hardly intelligible to the illiterate rural people.

(x) The most important problem is one of inadequate institutional credit. Small and marginal farmers still depend on local moneylenders and commission agents for meeting their urgent financial requirements. They are compelled to sell their limited produce to the moneylenders only at a predetermined low price.

In view of the several problems faced by the agriculturists and special characteristics of agricultural products, marketing of agricultural produce is really a complex problem. It is, therefore, essential that the state plays a more helpful role in marketing of agricultural products.

7. Marketable Surplus and Marketed Surplus

The surplus that a farmer, retains after meeting the domestic needs like requirements of food, seed, payment in kind, etc., is known as marketable surplus. In subsistence agriculture, the farmer generally produces only that much of produce which is required for his domestic use. If the farmer raises only food crops and uses them for domestic use, then there is no marketable surplus. If the farm producer produces something intended for marketing, then there is a possibility for marketable surplus. Commercial crops are mainly intended for marketing purpose. Therefore, it is not appropriate to use the concept of marketable surplus with reference to commercial crops.

Marketed surplus is different from marketable surplus. Marketed surplus is that amount of produce which is actually marketed or sold out of the total marketable surplus. The marketed surplus may be purchased by the needy in the rural areas as well as in the urban areas. Marketed surplus, sometimes may be equal to the marketable surplus. Or it may be either less or more than the marketable surplus. If the marketed surplus is equal to the marketable surplus, the economy is said to be smooth and perfect. Sometimes, the marketed surplus will be less than the marketable surplus. Farm producers, either because of expectation of higher prices in future or because of fear of crop failure, may keep more grain with themselves than what is required. In such a situation, the marketed surplus will be less than the marketable surplus.

Sometimes, what is marketed may be higher than the marketable surplus. This situation arises because of 'distress sales.' A farmer, either to clear of the debt or pay land tax or to meet some pressing cash requirements, may sell away his entire produce even without keeping any grain for minimum domestic requirements. In such circumstances, what is marketed is higher than what is marketable. In the case of small and marginal farmers, there are often distress sales.

8. Factors Influencing Marketable Surplus

The quantity of marketable surplus can be known if the produce is sold in organized markets. In rural areas, many farm producers sell away, in their respective villages, to the local merchant, village moneylender, etc. Sometimes, because of pressing cash requirements, small and marginal farmers may resort to distress sales. Therefore, all that is sold cannot be considered a marketable surplus. In the villages, there is also another practice. The small and marginal farmers may take some produce on loan basis from the neighbours to be returnable after harvest of their crop. In view of this, it becomes difficult to assess the marketable surplus accurately.

Generally, marketable surplus is more in the case of large size farming. The large-size farmer with adequate investment on modern inputs and through mechanization may be in a position to produce more. Marketable surplus, hence, may be higher. In the case of small-size holdings generally, crops intended for domestic use are raised and therefore, the marketable surplus is very negligible. However, in view of growing cash requirements, the tendency is to raise crops intended for sale.

The marketable surplus depends upon the domestic consumption requirements and the consumption habits of the farmers. In joint families, some persons may not work and contribute to higher output and yet remain as consumers. In view of this, the domestic consumption requirements may be more, leading to a fall in the marketable surplus.

9. Marketable Surplus and Price

Marketable surplus is mainly influenced by the price level. Ezekiel, Khatkhate, Mathur and Oison argued that there is an inverse relationship between price and marketable surplus. It means the marketable surplus is said to be higher at lower prices and lower at higher prices. The argument is that the farmers need only limited cash. When the price is high, only a small quantity is sold in order to get the required limited cash. Therefore, it is stated that at higher prices, the marketable surplus will be low. On the other hand, it is argued, that at higher prices, the marketable surplus will be low. On the other hand, it is argued that at higher prices, the producers will consume less and sell more and therefore, the marketable surplus will be more. Another explanation is also offered for the inverse relationship between price and marketable surplus. The small farmers generally face the problem of shortage of food. Even if the price levels increase there is no guarantee that they will sell more. In the case of small farmers, a higher level of production may result in a higher level of personal consumption.

V.M. Dandekar critically analyzed the argument of the inverse relationship between price and marketable surplus. He believed that under the conditions of limited cash requirements, it is possible that marketable surplus is low at higher prices and that it is more at lower prices. But this argument may hold good in the case of small farmers only and when the quantity they can sell is negligible. It is only the large-size farmers who sell more and they think of selling more at higher prices and not at lower prices. Therefore, V.M. Dandekar concluded that in the case of small farmers, there is inverse relationship, while there is a positive relationship between price and marketable surplus in the case of large-size farmers. However, it should be remembered that even the small farmers are in need of more cash. They also grow cash crops whose price levels are relatively high. It means that the small farmers may also be influenced by higher prices.

10. Recommendations of the National Commission on Agriculture (NCA)

For removing the defects and strengthening of rural marketing, among others, the NCA made the following recommendations:

(i) The markets should be very near to the villages with adequate facilities for grading, weighing and storage of all commodities.

(ii) The regulated markets should be strengthened in terms of adequate market yard, market functionaries, ware housing and storage facilities etc.,

(iii) There should be an agency which should take charge of calculators' produce, advance money for immediate needs, process the produce, arrange marketing for the next point and make final payment to them.

(iv) Branches of primary cooperative marketing societies should be established at all the regulated markets and all inputs needed by farmers for agricultural operations should be made available to them through primary cooperative marketing societies, Agro-Industries Corporation and sale depots of the private traders.

(v) Village roads should be improved by the Zilla Parishad and Public Works Departments.

(vi) The conventional bullock-cart needs to be designed so as to improve its efficiency.

(vii) Railways must provide easy and quick transport facilities to the rural areas, wherever possible.

(viii) Rural people must be provided with adequate storage facilities, apart from training the farmers in the methods of scientific storage and pest control.

(ix) The FCI should concrete on the construction and maintenance of suitable warehouse to store foodgrains needed for the public distribution system and buffer stock on a priority basis.

(x) Extension education in marketing should be improved through regulated markets, primary cooperative marketing societies and Farmer's Servicing societies.

11. State and Rural Marketing

The policies, programmes and actions of the government in its efforts to develop and modernize the rural marketing system are mainly in three directions:

(i) Institutionalizing of agricultural marketing by facilitating the formation of cooperative marketing societies

(ii) Regulation of markets for various agricultural products designed to minimize or eliminate unfair trade practices; and

(iii) Direct involvement of the State in the marketing of certain agricultural products.

In order to improve the marketing system along these three lines, certain steps have been taken as follows.

Encouraging Cooperative Marketing

Cooperative marketing societies provide several advantages:

1. Better bargaining power;
2. Reducing marketing costs;
3. Elimination of middlemen;
4. Storage facilities;
5. Prevention of loss in storage;
6. Grading facilities; and
7. Fair price to producer and consumer.

Legislation has been enacted by all the state governments regarding the formation and functioning of cooperative marketing societies. There are marketing cooperatives in a number of key rural areas like marketing of agricultural produce, marketing of agro-industry products, marketing of dairy, fishery and poultry products, consumer cooperatives at the retail level as part of the public distribution system, centralized marketing of cottage industry products, Khadi and handloom and handicrafts etc.

The rural cooperatives have the full potentialities of helping even the weaker sections in rural areas and this is evident from the Anand Milk Producers Union ltd., (AMUL). This unit which began in the village of Anand, Gujarat, in the 1940s has grown into Asia's most successful rural cooperative, now covering more than 800 village societies.

12. Regulated Markets

A regulated market is one which is established by the State Government and which functions under some rules and regulations. These markets are expected to solve many problems of agricultural marketing and improving efficiency. The aim of the regulated markets is to regulate purchase and sale of agricultural produce, create conditions for fair competition and ensure justice to the producers. The Royal Commission on Agriculture, 1998, recommended the establishment of regulated markets. Accordingly, in 1938, the Central Agricultural Marketing Department was established, Today, the above department is known as the Directorate of Marketing and Inspection. In almost all States, there are regulated markets now.

The State Government, through legislation, framed the rules in the matter of regulated markets. It is also officially decided what commodities are to be marketed and the area of a regulated market. After this step, a declaration is made for the establishment of a principal market yard as also its jurisdiction.

For the management of the regulated market, a market committee is constituted. The committee consists of representatives of growers, traders, local bodies, cooperative marketing societies and warehousing cooperation and also the governmental nominee. There is the Marketing Secretary, who is the chief executive and who exercise control over affairs of the market. The important functions of the regulated market are as follows.

(i) To maintain marketing yard and the building etc., property

(ii) To collect and disseminate market news

(iii) To ensure fair price to the agricultural producer

(iv) To settle disputes pertaining to sale and purchase in notified agricultural commodities.

The regulated markets undertake purchase and sale of commercial crops, food crops, fruits, vegetable etc.

The following are some of the advantages of the regulated markets:

(i) Through regulated markets, sales in the villages can be reduced and thereby, save the small producers from exploitation by the local moneylenders.

(ii) It is possible to ensure fair prices through regulated markets as there is accuracy in the weighment and measurement

(iii) The market charges are low in the regulated market. In spite of the many advantages of regulated markets, these markets are not highly useful to the rural producers.

13. Grading

Standardization and grading of the products are important in marketing of produce. Sometimes, standardization and grading are used as one, but they are different. Standardization refers to determination of a particular standard, based on certain qualities of the products. For example, based on the quality of the food grain, some stock of the food grains may be standardized. After standardization, goods are graded. Thus standardization precedes grading.

Grading is the process of dividing ungraded goods into different uniform lots based on certain characteristics such as quality, size, shape, colour and cleanliness. Grading assumes importance in agriculture since there are variations in quality in a given quantity of any produce. The importance of grading has increased due to commercialization of agriculture. The manufacturing industry requires specific raw materials and the consumers require better quality grains. In order to supply products according to the specifications indicated by the buyers, the products have to be necessarily graded.

Advantages of Grading

(i) Grading serves as an incentive to produce better quality products, and better products will help to get higher price.

(ii) Through grading, consumers will get good quality of products for the prices they pay.

(iii) Sales can be increased by grading of the products as the sellers need not give any special assurance about the quality of the produce, which is graded and, therefore, buyers easily think of purchasing the products.

(iv) It simplifies the financing of commodities while in storage. This is because the value of graded commodities can be easily calculated and the risks to price fluctuations can be safeguarded.

(v) It reduces wastage in marketing, particularly transport and storage. There is no need for weighing at every stage, where produce goes from hand to hand if it is graded.

(vi) Grading facilities easy export of products. A sample from the graded products is sent abroad and on this basis, the quality of the entire produce is decided. This facilitates export of products without physical inspection of the produce by the foreign buyer.

Grading in Rural Areas

The producers, particularly agricultural producers in rural areas, are not able to grade their produce according to quality, etc. Even where an attempt is made to grade the stocks, it is only a crude method. Very often, the farm producers leave the grading work to commission agents or traders or local money lenders. Under these circumstances, the agricultural producers are not able to get better prices for better quality products.

Even during the British period, attempts were made in the direction of grading. The Agricultural Produce (Grading and Marketing) Act was passed in 1937. This Act empowered the Government of India to prescribed grade standards for different agricultural and livestock commodities. This Act was amended in the later years and the scope of the Act was widened. As per this Act, the State Marketing Officer is competent to test the quality of the products. The tested products are stamped as "Agmark" (Agmark is the abbreviation for agricultural marketing) of the products quality duly certified by the marketing officer. The Agmark facility is now available to many agricultural and livestock products. For export purposes, certain products like wool, lac, sheep and goat-skin, tanned leather, cashew nuts, pepper, ginger and oilseeds, etc., must necessarily obtained Agmark. At present, the Directorate of Marketing and Inspection enforces compulsory quality control before export of agricultural commodities. Grading is now enforced for different items like ghee and vegetable oils, cream, butter, egg, rice, wheat atta, pulses, jaggery, honey, spices, cotton, etc.

The apex laboratory at Nagpur with 17 regional Agmark laboratories is actively engaged in quality control of agricultural products. The Indian Standards Institute, popularly known as ISI, formulates grades and standards for both agricultural and non-agricultural commodities.

In spite of these efforts, the problems of rural people in grading their produce remain without much of improvement. There must be agencies accessible to the rural areas in the matter of grading all the agricultural products. Grading requires storage facilities also. Therefore, in the matter of grading, storage problem becomes equally important.

14. Storage and Warehousing

Sometimes, the terms 'storage' and 'warehousing' are used in the same sense. Storage is a broader term, while warehousing is a part of storage. Warehousing is generally associated with trade and profit. Storage of goods is made possible by the network of warehouses, specially constructed for this purpose. The warehouses take the responsibility of the storage of goods. Therefore, a warehouse is defined as "an establishment of the storage or accumulation of goods." There can be either private or public warehouses. A private warehouse is under the direct control of the owner and it is solely used by him for storing his own produce. A public warehouse, on the other hand, is serving the general public. It operates under the rules and regulations of the government. The important functions of warehousing are:

(i) Storage,

(ii) Price Stabilization,

(iii) Risk bearing, and

(iv) Financing.

Warehousing in India

On the recommendations of the All-India Rural Credit Survey Report, the Central Ware housing Corporation was constituted. The functions of the Warehousing Corporation are:

(i) To acquire and build godowns and warehouses at suitable places;

(ii) To arrange facilities for transportation of agricultural produce to the warehouses;

(iii) The Central Warehousing Corporation subscribes share capital to the State Warehousing Corporation;

(iv) It acts as agent of the government for the purpose of purchase, sale, storage and distribution of certain commodities; and

(v) To carry out any other function as may be prescribed under the Act.

The Warehousing Corporation is expected to help in reducing wastage and losses in storage through appropriate steps. The Corporation is also engaged in training personnel to run warehouses. During the planning era, the warehousing branches have been increased, thereby increasing the storage capacity, meeting the storage requirements of individual farmers, FCI and the state Governments. Wherever warehousing facilities are available, the farmers are in a better position to store their produce. Further, by producing the warehousing receipts as security, they are able to get loans from the commercial banks, etc.

15. Market Information

Market information is essential for successful marketing. Market information includes "all facts, estimates, opinions and other information used in marketing decisions, which affect the marketing of goods." The information needed for decision-marking in marketing relates to prices of different products at different market centres, role of middlemen, selling methods and similar such relevant information. There are a large number of producers engaged in producing a variety of products. There are different stages in marketing like assembling, wholesaling and retailing. At each stage, a lot of information is required about various things. The success in the business depends upon the availability of timely, correct information. It is rightly said "good decisions are informed decisions." In the context of marketing information, two concepts, "market intelligence" and "Market News", are popularly used. Market intelligence refers to information relating to facts which are essentially of a historical character and is a record of what has happened in the past. The market news is concerned with current happenings in the market such as existing prices of goods in different markets. In the case of market news, the speed of communication is of utmost importance. This is because a person who first obtains the information of price, etc., is in advantageous position over others who receive the news late.

On the basis of adequate marketing information, a right decision can be taken regarding what to produce, how much to produce, where to sell and how much to sell, etc. The required marketing information may be obtained through letters, telephones, daily newspaper, trade journals, All India Radio, etc.

In recent years, the Directorate of Economics and Statistics, Ministry of Food and Agriculture, Government of India, has been playing a very useful role in the collection and spread of market information. This agency publishes some journals and reports containing market information. The important of them are Agricultural Situation in India, Bulletin of Agricultural Prices, Agricultural Prices in India and Bulletin on Food Statistics. Some of the State Governments also publish monthly magazines which contain detailed information about the agricultural marketing situation in their respective States. The main defect of the government publications is the time-lag in their publication which reduces their usefulness. But they have historical importance.

In spite of the different channels through which market information is spread, the rural are not fully benefited by them, because of large-scale illiteracy, sometimes reliable market information does not reach the rural areas. Information available from the businessmen, commission agents and neighbors may not be always correct. Even though the price boards are expected to be displayed at some of the shops and market centres, often these price boards remain blank. Further, information on retail markets is very negligible.

There is need for taking more appropriate and effective measures to collect and spread market information in the rural areas. More appropriate timings must be chosen for broadcasting the market news. All the village panchayats should have radio sets and through proper maintenance and care, they should be properly utilized for getting the market information also. The market news should be free from rumours, since it is harmful to all concerned in marketing. The staff in charge of collecting market information should be well trained in marketing and should work under some supervisory staff. The Government can think of marketing extension personal functioning at village level, timely and correct market information must be made available in the rural areas. Although some improvement has been made in the collection and dissemination of market information during the planning era in India, there is much more to be done in this direction to make rural marketing orderly and effective.

16. Price Fixation and Procurement by Public Agencies

The FCI, the Cotton Corporation of India, the Jute Corporation of India and the Civil Supplies Department of State Government and other public agencies make direct purchases from the farmers at stipulated prices. In view of this, a fair degree of certainty is ensured to farmers in the case of certain products.

17. Buffer Stock Operations

The State purchases at a minimum guaranteed price the food-grains whenever there is more of supply and stock them. In so doing, the likely fall in the price level due to increase in supply is checked by the government. The produce stocked is related for sale when there is a fall in the supply of food grains. Through this step, the government is able to check the rise in the price of foodgrains due to a fall in the supply. Purchasing when there is more supply and selling at a time when there is a fall in supply is known as buffer stock operations. Through buffer stock operations, the government is able to control price fluctuations and thereby help both the producer and the consumer.

18. Sustainable Rural Marketing (SRM)

Under the conditions of globalization and new economic policy in India, sustainable rural marketing assumes critical importance. So far, marketed rural products predominantly constituted food grains. Sugar, cotton, oilseeds, textiles, handlooms and handicrafts. In these areas, India has made really a mark. There are some other items in which, the emerging strength of rural India is already evident, namely, sericulture, marine and inland fisheries. In the light of global market, the area which is yet to be fully tapped is one of horticultural exports. B.S. Raghavan very rightly opines, "No other country grows such a wide range of fruits, vegetables and flowers and in such abundance, as India and yet, it has no record worth mentioning in horticultural exports. The rich variety of medical plants and herbs when processed and marketed can help India, take care of the health needs of its, population besides coming into its own as the world's pharmaceutical giant."[2] Promotion of sales of agricultural products demands marketable surplus, which means higher yield levels. Another equally important requirement is maintaining quality of our products. The quality of agricultural products in international market is already under cloud. This must be rectified. The need of the hour is guaranteed grade of high quality. Many studies reveal that Indian agriculture and its marketing process suffers under the conditions of inadequate institutional credit, poor processing facility and lack of minimum infrastructure for marketing even for domestic purpose. Now priority step needed is capturing global market for value-added horticultural products through networking of processing units. This apart quality of farm products and guaranteed grade need to be maintained to meet the expectations of the customers in the global market.

It is encouraging to note that the number of policy changes have been introduced during the planning era to promote agricultural exports and procurement by public Agencies increased significantly. These measures contributed to the increase in the exports of agriculture and allied products[3] from ₹ 37,266.5 crore in 2003-04, to ₹ 77,769.7 crore in 2007-08 In All India the procure met of rice was 129.68 lakh tonnes 1996-97. and it increased to 260.56 lakh tones in 2007-08. Regarding wheat procurement, is increased from 81.58 lakh tonnes in 1996-97, to 222.25 lakh tonnes in 2008-09. Procurement of cotton increased (from 430 thousand bales of 170 kgs each) in 1998-99 to 979 in 2007-08. Raw jute from 54.50 (thousand bales of 180 kgs each) in 1998-99, to 755.92 in 2007-08

Since Cooperative Marketing and Processing Societies are contributing for rural marketing, they have to be strengthened. Operation Flood programme, where Production, processing and marketing of milk are integrated, is of immense use in marketing of milk and milk products through well designed structure. Under this programme, the dairy farmers in villages are organized into village level cooperative societies called the Diary Cooperative Societies (DCS). In turn the DCS in a district form a milk producers cooperative union at the district level. These unions take the responsibility of collecting milk form DCSs, processing it in their dairy and marketing it. Further, the unions also take the responsibility of supply of necessary inputs required for milk production to the diary farmers, These inputs include feed, fodder slips, health care, artificial insemination, purchase of milk at an assured price and other related services. The unions at district level form a Milk Producers' Cooperative Federation at the State level. The Federation at State level render necessary help to the unions and in turn to the DCSs in the process of marketing of milk and milk products. These types of organizations are popularly known as "Amul" pattern societies. This type of arrangement is in vogue in different States in the country and it is estimated that about eight million litres of milk is collected every day through these societies[4]. The societies formed on these lines are selling milk and milk products on brand names — Amul in Gujarat, Nandini in Karnataka, "Aavin" in Tamil Nadu and Vijaya in Andhra Pradesh.

The success in dairy project in Gujarat is the outcome of devoted efforts of Dr. Kurien. Encouraged by this experiment, on similar lines some new projects have been taken up. The important them are Horticultural Producers' Cooperative Marketing and Processing Society Limited (HOPCOMS) in Bengaluru, Fresh Marketing Society in Andhra Pradesh etc. The Fresh Marketing Society in Andhra Pradesh ensures the consumers that the vegetables sold to them are always fresh. Under the present policy of liberalization, private entrepreneurs are encouraged to set up food processing industries. This encouraged some multinationals to enter into the field. Entry of Pepsi has given a boost to tomato cultivation in Punjab and other places. Mahagrapes in Pune is another private project engaged in packing and precooling of grapes.

With advanced technology, molasses, a byproduct from sugar industry is being distilled into alcohol and now it attains high value. Similarly, rice bran which was normally fed to cattle is now being used to extract rice bran oil and the deoiled cake is fed to the cattle. Paddy straw, husk, cotton stalks are now being used for manufacturing hard boards. But as Gopalaswamy opines, "the impact of the development of modern marketing tools and techniques has been very minimal on agricultural produce marketing."[5]

Marketing of rural/cottage industry/artisan products becomes equally important in strengthening rural marketing. During the planning era, several steps have been taken for the betterment of this sector in term of technology upgradation, supply of raw materials, establishments of work-sheds, domestic markets through retail outlets and export marketing wherever possible through cooperative marketing societies. Some of these societies try to promote a brand image in marketing within the country as well as outside to their products like COOPTEX of Tamil Nadu, Priyadarshini of Karnataka, APCO FABRICS of Andhra Pradesh, WEAVER of Haryana and others. The marketing efforts put forth by these organizations are said to be quite commendable.[6] But yet, marketing of products of the rural industries is not that comfortable under the conditions of high cost of production and consequent high prices and growing competition from the major industries.

To facilities rural marketable surplus, modern inputs need to be made available and hence promotion of sales of such inputs manufactured in the urban sector is essential. The urban manufacturer is yet to fully tap the rural markets with wide coverage with wide coverage. In view of vastness and heterogeneous nature and illiterate customers T.P. Gopalaswamy rightly opinions the need for rural market segmentation and appropriate strategies for sales promotion.[7] Rural market segmentation may be based on: (a) extent of land holding, (b) extent of irrigation, (c) cropping pattern, (d) literacy level and progressiveness of the people, (e) proximity to towns/urban centres and (f) facility for supplementary source of income.

Farm producers with large size holding, assured irrigation and supplementary of occupation are certainly potential enough to be tapped for the sale of consumables like fertilizers, pesticides, etc. and consumer durables like TV, Radios, Transistors, Refrigerators etc. The experiences of companies operating in the rural market reveal that rural people generally prefer small unit packings, low priced packings, new product designs, sturdy products, utility-oriented branded goods. Keeping in view the hierarchy of markets. For the rural consumers, multilevel market centres like shandies or village fairs, mandies of towns/taluk/mandal levels, wholesale markets in major cities, etc. need to be fully utilized in sales promotion. Apart from print and electronic media, mass media such as well paintings, mobile publicity with audio and video equipment, posters, banners, and gift schemes need to be fully used. It is well noticed that the "word of mouth'' information is very important in rural areas. The rural buyers are often guided by the experiences of others in using consumables or consumer durables. Hence "Opinion leader" supposed to be knowledgeable person is key person influencing market transactions. The "Opinion Leader" may be landlord or progressive farmer or teacher or educated persons in the village or village level extension personnel. This source is also is to be tapped in furthering marketing in rural areas. Apart from appropriate media, appropriate timing of these campaigns is more important. Most suitable and appropriate time would be peak harvest season when the farmer would be selling crops and be in possession of disposable incomes.

In spite of the phenomenal improvement in different aspects of rural marketing, the rural poor still face certain problems in both selling their products and also purchasing their requirements. Some remote villages have no easy access to the market centres. Rural marketing infrastructure development needs priority in term of Rural roads and transport, warehouses, cold storage and assured remunerative price. Electronic media should be made helpful in the promotion of sales of rural artisans. If rural market has to expand and grow it is necessary to pay adequate attention to both marketing of rural products as well as marketing of urban products in the rural areas. Such an approach will be to the advantage of both rural and urban sectors. More importantly, with efficient marketing system disposable incomes of rural people will increase and this in turn generate market for manufactured products. In short, marketability of rural produce has to be improved significantly to ensure strengthening market for urban products and services, leading to sustainable development of the Indian economy.

Eleventh Five year plan report observes[8] the agricultural sector needs well functioning markets to drive growth, employment and economic prosperity in rural areas. The markets lack even basic infrastructure at many places. At present only one fourth of the markets have common drying yards, and godown. At present, cold storage facility exists only in 9% of the markets, where perishable commodities are brought for sale. The basic facilities such as loading and unloading facilities, weighting equipment, grading facilities are not available in all market yards. It is to be noted that a food processing unit (Food park) is coming up at Mogili village in Chittoor district in about 140 acres. This is the second of its kind in country sponsored by the Central Government. The first one functioning in Hardawar. The Central government proposes to establish ten such food parks in the country.

In the context of market regulation and development all state governments should take the following steps:[9] (i) To hold regular election to the Market Committees and bring professionalism in the functioning of regulated markets, (ii) To plough back the market fee for development of marketing facilities, (iii) To extend greater flexibility to stakeholders, sellers as well as buyers to interact in the markets, (iv) To promote facility of grading, standardization, packaging, etc. in the market area, (v) To ensure transparency in auction system, prompt payments to farmers, dissemination of market intelligence, and speedier transactions in the market, (vi) To improve weighting system by installing bulk weighment system.

It is noticed that post-harvest losses are about 20%-30% is different crops. It is mainly due to inadequate infrastructure development for post-harvest management (PHM) including pre-cooling. Demand for horticulture products will be sustained by development in agro-processing. There is a rising demand for new products such as dried powder fruit-based milk-mix, juice punches, banana chips and fingers, mango nectar, fruit wines, milk powder etc. Through agro-processing facility, marketing scope can be extended.

Conclusion

Rural marketing must necessarily cover not only marketing of products and also marketing of rural/cottage industry/artisan products. During the planning era, several steps have been taken for the betterment of this sector in terms of technology upgradation, supply of quality raw material strengthening of cooperative marketing societies, provision of storage facility food processing assured credit, and most importantly steps to assure remunerate price for the rural produce of all categories. Some state governments have taken certain steps in improving the marketing facilities like Co-optex of Tamil Nadu, Priyadarshini of Karnataka, Apco Fabrics of Andhra Pradesh, WEAVER of Haryana and other schemes. The marketing efforts must be further strengthened. But yet marking of products of the rural industries is not that comfortable under the conditions of poor infrastructure, high cost of production, poor technology, growing competition from the major industries as well as from international markets under the new economic policy. As Eleventh Plan stated, inadequacy of credit, absence of adequate number of cold storage, facilities and inadequate infrastructure pose problems in rural marketing. The producers of both farm as well as micro enterprises are not able to get remunerative price for their products. Rural marketing infrastructure development needs priority in terms of rural roads and transport, warehouses, cold storage, and assured remunerative price. The recommendations of Swaminathan Committee must be honoured and implemented with all seriousness ensuring remunerative price to the farm producers. Electronic media should come forward in the promotion of sales of rural products both farm and of village industries and rural artisans. More importantly, with efficient marketing system disposable incomes of rural people will increase and this in turn, generate market for manufactured products. In short, marketability of rural produce has to be improved significantly to ensure strengthening of market for urban products and services as well leading to sustainable development of the Indian economy.

REFERENCES

1. Reddy, Venkata K., *Agriculture and Rural Development,* Himalaya Publishing House, 2001, pp. 171-172.
2. Raghavan B.S., "Eyeing Global Markets" in *The Hindu – Survey of Indian Agriculture,* 1994, p. 13.
3. *Agricultural Statistics 2008,* pp. 233-35.
4. Gopalaswamy T.P., *Rural Marketing – Environment, Problems and Strategies,* Whether Publications, New Delhi, p. 94.
5. *Ibid.,* p. 98.
6. *Ibid.,* pp. 100-101.
7. *Ibid.,* pp. 104-131.
8. *Eleventh Five Year Plan Report,* Vol. III on Agriculture, p. 22.
9. *Ibid.,* p. 23.

19 APPROPRIATE TECHNOLOGY FOR RURAL DEVELOPMENT

1. Introduction

In the process of rural development, science and technology must necessarily play a crucial role. Rural India has abundant human resources, mostly of unskilled nature. Also, rural people are tradition-bound and cautious in adopting new practices. They have the problem of lack of capital and low capacity to bear the risk of any loss. Therefore, there must be a realistic approach in prescribing a particular technology without being caught up with the jargons of high technology, intermediate technology or soft technology. Technology required must be one which is resource-based, need-based, within the available means and which suits well with the aspirations, environment and culture of the people.

2. Appropriate Technology Defined

The word "technology" means a different thing to different people. Technology refers to a particular type of technique or method of doing something with special skill or knowledge. The knowledge used in doing a thing may not necessarily be one written in a book based on scientific theory. A person may acquire knowledge with practice or his own experience in a particular work. It is stated that "Technology is not primarily public knowledge, available in written form and easily transmissible from one country to another, but know-how, learned in the doing by trial and error, the practitioner's skill rather than the scientist's theory."[1] A technology is always embodied in a set of factors of production used at any time. The need for technical change based both on application of growing knowledge of science and scientific theory as well as on certain skill or art evolved out of experience is well recognized.

Alfred Marshall rated knowledge very high and considered it the most powerful engine of production.[2] In the definition of the law of diminishing returns, the condition "unless it happens to coincide with an improvement in the art of agriculture" and his use of "increasing returns" clearly highlight the importance given by Marshall to science and technology.

Schumpeter attributes higher growth mainly to sources other than increase in the labour force and increase in the stock of traditional forms of capital.[3] Another modern economist speaks of the importance of increase in 'useful' knowledge, as a produced means of production.[4]

Speaking about the importance of science with reference to agricultural growth, T.W. Schultz rightly observes: "The man who farms as his forefathers did, cannot produce much food no matter how rich the land or how hard he works. The farmer who has access to and knows how to use what science knows about soils, plants, animals and machines can produce an abundance of food though the land be poor.[5] T.W. Schultz argues that there is no death of allocative efficiency in the agriculture of poor countries. The resources at the disposal of the farmers, through their long experience, are efficiently used and Schultz argues that "No appreciable increase in agricultural production is to be achieved by reallocating the factors at the disposal of farmers who are bound by traditional agricultures."[6] The implication is that there is a need for an appropriate technology. A technology is always embodied in particular factors. A technique of production is an integral part of one or more factors.

Modern economists divide the production agents into two parts-one consisting of "land, labour and capital (goods)" and the other of "technological change". In the opinion of T.W. Schultz, 'Technological change is a list of short hand for an array of (few) factors of production that have been omitted in the specification of the factors…. technical change is in essence a consequence of either adding or dropping or changing at least one factor of production."[7] For example, using an iron plough instead of a wooden plough implies a technical change.

Technical change is not a ready-made product that can be copied from development countries and introduced into the productive process in the poor countries. A body of useful knowledge developed in advanced countries may be used to develop similar or superior new factors appropriate to the biological and other conditions that are specific to the agriculture and other productive fields of poor countries. What is important is that appropriate technology relevant to Indian condition must be evolved and made available at a price within each poor man's reach. A technology becomes appropriate when a given technology is feasible and desirable in a given situation.

The Western countries have attained higher levels of economic development by using science and scientific knowledge to achieve higher levels of labour productivity and aggregate output by intensive capital use. Technology has become an important factor of production an addition to land, labour and capital. Physical natural resources like cropped land, raw material etc., are limited in supply. But through science and technology, these limited resources are used effectively and higher levels of productivity per unit of resource are achieved. Because of large increase in output due to a significant rise in productivity levels, it has been possible to cope with the demands of a rising population. Science has helped man to control nature and environment and produce more efficiently. Scientific discoveries and innovations from time to time have facilitated and increased the ability of man to increase wealth, income and consequently, achieve better standards of living. No developing country can achieve real economic progress without the widespread use of modern science and technology. In a dynamic society, with ever-growing population and fast changing and increasing wants, there is a growing demand for a variety of goods and services. Further, there is an increasing competition also in the local, national and international markets among different producers of multifarious goods and services. Science must help discover ways and means of producing better quality products at relatively cheaper costs and only then profits can arise. Science must help to find out innovative techniques leading in competitive markets.

It is not investment alone, in the required quantity, that matters. Innovation is most important. Schumpeter speaks of five kinds of innovations:

(i) Producing a new good of one of different quality;

(ii) Producing a good in a new way which has not been tried by anyone;

(iii) Entering a new market which a particular industry has never previously penetrated;

(iv) Taking raw or partly processed materials from an existing source not formerly used by an industry; and

(v) Organizing production in a new form such as creating or breaking a monopoly. In applying innovative techniques, the rural realities in India must necessarily be borne in mind.

In rural areas there are economic imbalances leading to a lopsided development which may be traced to variations in natural endowment growing misdistribution of income, sources of income and employment opportunity, etc. Beneficiaries of technological change in the initial stages are understandably comparatively small pockets with relatively better economic strength and support. The economic imbalances as such cannot be attributed to technology. It is rightly observed that "For the removal of poverty and unemployment, the lion in the path is not technology, but an improvident choice of technology in highly unequal societies characterized by economic and socio-cultural extremes.[8] In the economic context, technology means "translating knowledge and inventions based upon it into devices, equipment or practices by which qualitative changes occur in the methods or process of production and in "factors of production" and "services" as the result of which new products are created of new ways of producing old products are developed, and by which unit costs are reduced (on a larger output with the same expenditure of labour and capital or the same output with a smaller expenditure).[9]

3. Appropriate Technology in Rural Development

There is need for an appropriate technology in different aspects of rural development as discussed below.

Appropriate Technology in Agriculture

Agricultural situations vary from region to region. There are areas with assured irrigation and abundant capital but with scare labour. In such situations, widespread use of high-yielding variety technology of capital-intensive nature becomes appropriate. The affluent section of the farm community in such areas substitute capital for labour. However, a major part of the cropped area in the country is rain dependent and the problem is one of both land and capital scarcity with lot of uncertainty associated with farm production. In such situations, land augmenting technologies with labour use become necessary.

4. Technology in Irrigation

Extension of irrigation through minor irrigation schemes of different kinds is by itself an appropriate technology for agricultural development. The storage dams and the main canal system, large pumping stations and deep tube wells may be viewed as capital-intensive projects. But, organizing farmers for construction and maintenance of on-farm small water distribution system and for equitable distribution of water and use of small low-lift pumps, shallow tube wells, hand or pedaled pumps, etc., become appropriate rural development activities with regard to lift irrigation. One important example of use of appropriate technology has been bamboo tube wells and coir and cavity borings. It is said that in some parts of Bihar, they created a revolution in the use of subsoil water.[10] Drip and sprinkle method of watering is now gaining popularity as an appropriate technology in irrigation.

5. Technology of Dry Farming

Dry farming technology has to be improved covering ways and means of capturing and retaining as much of rainfall as possible, conservation of natural moisture, conjuctive use of water, evolving drought-resistant strain varieties to give adequate nutrition to the plant in dry areas. Depending upon the soil type, a suitable tillage and moisture conservation technology has to be introduced. Deep ploughing in some cases and minimum cultivation in others and the sowing of Khariff crops in ridges so that moisture for the rabi crops can be captured in furrows make a great difference to crop growth. As many as possible, low cost water reservoir have to be constructed for the shortage of rain water. The new water harvesting technology like the techniques of sprinkler and trickle irrigation may be undertaken. Many research centres have succeeded in evolving drought-resistant variety of seed and the results have to be carried from the laboratories to the lands. The nutrition requirements for different crops grown under wet as well as dry conditions are worked out and this knowledge must reach the farmers. Through necessary infrastructure, demonstration and training, the different dry farming techniques have to be carried to the farmers. The extension of Green Revolution to dry arid tracts with new varieties of seeds that can thrive in rain fed areas, is necessary with all the follow-up measures.[11]

6. Appropriate Land Use

The fundamental problem of agricultural planning is to establish a pattern of land use between forest, pasture and crop land which will converse the fertility of land. An important aspect here is the integration of crop production and animal husbandry into some type of mixed farming. In backward regions, considerable extent of land may remain fallow and it may not be of much use for remunerative cultivation. Conversion of fallow lands into a productive asset is a difficult task but "putting these lands under permanent pastures would be an ideal from the point of view of land use and control of erosion."[12] The 'wasteland' can be gainfully used for growing fodder. The increase in the fodder supply would facilitate live stock development and thus development of dairying.

7. New Farm Practices

In recent years, has been a significant contribution of agricultural research in several directions. High-yielding varieties of seed suitable for different situations have been evolved. A number of improved agro-economic practices relating to various crops have been developed. Fertilizer dosage rates, including the mix of chemical fertilizer and farm yard manure, have been worked out. Research on weed control and control of disease and pests has been significant. Research on utilization of by-products such as jute coir, coconut husks, tobacco waste, cotton waste, etc., has yielded good results. The useful results of the research centres have to be transferred to the fields and for this purpose, an extension programme has to be made effective.

8. Appropriate Agricultural Tools

Agricultural engineers and experienced farmers have developed certain tools and implements that are well suited to the prevailing conditions of soil, topography, etc. Use of iron ploughs, redesigning of seedling ploughs and weeding instruments, use of small size tractors are some of the instances of appropriate technology.

With the advent of the high yielding variety of seed, whose maturity period is short, there is the possibility of raising more than two crops a year. Therefore, the different farm operations must be carried on very quickly in every season so as to keep the land ready for raising the next crop. Now the farm producer is in need of more labour and in peak seasons, acute scarcity of labour is faced. As T.W. Schultz has pointed out that the Indian farmers are efficient so far as utilization of resources at their disposal is concerned. With the traditional inputs or age old technology at their disposal, an equilibrium point is reached at low levels of production. If production levels are to be raised, appropriate technology in terms of improved seed, improved tools and implements, etc., must necessarily be used.

While tractors, power tillers and other farm machinery may be capital-intensive, improvement of animal-drawn or hand implements would become appropriate technology.

Modern drying and storage compounds for handling grains going into commercial or government godowns are no doubt necessary. But schemes like on-farm drying and storage to reduce loss of grain of millions of small farms, where farmers store grain for home consumption, become very appropriate. An adequate number of fertilizer factories and regional fertilizer storages are required. Equally important steps of rural development activities are transit storage network in rural areas, marketing distribution system to ensure receipt of required fertilizers by small farmers at fair prices and of reliable quality.

Regional seed farms and seed laboratories for evoking new improved seeds become necessary. At the same time, necessary facilities for small farmers to maintain viability of their own seeds and viable extension system regarding all aspects of rural development reaching small farmers must be provided.

Since rural development is primarily meant for the development of weaker sections, institutional credit for seasonal production for small farmers and tenant cultivators, credit for low-lift pumps, tube wells, sprayers, credit for processing and storage equipment, etc., become most essential.

9. Recycling of Wastes and Oil Technology

Recycling of a large variety of agricultural wastes is necessary. It is said that a million tonnes of minor oil seeds like neem, mahua, kusum seed, etc., are wasted every year. Only a negligible part of these are used for crushing. It should be possible to collect and crush at least 0.3 to 0.4 million tones of oil seeds and fats extracted for use in production of soaps, medicines, manures, etc. Groundnut cakes, after extraction of oil, can be used as protein-rich-food. Rice bran is presently used as cattle feed and only a fraction of it is being processed for extraction of industrial grade oil by solvent extraction process. Further, rice husk, instead of using for boilers, can be profitably used for manufacturing silicon tetrachloride or activised carbon and the cash ash can be used for production of building materials.[13] Again, a large quantity of baggase is being burnt as fuel in the boilers while manufacturing jaggery and sugar. This material can as well be used profitably for the manufacture of paper dolls, etc. This would reduce pressure on the limited forest resources of the country and provide employment to many in the rural areas. Through appropriate technology, more and more uses of the variety of agricultural waste material can be found as they are our renewable assets.

The Oil Technological Research Institute, Anantapur, Andhra Pradesh, under the leadership of S.D. Thirumala Rao, has developed a very useful technology in extracting oil from many seeds and agricultural wastes.[14] The research findings are helpful in providing appropriate technology in starting small scale industries with locally available raw material and other resources. If the technology is utilized, oils from different sources can be produced in small units which have rich and ready local as well as foreign markets. Among others, the research institute has developed a technology for extraction of oil from ambadi-seed, also called gogu-seed. Andhra Pradesh accounts for 80 per cent of total production of 75,000 tonnes of the seed in the country. Gogu-seed oil is edible in nature.

It can also be used for linoleum, paints and varnishes. The cake is a good cattle feed besides being a fertilizer. Andhra Pradesh has all the potentialities of developing this industry making use of available technology. Again, Andhra Pradesh produces about 58,000 tones of castor seed which accounts for 53 per cent of the total production in India. Appropriate technology is available for the development of this industry and it is necessary because castor oil is used in a variety of ways — in the manufacture of medicines, dehydrated castor oil, urethane foams, synthetic fibers, specialty lubricants etc. Appropriate technology is available for exploitation of cotton seed for production of edible oil. Washed cottonseed oil is being extensively used in the manufacture of vanaspathi. Fully refined cottonseed oil is a good edible oil. Again, technology is available to extract oil from Davana (*Artemisia palleus* well), an important aromatic plant. This Davana oil has a very rich and ready foreign market as the oil is used in several costly perfumes and cosmetics. There is technology to use rice bran for extraction of edible oil. Fresh bran has to be immediately stabilized to prevent its deterioration. By extraction of fresh bran or stabilized bran, crude rice bran oil with low free fatty acid and fortified with vitamins A and D to yield high grade rice bran oil.[15]

Again, technology is available for extraction of oil from Pudina. The leaves on distillation yield an essential oil known as "mint oil" or "Japanese peppermint oil". The oil is used extensively in flavouring pharmaceutical preparations, food products, confectionaries, liquors and in medicine. The oil has mild antiseptic and anesthetic properties.[16]

10. Animal Husbandry

Artificial insemination technology is an important factor in improving the quality of the cattle wealth. In the filed of dairy, economical methods of manufacture of milk products under rural conditions have been developed. The storage of large quantities of milk is now possible by the setting up of chilling plants in rural areas. The modern technology in the field of dairy has already resulted in white revolution in selected places. For transformation of rural economy, dairying technology must be popularized in different areas.

In the field of poultry development, balanced poultry feeds have been evolved and indigenous strains of birds have been developed with high rate of egg production. Methods have also been found for decertification of eggs with a view to increasing their keeping quality. Based on scientific study, appropriate sheds are built with material available in rural areas for housing poultry.

Further, in the field of fisheries, new techniques have been evolved to develop fishing industry even at individual farmer houses. The concept of social forestry is gaining momentum and appropriate plants are evolved. In the filed of afforestation and soil conservation, considerable research has been done and useful knowledge is available. It is necessary that effective steps are taken to carry the results of research in different directions of development to the doors of the rural people and help the rural poor in making required investment so as to enable them to use the available technology in a fitting manner.

11. Biogas Technology

Biogas technology can play a crucial role in meeting the energy requirements of rural areas in different directions. Through biogas plants, both at individual family level and as community gas plants, domestic requirements of energy for heating, cooking, lighting, etc., can be easily met and in this way huge quantity of conventional energy like kerosene, electricity, etc., can be saved. Further, indiscriminate falling of trees, leading to deforestation and consequent environmental problems, can be avoided through biogas technology.

A study at the Institute of Science, Bengaluru, reveals that about 26,000 family size biogas plants will produce as much fertilizer as a single 140 million coal-based plant. In addition, the construction of biogas plants will cost 14 million less and will generate 130 times as much employment.[17] Since there is wide scope for setting up a large number of biogas plants, fertilizer production can be undertaken in the rural areas through biogas technology. Fertilizer production at village level would solve problems like packing, transport and storage and facilitate supply of fertilizer at a cheaper rate.

In the absence of the minimum number of two to three cattle heads required, nearly 75 per cent of rural families cannot have family-size biogas plants. The answer to this problem is one of installing community biogas plants using cow-dung collected from individual farmers in small quantities and various organic wastes like crop residues, etc.

There is need for a separate agency in each State with apex agencies at district levels for the promotion of biogas technology. The agency must take up the responsibility of encouraging the villagers to set up biogas plants with the necessary technical guidance and supervision. Further, periodical inventory may be made to identify the non-working plants, reasons for failure, the required repairs and to take immediate steps to set right the defective plants. Agricultural universities and dairy development corporations and voluntary organizations may be involved actively in the promotion of biogas technology,

12. Micro Hydel Schemes

Successful attempts are being made in the installation and running of three-to-ten kilowatt hydro-electric generating plants to electrify villages at a low cost in far-flung hilly areas. Wherever it is possible to harness water resources of small rivulets and streams and a waterfall, the use of small hydel power generating sets must be encouraged. Such micro hydel scheme would give impetus to local talent to electric gadgets which will be locally used. This will help to avoid loss of energy on transmission lines and supply energy at a low cost.

13. Harnessing of Solar Energy

The National Committee on Science and Technology emphasized the need for using solar energy. Technology has been developed to make economically feasible equipment for cooking and crop-drying purpose through solar energy. In the rural context, there is need for developing technology for harnessing solar energy for lifting of water. Once production of electricity from solar energy becomes feasible and economically viable, it should be possible to electrify all villages in the country.

14. Bullock Cart Transport

Bullock carts are a major transport means in rural areas. The bullock carts are also used in towns and cities to transport goods from railway stations to the doors of consumers. The widespread use of bullock carts may be attributed to their cheapness, ready availability, less cost in making and repairing. Suitability for rough roads, etc. It is said that nearly half of the Indian villages do not have proper roads for use of trucks and tractors. The loads available in villages are too small and production centres are scattered. The distances to be covered in transporting loads are also normally too short. Often, the time involved in loading and unloading is more than the journey time. All these factors make transport by trucks and tractors most uneconomical and bullock cart is rightly described as an irreplaceable factor in rural transport.

It is possible to make effective use of animal energy if there is improvement in the design of implements yoked to these work animals. The energy transmission losses in contemporary designs of animal-drawn agricultural machinery are considerable. Apart from design defects, the fitment between the power source and the implements is defective. Through nominal improvement in design of the bullock carts and ploughs, it is said, the present animal energy capacity can be doubled or trebled without imposing additional burden on the animal. Due to lack of proper fitment between animal and the cart or plough, the skill tissues of the neck of the animal are damaged leading sometimes to cancer and finally death of the animal. The animal, as at present yoked to the cart, not only hauls a load horizontally but also carries some of it vertically. This cortical load makes for insufficient utilization of bio energy.[18] There is need for manufacture of improved animal drawn carts, accessories, implements and machinery suitable for different areas.

15. Inland Water Transport and Coastal Shipping

The country has got 5,000 kilometres of coastal shipping and another 5,000 to 6,000 kilometres of navigable inland water ways which can relieve the pressure on rail and road transport system and help the consumers to save due to cheaper transport costs. Here again, the manufacture of appropriate vessels adaptable to the type of commodity and the characteristics of waterways is essential to increase the scope and to intensify the operational efficiency of this type of transport.

16. Technology in Rural Housing

The housing problem is not one of providing mere shelter alone. It covers a variety of needs, both physical and social, of the group of people to whom shelter is to be provided. As a part of the housing problem, attention has to be paid for the provision of development of public utilities, public health services and community facilities for providing a better and healthier environment.

It is a well-known fact that there is a considerable shortage of rural housing in India. Furthermore, in many respects, the housing conditions in rural areas are far from satisfactory. A vast majority of house in the rural areas are built employing indigenous and traditional building materials such as mud or clay, bamboo, palm leaved, grass, thatch, etc. The use of these materials results in non-durable and unsafe houses which require of fire. In may cases, for want of a separate kitchen, cooking fires in the living room fill the house with smoke. The areas surrounding the houses are often polluted owing to lack of adequate drainage and sanitation. In many villages, there are no latrines and in cases there they exist, they are quite inadequate. According to a National Sample Survey report, in India, about 94 per cent of the rural households lack built-in latrines and 97 per cent have no bathrooms.[19] The deterioration in housing conditions in rural areas in India can be attributed, among others, to the high density of population, uneven distribution of population and a fast rate of migration of rural people to urban areas because of poor economic conditions prevalent in rural areas. The process of migration depletes rural areas of young working hands, skilled labour and also leadership. This has a negative effect on rural development.[20]

There is pressing need for appropriate technology in reducing the cost of rural housing. New techniques in making good quality bricks and tiles must be taught to the rural people. New innovations in the use of bamboo strips as reinforcement for roofing have been successfully tries in some places. It is necessary to pool all this knowledge for the benefit of the rural masses. The known technology in the filed must be popularized through appropriate media, particularly through exhibitions and construction of model houses in different states as seeing is believing. Goswami Gajenam Puri, former Engineer-in-Chief of Madhya Pradesh, assisted by two others had developed a technology with which a decent working house of 220 sq.ft. consisting of one living room, a kitchen, a small both-cum-toilet costing just ₹ 6,000 only. It is said there is scope for considerable reduction in costs in the rural setting with the labour contribution of the beneficiary. The house is designed with hollow walls, hollow clay plate roof, reduced wall and roof loads and with less of foundation costs. Mr. Puri argues that with “do-it-for-yourself” technology, the cost would be stills less.[21] A lot of improvements have been suggested by the National Buildings Organization and the Central Building Research Institute in the manufacturer of bricks at a low cost in a rural situation through improved skills. There are designs where a tile roof house of 30 to 40 sq. m. plinth area with use of lime and economical building material can be built for about ₹ 3,000. Different type of house designs suitable for different climatic conditions are available with the National Buildings Organization and this knowledge should reach rural areas. It is needless to state that well-designed and well-constructed houses play an important role in improving the environmental conditions of the rural areas.

17. Small and Cottage Industries

Improved technology has been developed with the aim of making small and cottage industries economically viable. Improved methods are now available in oil extraction, in crushing sugarcane, in manufacturing gur and khandsari, in the manufacturer of leather products, etc. The Khadi and Village Industries Commission and the Handicraft Board have been instrumental in improving designs and quality of production through appropriate technology. The known technology must be popularized and the rural poor must be helped in all directions, including the provision of institutional credit, to make use of the technology and derive the benefits. Technology is available for mini groundnut oil mill and there is scope for the setting of many more small oil mills as there is good demand for groundnut oil and cake locally as well as from vanaspati manufacturers.

The Council for Advancement of Rural Technology (CART), set up by the Union Government in 1981, is engaged in the development and dissemination of technology appropriate for rural industries. This would be done by the twin strategies of descaling technology from the organized small-scale sector as well as modernizing the traditional technologies. The CART would arrange to collect field problems, serve as liaison with research institutions, approve of new technology and arrange for their transfer to the rural artisans and village industrial units.

18. Appropriate Education and Information Technology

To facilitate the rural illiterate in adopting appropriate technology in different areas of their concern, appropriate education and training for them becomes crucial. The education and training for rural masses must be of such type which would help equip them with a variety of skills appropriate for their vocation and day-to-day life. Rural development personnel such as Mandal Development officer, Village Development officer and Extension Agents should be first trained well before they are drafted for different tasks of rural development.

Information Technology (IT) has become an important component of technological infrastructure, very much needed for development. It deals with all those critical areas concerning dissemination of knowledge like generation, transmission and utilization of information with techniques of informatics, statistics, office automation, data processing. Management Information System (MIS), Design Support System (DSS), Telecommunications, broadcasting and computer networking. The Ninth Five Year Plan,[22] proposes to strengthen IT as it has immense potential for employment generation to the extent of 25 per cent of the labour force in the long run. It is pointed out that the impact of IT will be especially predominant in the social sectors like health, education, judiciary and rural development. With the goal of making India, a global IT super power, a National Task Force on Information Technology and Software Development was setup in 1998, with Shri Jaswanth Singh as Chairman. The Task Force made concrete suggestions for development, application and export of software and Data Communication and development and export of Hardware. The recommendations submitted by the Task Force in three phases cover a wide spectrum of issues relating to Telecom, Finance, Banking, Revenue, Commerce, Electronics, HRD, Defense and Rural Development. It is encouraging to note "Rural Development" figures in the recommendations of the Task Force. Some of the recommendations are being implemented in a phased manner.

Conclusion

Keeping in view the present situation of rural areas and rural poor, a realistic plan of action has to be designed in the strategy of application of appropriate technology. In designing and developing rural infrastructure in terms of rural roads, transport, rural housing, primary health etc., appropriate technology must be identified and rural people must be convinced of the relevance of the proposed technology and that way popularize the technology.

REFERENCES

1. Cairn Cross, A.K., "The Role of Technology and Natural Resources in the Development Process" in *Reflections on Economic Development and Social Change,* (Eds.) C.H. Hanumantha Rao and P.C. Joshi, New Delhi, Allied Publishers, 1979, p. 64.
2. Marshall, Alfred, *Principles of Economics,* London, Macmillan and Co., 8th Edn., 190, p. 138.
3. Schumpeter, Joseph A., *The Theory of Economic Development,* Cambridge, Harvard University Press, 1951, p. 68.
4. Knoght, Frank H., "Diminishing Returns from Investment", *Journal of Political Economy,* 52, March 1944, pp. 26-47.
5. Schultz, T.W., *Transforming Traditional Agriculture,* Ludhiana, Lyall Book Depot, First Indian Edition, 1970, p. 3.
6. *Ibid.,* p. 39.
7. *Ibid.,* pp. 132-33.
8. Ganguli, B.N., "The Economics of Technology, Output and Employment in Poor Countries" in *Reflections on Economic Development and Social Change,* (Eds.) C.H. Hanumantha Rao and P.C. Joshi, New Delhi, Allied Publishers, 1979, p. 43.
9. *Ibid.,* p. 41.
10. Arora, R.C., *Integrated Rural Development,* New Delhi, S. Chand and Company Ltd., Second Revised Edition, 1986, p. 124.
11. Reddy, K. Venkata, *Agricultural Productivity in Andhra Pradesh,* S.V. University, Tirupati, 1977, p. 209.
12. Aurora, D., "Drought-prone Areas: Some Development Issues", *Indian Institute of Public Administration,* New Delhi, 1977.
13. Arora, R.C., Integrated Rural Development, *op. cit.,* p. 125.
14. Oil Technological Research Institute, Anantapur, *Profiles in Oil Technology, 1977*. The profile contains employment potential, estimates of investment and probable returns from different enterprises in the extraction of oil.
15. *Ibid.,* p. 21.
16. *Ibid.,* p. 58.
17. Bihari, L. Patel, "Biogas Technology and Rural India", in *Kurukshetra,* Vol. XXXIV, No. 6, March 1986, p. 14.
18. Ramaswaamy, N.S., "How Valuable is Animal Energy?, *op. cit.,* p. 8.
19. *National Sample Survey, Report,* No. 67, Tables and Notes on Housing Conditions in India.
20. Mathur, G.C., "Improvement of Rural Housing and Environmental Conditions in Developing Countries" in *Rural Habitat Transformation in Worlds Frontiers,* (Eds.) R.L. Singh and Rana P.B. Singh, 24 IGC, Tokyo Publications, 1980, pp. 540-41.
21. Narayanan, K.P., "Technology at the Doorstep" in *The Hindu,* Oct. 11, 1986, p. 18.
22. *Ninth Five Year Plan,* Vol. II, p. 923.

20 HUMAN RESOURCE DEVELOPMENT/HUMAN DEVELOPMENT INDEX

1. Introduction

Of all the resources, human resource is crucial in the process of development. Development required a systematic effort of coordinating different resources, skills and ability of human resources. As John P. Lewis said "In Agriculture and the rural economy, the development problem is primarily an organizational one. It is tougher, not easier for this reason".

If development is equated to economic growth, then development is measured in terms of Gross National Product and per capita income; completely ignoring the role of non-economic factors in the process of development. Development has many dimensions and includes qualitative changes in social, economic, cultural, environmental aspects. Development is a continuous and unending process of improving the quality of every individual through necessary skills and that way develop the society as a whole.

Development[1] ultimately means development of human resource and therefore, it is to be judged by what it does to man. In the rural areas, a good number of people for over several years lived a life of dependency or almost complete slavery. Because of object poverty and consequent under development or social stagnation people lost faith in themselves and in their potentialities for development and they remain without active participation in social, economic, cultural and political life. It is important that they come out of this apathy and skepticism. They have to be motivated to think freely and act wisely. Human resource development is meaningful when every individual in the society has the necessary skills and ability to think for himself, set right himself and has the mind of cordial relationship with all numbers in the society. Such cordial relationship is required among different sections of society and that speaks the quality of human resource.

Mahatma Gandhiji long back said that development should not be measured only in terms of Gross National Product and per capita income alone. Other criteria such as perfect relationship among different members of a family; cordial relationships in between social groups in the society; development of different Arts and Spiritual growth. Gandhi gave importance to ethical aspect of life. On one occasion, he made reference to Pancha Papalu — five sins (in Chapter I of this book). In Gandhiji's line of thinking, it is not acquisition of wealth which matters, but the central point is right conduct and proper social relations facilitating human resource development.

2. Human Resource Development Defined

The United Nation Development report of 1997[2] has defined human development as the process of enlarging peoples choices. The choices include healthy life, appropriate education, decent standard of living, political freedom, assured human rights to live with self respect. Thus, the Report states that Human Development is a process of widening people's choices as well as raising the level of well-being achieved. In this way, here, the concept of human development puts people at Central stage, as envisaged by Gandhiji long back.

In the words of Mahabub-UlHaq,[3] "the defining difference between the economic growth and the human development is that the first focuses exclusively on the expansion of only one choice — income, while the second embraces the enlargement of all human choices — whether, economic, social, cultural or political." It is argued that with economic growth and increase in the levels of income, it may be possible to enjoy all choices. Mahabub UlHaq replies this line of argument with different reasons. First income may be unevenly distributed within a society and some may not have access to income and if they have access, it may be very negligible. Thus, economic growth may not "trickle down". Second, priorities chosen by the society or political set up may not allow the income expansion facilitating human options. Further, accumulation of wealth may not be necessary for the fulfillment of different human choices. In fact, some choices do not require any wealth at all. For instance, "a society does not have to be wealthy to represent the rights of each member. A nation does not have to be affluent to treat women and men equally. Valuable social and cultural traditions can be and are maintained at all levels of income."[4] These are human choices that extend far beyond economic well-being. Knowledge, health, a clean physical environment, political freedom and simple pleasure of life are not dependent on income. Haq very rightly warns, "unless societies recognize that their real wealth is their people, an excessive obsession with creating material wealth can obscure the goal of enriching human life."[5]

Paul Streeten categorically states that human development is necessary for more than one reason.[6]

(i) Human development is the 'end', while economic growth is only a means to this end.

(ii) Human development in terms of a well-nourished, healthy, educated, skilled, alert labour force, is certainly a means to higher productivity,

(iii) Human resource development in terms of improvement in education, particularly of girls, better health facilities and reduction in infant mortality leads to lowering the family size by reducing human reproduction;

(iv) Human development facilities, good physical environment. Paul Streeten basing on research studies, states that deforestation, desertification and soil erosion decline when poverty declines with human development,

(v) Human development and reduced poverty would contribute for a healthy civil society, orderly democracy and greater social stability.

The above analysis cleanly indicates that human development paradigm embraces different aspects of society and not just economic growth. The social, political and cultural factors are given as much importance as the economic factor. Further, while people are regarded as end of development the means are not ignored. However, the character and scope of economic growth are measured in terms of enriching the lives of people. People do not just remain the instruments of production, but acquire the centric stage.

3. Essential Elements of Human Resource Development

Mahabub-Ul-Haq indicates four essential components in the human resource development paradigm: equity, sustainability, productivity and empowerment.

Equity: If development is to enlarge people's choices, "people opportunities enjoy equitable access to opportunities. Equity in access to opportunities demands a fundamental restructuring of power and effecting changes on the following lines: (1) Change in the distribution of productive assets through land reforms, (2) Major restructuring in the distribution of income through progressive fiscal measures aimed at transferring income from the rich to the poor, (3) Overhauling of the credit system aimed at meeting the credit needs of the poor, (4) Equalization of political opportunities through reforms in voting system aimed at preventing domination of political power of a feudal minority, (5) Undertaking steps to remove social and legal barriers that limits the access of women, minorities ethnic minorities to some of the key political and economic opportunities.

Sustainability: Sustainability must become an essential component of human development, in the sense that the next generation must have the right to enjoy the same well-being as the present generation enjoy. It simply means, the present generation must so utilize the basic resources like land, water, forest, that these are passed on to the future generation safely for their development. Further, sustainability demands that wide disparities within and between nations must necessarily be reduced and appropriate steps must be taken to reduce them. This is for the simple reason that high degree of inequality of incomes would lead to social tension and is inherently unsustainable, politically economically and environmentally.

Productivity: Productivity is an essential component of human development which required investment in people aiming at their maximum potential. Mahabub Ul Haq, places human development as central issue and therefore it is better to treat productivity as one part of human development paradigm, with equal importance given to equality, sustainability and empowerment.

4. Empowerment

Human development envisages full empowerment of the people. Empowerment means that people are in a position to exercise choices of their own with free will. It implies a political democracy in which people can influence decision about their lives. It requires economic liberalism so that people are free from excessive economic controls and regulations. It means decentralization of power so that real governance is brought to the doorstep of every person. It also means that all members of civil society, particularly non governmental organizations, participate fully in making and implementing decisions."[7] Understandably, empowerment implies empowering both women and men, so that they can compete on an equal footing; also empowerment requires investment in education, and health of people; again empowerment requires clearly an environment where everyone has access to credit and other inputs of importance.

5. Human Resources and Rural Workforce

Human resources are most important in all the developmental programmes. The term 'human resource' comprehends more or less the entire population of a country. It covers not only the skilled manpower, but also, the unskilled labour as well, indeed all persons who could be put to productive work. The development of human resources implies the development of both the physical and mental aspects of the productive capacity of workers. The physical aspect of the productive capacity can be developed by means of nutrition, health services, etc. and mental aspect by education and training.

Rural workforce consists of farm workers including cultivators and agricultural labourers, rural artisans, like blacksmiths, carpenters, potters, shoe makers, weavers and those engaged in trade and commerce, rural transport and communication and other service establishments. Among these categories, farm workers are predominant. In 2001, as many as 45.6 per cent of the workers were in the agricultural sector. Most of the farm and non-farm workers are illiterate and their level of efficiency is low. There is underutilization of labour force in rural areas. In a situation of under utilization of labour, the earnings of many are too low to meet even the requirements of subsistence living. Such a situation affects the rate of saving and investment. Further, the demand for different industrial products is also affected. Underutilization of labour leads to growing impoverishment and creation of an atmosphere of discontent and strife, hardly conducive to rural development.

Human resources have two aspects, namely, quantitative and qualitative aspects. India is rich so far as quantitative aspect is concerned as more than 600 million live in rural areas. At the present population growth rate of about 1.8 per cent per annum, nearly 20 million are added every year to the already existing huge size of population. The most disappointing feature of manpower resources is with regard to the qualitative aspect. As per 2001, census, only 40 per cent of the total population is literate in all-India. The literacy rate in rural areas is still lower at 35 per cent only. The qualitative aspect of human resources can be better judged from the angle of motivation for development and technical skills and ability of labour to handle new technology.

In India, agriculture is more or less a monopoly of the illiterate, ignorant and generally poverty stricken people. The basic problem of agriculture is its low productivity or low yield per hectare. The land available for cultivation is almost exhausted with little scope for extensive cultivation. New land resources must be found in the form of achieving higher yields per hectare. From the land that is already under the plough increasing agricultural productivity or yield per hectare depends not merely on physical inputs such as good seed, fertilizer and credit, but also on human factor. Assuming that all physical factors are available for growing crops, proper utilization of them to the maximum benefit depends primarily on one factor, namely, the human factor, i.e., the farmer. The farmer must have the necessary know-how and the skill to make farming a profitable proposition with higher net returns per hectare. In India, farming is a primary occupation to a vast majority of people. Yet, farm production is not sufficient to meet the growing demand of the fast increasing population with minimum nutritional standards. The average yield per hectare crops in India is far below that of the developed countries even though the performance of a few farmers in some region is quite satisfactory. The weakness of agriculture in India can be attributed to many factors, of which the weakness of human factor seems to be the major. For the betterment of farming the farmer himself must be improved. The farmer[8] must have necessary knowledge and skill to modernize agriculture. Modernization and illiteracy go ill together. The transformation of traditional agriculture into modern farming leading higher yields per hectare is possible when the man behind the plough is modernized. The illiteracy of the farmer and his low level of skill and technology clearly reveal that human factor is a neglected input so far as motivation, outlook and training for increased production are concerned. Poor soils, low per capita acreage and scarcity of capital alone are not responsible for lower growth of agriculture productivity. The higher yields in Japan, Italy, Greece and Australia with less per capita acreage than what it is in India, can be traced to the quality of the material used and again to the capabilities of farm people. T.W. Schultz,[9] analyzing the causes for differences in the yield levels in various countries, rightly remarks "the differences in land are least important, differences in the quality of material capital are substantial importance and differences in the capabilities of farm people are the most important in the amount and rate of increase of agricultural productivity. Thus, the importance of human resource is well analyzed by T.W. Schultz.

6. Extension Education and Human Resource Development

An urge for technological change in modernization of agriculture in developing countries in the midst of poor education and widespread illiteracy necessitates an extensive method of educating the farmer in his farming activities, which is known as extension, education, after the American pattern. It is the rural adult education designed to build rural people in their many-sided activities with special reference to improvement of agriculture. "The task of extension work is to help rural families, apply science to day-by-day routine of farming, home making and other aspects of rural living.[10] The programmes of Extension Agency are normally oriented towards the production problems of farmers. The extension education is interested in carrying to the doors of the farmers the results of research centres and demonstrate it to the farmer on his field. The "result-demonstration" is a corner stone in extension education. When an improved variety, say of paddy, is developed, the extension personnel work closely with the farmers in determining whether it is adaptable to local farming conditions. The new variety is grown in one plot and the traditional variety in the other. The farmer and his neighbors themselves judge whether the new variety is superior. If proved superior, the farmers adopt new varieties quite willingly and better yields. Extension not only takes the findings of science to the farm, but also takes the problems of the farmer to the research laboratories and helps to formulate new types of research activities designed to meet the growing problems facing the farm people. Extension education is again interested in training and guiding the farmers and thus facilitate innovation and scientific management so as to achieve higher productivity. Practices that have been proved sound by research, and local experience are extended through demonstrations, group meetings, bulletins, radio, press and other media. The strength of the extension lies in gaining the confidence of farm people by bringing to them tried and improved methods and practices, the value of which farm people are encouraged to test for themselves and get better yields. Thus, extension education has manifold functions. It is through more systematic application of science to farming that increasing productivity can be realized and the aim of extension is to carry science to farming. "Science, wherever it has gone around the world, is responsible for progress in agriculture and rural welfare. Bringing that science to the door steps of more and more people is the job of extension".[11]

The basic principle in extension work, when approved as a stimulating process, is that any programme must be in harmony with culture of the people. At any time and in any place, there is well established social life, men have their own habits and organizations. They have religious beliefs, values and practices. They have their own machines, skills and methods to make a living. The pattern of life of the people as determined by the society as a whole is well settled and deep rooted and any change in the existing set-up may not be very easy. The extension agent must clearly understand the existing social pattern and try to introduce change in a manner most acceptable to the people. A dictated programme with a force behind it would only face resistance and the desired programme must, therefore, be in harmony with the culture of the people. The knowledge that extension imparts must be viewed with sympathy and understanding. Once people are convinced that extension service is their own affair and for their good, success is certain. It is necessary that the persons in charge of extension service must become one with the rural folk and give them the required services in a suitable manner and at an appropriate time. The response or reaction of the farmers must be carefully examined. In a convincing way the message of modern farm technology must be delivered to the farming community. Extension workers must continuously strive to bring about changes in the behaviour of the cultivators and their families. They must influence the cultivators to effect improvement over traditional methods of cultivation by modern farm technology. They must impress the cultivators the advantages of better seed, chemical fertilizers and the usefulness of insecticides and pesticides etc.

Every extension worker is a teacher. Unlike the teacher of the elementary or secondary school in a village, who teaches in a classroom a particular subjects, the extension workers teaches the farmer at any place on any subject. The extension worker has no fixed classroom. His classroom is wherever he meets people. It may be a field, a home or under the shade of a tree. His pupils range from boys or girls to grey-haired men who are village leaders. He teaches any problem of immediate importance in farming or rural living. The success of his teaching and its impact on the farmers and on farming depends on various factors, the important factor being the educational background of the farmer.[12]

The primary function of the extension agency is first to interpret to the farm people in their terms and at their levels of comprehension such authentic and scientific information as may be relevant to and feasible in local conditions and secondly, to support this field education by timely and adequate supplies and services. Under the present arrangement, the key officer at the village incharge of the extension service is the village development officer (VDO). The VDO, often a matriculate and with elementary training in agriculture, is asked to look after nearly 1500 farmers and give them guidance not only in agriculture but also in animal husbandry, cooperative organization, social service, etc. Such a type of extension service is most inadequate and cannot serve the purpose. In countries, which have a progressive agriculture, the agricultural extension service is more intensive. For instance, in Scotland only 200 farmers are served by one extension agent, in Holland about 400 and in Japan between 500 and 600. Further, the private manufacturers of fertilizers, pesticides, agricultural implements, etc., also provide extension services to the farmers. In China, the area allotted to each extension worker is about one-fourth in India.[13] It is to be realized that unless there is enough and fully qualified extension staff to attend to agricultural programmes, there is no hope of making much progress in the farm production. In view of the fact that the agricultural programme is the most important single programme of village upliftment in India, it is necessary that all Gramsevaks should be graduates in agricultural science and paid accordingly.

7. Human Resource Development in Rural India

There is a pressing need for increasing the productivity of labour engaged in both farm and non farm sectors of rural economy and raising their levels of income. The non-farm agricultural sector needs the support of a strong industrial base for its development. There is, therefore, need for integrated development of both agricultural and non-agricultural sectors. This strategy will throw up new challenges to the policy-makers in the field of manpower planning, employment and education. Among others, the following steps become necessary to develop human resources in rural areas: (i) There is a need for technical literacy for the farmers through appropriate and effective extension services. (ii) Training facilities have to be provided to the farmers at selected local points so as to cover a large number of farmers and farm workers. (iii) Adequate number of workshops at suitable places should be set up so as to help the farming community to have the services in the matters of repair of agricultural tools and implements, coil-building of electric motors etc. The Regional Rural Banks (RRBs) should take initiative in evolving schemes for encouraging private entrepreneurship in the rural areas. The RRBs may initiate different self-employment schemes in rural areas so as to provide technical service, including consultancy service in modern farming, etc. (iv) The Krishi Vigyan Kendras now sponsored by the ICAR need to be strengthened and widened, covering all categories of farmers of both wet and dry crop cultivation. (v) In view of widespread illiteracy, poor economic conditions and underutilization of manpower resources, informal and on-the-job training facilities have to be provided for skill-generation in the rural areas. (vi) In the rural schools, where a majority of students are drawn from the farming community, basic agricultural education must be made compulsory. Even if some discontinue with the X class, what they learnt about agricultural services must be useful in their farm operations. At 10 + 2 stage, there can be a separate stream of education with subjects relevant for rural development, catering to the skill requirements of the farm and allied sectors and non-farm sectors. (vii) Vocational education to the rural adults is more important for achieving the immediate results, while giving appropriate and socially relevant education to the rural youth, is necessary in the long run. The skeptical and traditional-minded elders, sometimes, may not accept second hand advice, that, too, from their children. Therefore, the education of the adults through extension and appropriate vocational training programme is essential. The present-Farmer's Training Programmes are found to be quite inadequate in terms of training imparted and their coverage.

8. Rural Artisans

Traditionally, rural artisans constituted the backbone of rural economy as they effectively provided the needed servicing facilities for the traditional farm equipment and other assets, possessed by the villagers. Rural artisans included carpenters, blacksmiths, potters, weavers, cobblers, toy-makers, basket makers etc. It was the artisan who contributed considerably to the development of techniques of making a bullock-cart, plough, agricultural tools, pottery etc. These techniques emanating from the indigenous skill were duly accepted by the rural masses, because the artisan had fully involved himself in the development of native technologies. He received encouragement from fellow villagers who had confidence in him and his ability to do the required job. However, in the present context, they are a class decaying for a variety of reasons: (i) tendency of the farmers to go for improved agricultural implements and incapability of the rural artisans to provide servicing and repairing facilities to such modern agricultural implements; (ii) villagers are becoming increasingly modernized with increased incomes and purchasing power. This had adversely affected the demand for the products produced by rural artisans and (iii) illiteracy and lack of other skills is rendering the rural artisans increasingly irrelevant to the rural scene.

Efforts should also be made on a priority basis to resettle the rural artisans and render them functionally effective. They may be motivated to take up new activities with necessary financial help and required training. The rural artisans may be involved in setting up of repair workshops in the villages. They may be given suitable training to acquire different skills such as tractor mechanic, tubewell mechanic, gobar-gas plant mechanic, electrical mechanic, etc., efforts should be directed towards designing the training courses for such composite skills to rural people. Such courses should be linked to the local needs and should emphasize more on work practice than theory. Necessary institutional support in terms of financial assistance, training, marketing, etc., must be provided to encourage the different sections of rural population to come forward to start the rural industrial units and manage them efficiently. The existing facilities for institutional training through TRYSEM, IRDP, Industrial Training Institutes, Polytechnics etc., should be reviewed and redesigned with the objective of meeting the needs of rural people.

9. Human Development Index (HDI)

The HDI is a statistical analysis as an index/glossary to rank countries on the level of "human development" and classify countries as developed (high development), developing (middle development) and underdeveloped (low development) countries. This exercise is attempted based on statistics for Life Expectancy, Education, Standard of living and GDP collected at the national level using a formula.[14]

The origins of the HDI are found in the United National Development Programmes (UNDP), and Human Development Reports (HDRs). There were devised and launched by an Economist of Pakistan, Mahbub Ul Haq in 1990, fully assisted by development economists like Paul Streeten, Francis Stewart, Gustav Ranis, Keith Griffin, Sudhir Anand, Meghnad Desai and Amartya Sen. Sen, though expressed doubts to capture the full complexibility of human capabilities in a single index, but finally agreed to proceed with the exercise which may be of great use for policy makers ensuring human development as a central point in economic growth. Since 1990, the HDI has been used by the United Nations Development Programme for its annual Human Development Reports.

The HDI combines three dimensions: (1) Life expectancy at birth, as an index of population health and longevity. (2) Knowledge and education as measured by the adult literacy rate (with two-thirds weighting) and the combined primary, secondary and tertiary gross enrollment ratio (with one-third weighting). (3) Standard of living as measured by the natural logarithm of gross domestic product per capita as purchasing power parity.

Conclusion

Human resource needs to be developed as a central point of development exercise in terms of necessary skills, knowledge, health services and similar such crucial inputs. Along with this, human resource needs to be strengthened ethically coupled with work culture, only then full-fledged growth of human resource is possible. It is well established that human resource development has been mainly responsible for Japan's quick recovery and faster growth, after its devastations in the Second World War. With human resource development only, faster growth and development on right lines can be achieved.

REFERENCES

1. Reddy K. Venkata, *Agricultural and Rural Development* (A Gandhian Perspective) Himalaya Publications, 2001, p. 15.
2. United National Development Programme, *Human Development Report 1997*, Gyord University Press, 1997, p. 15.
3. Mahbub Ul Haq, *Reflection on Human Development*, Casefold University Press 1997, p. 15.
4. *Ibid.*, p. 14.
5. *Ibid.*, pp. 14-15.
6. Paul Streeten's "Foreword" to Mahbub Ul Haq's Book, pp. IX-X.
7. Mahbub Ul Haq, *op.cit.*, pp. 16-20.
8. *Ibid.*, p. 20.
9. T.W. Schultz, *Transformation of Traditional Agriculture*, Quoted from FAO Agricultural Commissioner Projections, p. 16.
10. Brunner, Edmund Des, *The Farmers of the Works – The Development of Agricultural Extension*, New York Columbia University Press, Fourth Edition, 1954, p. 1.
11. *Ibid.*, p. 7.
12. Reddy, K. Venkata, "Education, Extension and Agricultural Productivity", in *Khadigramodyog*, March 1970, p. 7.
13. Sen, S.R., the *Strategy for Agricultural Development*, Bombay, Asia Publishing House, Second Edition, 1966, p. 15.
14. For details of the Human Index formula with illustration, see *Indian Economy* by Misra and Puri, Himalaya Publications, 2009, pp. 24-25.

PART IV

SCHEMES IN DEVELOPMENT

21. Poverty Alleviation and Employment Generation Programmes
22. Social Security Schemes in India
23. Rural Housing
24. Health Care for All
25. Education for All
26. Dr. A.P.J. Abdul Kalam's PURA Mission for Rural Reconstruction

21 POVERTY ALLEVIATION AND EMPLOYMENT GENERATION PROGRAMMES

1. Introduction

Poverty is a burning problem in India. India is a big country in the world with abundant natural and human resources. However, India is a nation still with over 300 million poor people. Poverty, illiteracy, ill-health and civic inertia normally go together. Majority of the poor are illiterate, unskilled, mainly depending on agriculture for their livelihood. The poverty, illiteracy, ill health and civic inertia are all concentrated in rural areas. With migration of rural poor to the urban centres, the poverty associated problems are all migrated to the urban centres. The urban slums welcome the rural poor and such situation adds to environmental problem in urban centres.

2. Poverty Defined

The term poverty is commonly understood as the condition of having low income and consequently denied of minimum needs and services for human existence. Poverty may be due to less access to income, resources and opportunities. Absolute poverty is a situation of not having the resources, capabilities and opportunities to meet the basic needs for living. Poverty conditions exist in different parts of the country. In fact, almost by all the countries there is some degree of poverty. Although poverty is generally considered to be undesirable because of the pain and suffering that may accompany it, in purely spiritual context, poverty may be seen as a virtue. Renunciation of all kinds of desires and worldly pleasures may give happiness to some people and in this context, poverty may be a welcome featured. Poverty is a relative term. The list of basic needs and minimum level of consumption will vary from man to man depending upon general level of economic development and socio-cultural factors. Unemployment and underemployment are the chronic problems leading to poor incomes in the rural areas. They remain untackled effectively inspite of series of schemes during the planning era. There is high degree of poverty in rural areas in the midst of surplus manpower and other resources. It is encouraging to note that the poverty ratio has been on the decline from time to time. In 1973, the poor constituted 54.93% of total population and it has come down to 27.5% in 2004.

The percentage of population below poverty line varies between rural and urban. Rural poverty continues to be always higher than the Urban poverty. Table 21.1 indicates the degree of variation in poverty between rural and urban.

Table 21.1

Percentage of people below poverty line in India (1973-2004)

Years	Rural	Urban	Combined
1973	56.5	49.0	54.9
1983	45.7	40.8	44.5
1993	37.3	32.3	36.0
2004	28.3	25.7	27.5

Source: Planning Commission.

It is evident from Table 21.1 that throughout the period under reference the percentage of people below the poverty line in the rural areas has been consistently higher than that in urban centres.

According to the Planning Commission (Eleventh Plan draft), a monthly consumption expenditure of ₹ 356 per head in rural areas and ₹ 539 per head in urban areas would be required to meet the minimum basic needs of a person. Such of those below these levels of expenditure are considered as those who live below poverty line. However, it is agreed that the monthly per capita consumption expenditure of ₹ 356 for rural areas and ₹ 539 for urban areas in India is too low. But yet, this indicates poverty situation in India. No doubt, there are difficulties in measuring the poverty. Rural poverty is visible, by and large, in the form of poor dietary conditions with malnutrition, primitive shelter, poor health, low level of literacy, high propensity to consume with very low level of savings, lower ratio of capital to labour and unproductive agriculture. Though rural India emerges as a developed country, still many rural people suffer for want of basic amenities such as safe drinking water, reasonable food and housing, primary health and education facilities, These basic requirements of life can easily be assured if gainful employment with a living wage is provided to all unemployed and underemployed in rural areas.

3. Measurement of Poverty

There are difficulties in measuring poverty in rural areas with a large proportion of self-employed population. Further, in rural areas, workers cannot be easily classified into employed and unemployed since almost everyone is employed for some part of the year and unemployed for some other part. Again, some households need additional employment for supplementary income while others need employment for regular earnings. The careful and painstaking attempts of the studies of Datenwada Committee, Dandekar, Neelakanthrath and G. Parthsarathy in analyzing the nature and magnitude of rural employment clearly pinpoint all pervading problems of rural poverty and unemployment.[1] The composition of the poor has been changing and rural poverty is getting concentrated in agricultural labour and rural artisan households and urban poverty in casual labour households. Agricultural households accounted for 41% of rural poor both in 1993-94, as well as in 2004-05. The share of the self-employed in agriculture, among the rural poor had fallen from 32% in 1993-94 to 21.6% in 2004-05. The casual labour households, accounted for 62.6% in 1993-94 and 56.5% in 2004-05[2] in urban areas.

Among social groups, SCs, STs and backward castes accounted for 80% of the rural poor in 2004-05, considerably more than their share in rural population.[3] In 2004-05, while the poor in total rural population was 28.3%, among SCs it was 36.8%. In urban areas, while the overall poverty was 25.7%, among SCs it was higher than in rural areas at nearly 40%.[4]

The proportion of STs population among the rural population living in poverty is high. It is about 15% in 2004-05, double that of their share in the total population of India. For rural population the incidence of poverty among STs had fallen from 51.94% in 1993-94 to 47.3% in 2004-05, where as it had fallen from 37.3% to 27.5% in the total population. In 2004-05, the incidence of poverty among STs has barely fallen compared to a decade earlier.[5]

The share of poor in the total urban population in 2004-05, was 25.7% but 33% of STs in urban areas were estimated to be poor. Thus, it is evident that STs with 47.3% in poverty is higher than that of SCs with 36.8% in poverty in rural areas. It is evident that both SCs and STs in rural and urban areas, under the category of poverty, are higher than the overall poverty percentage. At the same time, the proportion of ST population living below poverty line is higher than that of SC population.

Some states have been particularly successful in reducing the proportion of poor in the total population. In 2004-05, the states with lowest poverty ratio were Jammu & Kashmir (5.4%), Punjab (8.4%), Himachal Pradesh (10%), Haryana (14%), Kerala (15%), Andhra Pradesh (15.8%) and Gujarat (16.8%). On the other land, in some states poverty percentage is recorded to be higher like Orissa (46.4%), Bihar (41.4%), Madhya Pradesh (38.3%) and Uttar Pradesh (32.8%). In the States that were formed recently, the poverty ratio is recorded to be higher as Chhattisgarh (40.9%), Jharkhand (40.3%), and Uttarakhand (39.6%).

Deprivation of SCs and STs in the form of inequalities of wealth is very pronounced. Most SCs live in rural areas, where land is the main asset. In the overall population, 60% of the rural households were cultivator households. But SC households are more disadvantaged among rural households; only 47% were cultivators and a quarter of the SC households were agricultural labour households. About 6% to 7% of the SC house holds were artisan households.[6]

4. Poverty Among Women

The incidence of poverty in terms of income, among females tended to be marginally higher in both rural, and urban areas. The per centage of female persons living in poor households was 28% in rural and 26% in urban in 1993-94, and 29% and 23% respectively in 2004-05. It is evident that the percentage of rural female poverty increased from 26 to 29. However, the poverty ratio in the case of urban female population declined from 26 to 23. The lower percentage of female persons among the poor despite higher female poverty ratio was due to an adverse sex ratio- which itself is a reflection of the discrimination that women and girls face over their life cycle.[7] It is also stated that SC and ST category of girls are the worst off in terms of educational outcome indicators. For instance, in the critical age group of 15-49 years, when women are in their reproductive period, 73% of SC women, 79% of ST women and 61% of OBC women are illiterate. 61% of Muslim women are also found to be illiterate. This percentage is relatively low because the rate of urbanization among Muslims is higher than the other social groups. Some 42% of SC women and 46% of ST women were malnourished. Again 30% of the 'other category' women were also malnourished.

5. Poverty Among Children

Child poverty is widespread in India both in rural and urban areas. The percentage of children below 15 years living below poverty line (BPL) households constituted 39 in rural and 41 in urban areas in 1993-94 and this situation changed to 44 in rural areas and 32 in urban areas in 2004-05. It means children in poverty households in rural areas increased, while it decreased in the care of urban. The high and rising level of child poverty is now only linked to a high incidence of child malnutrition, but also under mines their future capabilities and adversely affects equality of opportunity.[8]

The globalization and urbanization necessitate steps for development. Those with education, technical know-how, skills took advantage and progressed. Such of those lagging behind in acquiring the skills and information become poor and poorer. The impact of globalization as revealed by some studies is one of widening the gap between the haves and have-nots. The UNO very rightly has set the millennium goals for the promotion of over all development of the individuals and countries. They include: (1) Eradicate extreme poverty and hunger by reducing by half, both the number of people living on less than a Dollar a day and the proportion of people suffering from hunger, (2) Achieve universal primary education ensuring that all boys and girls complete a full course of primary schooling, (3) Promote gender equality and empower women, (4) Reduce child mortality, (5) Improve maternal health, (6) Combat HIV/AIDS, malaria and other diseases, (7) Ensure environmental sustainability and (8) Develop a global partnership for development.

It is encouraging that the government is proceeding on these lines. Poverty in India being a chronic nature for various reasons historical and socio-economic, the government is taking steps to identify the source of poverty and areas of intensity of poverty and initiate appropriate steps. The objective of 'inclusive growth' policy of the Eleventh Five year plan is one of addressing different issues centered round poverty. Regions which have large number of chronic poor people are identified such as: (1) Tribals in forests mainly in central and eastern parts of the country. (2) Also, semi-arid regions where agriculture is mainly rain fed located in states of western and southern India. Even in areas where agriculture is irrigated, poverty is located in situations where the land-man ratio is the lowest. In such regions it is rightly observed that "without effective land reforms and agricultural services poverty reduction is not possible". (3). Dependence on casual labour has grown and a large proportion of the chronic poor are depending on wage labour. Most of these are SC and STs. While male casual laborers have increasingly worked in the non-farm sector, women labour have increasingly been concentrated in agriculture. Most of ST and 40% of SC casual workers are poor, the landless casual workers being the poorest. On an average, women are poorer than men casual labourers. (4) At least, 18 million rural people do not have a shelter over their heads.[9] This situation calls for adequate employment opportunities to the rural poor. It is only to meet this requirement, NREGP is implemented and in that way eradicate poverty. Empowerment of workers and creation of durable assets is required to eradicate poverty. This required linkages of NREGP with other programmes like National Rural Health Mission (NRHM), National Mission for Literacy and Elementary Education and other livelihood and infrastructure initiatives.

6. Steps to Tackle poverty

No doubt, eradication of poverty and unemployment had figured in the list of objectives of our series of Five Year Plans from the First Five Year down to the Eleventh Five Year Plan currently in operation. In fact several measures have been taken to eradicate poverty and unemployment. However, the steps taken so far, are of *ad hoc* nature and hence, the success achieved in eradication of poverty is not encouraging as expected. It was taken for granted that in the process of achieving certain physical targets of agricultural and industrial production etc., in each plan period, the employment objective would take care of itself. Moreover, eradication of unemployment had never been a primary objective of our planning. While the aspiration was one of high rate of development, with adequate employment, the experience has been only inadequate development without adequate employment. The fruits of development have been pocketed by selected rich few. The rural poor have been bypassed both in terms of growth and distribution of gains of growth. MacNamara, former President of the World Bank, rightly observed, "Growth is not equitably reaching the poor and the poor are not significantly contributing to growth". The problem of surplus population subsisting on land and the consequential increase in rural unemployment and underemployment leading to poverty in India has been stressed by several competent authorities since the report of the Famine Commission of 1880. According to the 1981, census the total unemployed in rural areas were 27.06 million. This figure increased to 34.73 million in 2004-05. In 2004-05, the rural unemployed persons worked out to 25.09 million. Thus, from time to time number of unemployed has been increasing.

7. Trend Towards more number of Agricultural Labour

The most disturbing factor in recent years, has been faster increase of agricultural labourers and gradual decline of number of cultivators. In 1951, total population in India was 361.1 million, out of this, rural population was 298.6 million or 82.7% of total population of this, agricultural labourers were 27.3 million or 28.1%. As per 2001, census, the total population was 1028.7 million. The rural population was 742.6 million or 72.2%. In this year, the agricultural laborers were 106.8 or 45.6%. The cultivators were 127.3 million or 54.4%. Thus, in 1951s the agricultural labourers constituted 28.1% of total agricultural workers. This percentage increased to 45.6% in 2001. It is evident that the magnitude of agricultural labourers has been increasing from time to time. At the same time, the cultivators in 1951, constituted 71.9% of total agricultural workers. This percentage declined to 54.4% in 2001. The trend of population growth and agricultural workers is presented in the following Table 21.2

Table 21.2

Population and Agricultural Workers

Year	Total Population	Average Annual Exponential Growth Rate (%)	Agricultural Workers			
			Rural Population	Cultivators	Agricultural Labourers	Total
1	2	3	4	5	6	7
1951	361.1	1.25	298.6 (82.7)	69.9 (71.9)	27.3 (28.1)	97.2 (100)
1961	439.2	1.96	360.3 (82.0)	99.6 (76.0)	31.5 (24.0)	131.1 (100)
1971	548.2	2.22	439.0 (80.1)	78.2 (62.2)	47.5 (37.8)	125.7 (100)
1981	683.3	2.20	523.9 (76.7)	92.5 (62.5)	55.5 (37.5)	148.0 (100)
1991	846.4	2.14	628.9 (74.3)	110.7 (59.7)	74.6 (40.3)	185.3 (100)
2001	1028.7	1.95	742.6 (72.2)	127.3 (54.4)	106.8 (45.6)	234.1 (100)

Source: Directorate of Economic and Statistics Department of Agriculture & Cooperation, Ministry of Agriculture, Govt. of India. ***Agricultural Statistics 2008***, p. 37.

It is evident from Table 21.2 that agricultural labour has been increasing from time to time. This is due to several reasons. Due to division and subdivision of land holdings from generation to generation, and consequent creation of tiny holdings and growing indebtedness, cultivators with small holdings are gradually converted into landless labour. Again due to the decline of demand for products of village artisans, carpenters, potters, weavers etc., the artisans are converted into agricultural labourers. In the absence of ownership of physical assets like land, etc., the rural poor aspire for more children whom they consider as their assets and on whom they have to depend during the old age. The family planning programme is yet to have significant impact on the rural poor.[10]

8. Poverty Alleviation Programmes of Past and Ongoing

From time to time, several programmes have been initiated to tackle the problem of unemployment and poverty. To name, a few of such programmes:

1. Rural Works Programme of 1970-71, which was later named as Drought Prone Area Programme
2. Desert Development Programme (DDP) of 1977-78.

The ongoing programmes are as follows:

Pradhanmanthri Gram Sadak Yojana (PMGSY)

Pradhanmanthri Gram Sadak Yojana (PMGSY) was launched on December 25, 2000. It is 100% centrally-sponsored scheme with the primary objective to provide all-weather connectivity to all the eligible unconnected habitations in the rural areas. The Programme is funded mainly from the earnings of diesel cess from the Central Road Fund. Under this scheme, upto March 2009, a total of about 2,14,281. 45 kms of road works has been completed with total expenditure of ₹ 46,807.21 crore.

Indira Awas Yojana (IAY)

This scheme addresses housing shortage as an important component of poverty alleviation in rural India. The objective of IAY is to provide financial assistance for construction and upgradation of houses to BPL households belonging to the Scheduled Castes and Scheduled Tribes, freed bonded labourers, non-SC/ST rural households, widows and physically handicapped persons living in the rural areas. The scheme is funded on a cost-sharing basis of 75:25 between the centre and the states. However, in the case of North-eastern states the funding pattern has recently been revised to 90:10. Till March 31st, 2009, as many as 21.05 lakh houses had been constructed.

Swarna Jayanthi Shahari Rozgar Yojana (SJSRY)

In Dec. 1997, this scheme came into force. This scheme is composed of Urban Self-employment Programme (USEP) and Urban Wage Employment Programme (UWEP). The aim is urban poverty alleviation. During 2008-09, an amount of ₹ 540.67 crore has been spent on this programme and as many as 9,47,390 urban poor were assisted to set up individual/group micro enterprises and 14,84,209 urban poor were imparted skill training under SJSRY.

Swarna Jayanthi Grama Swarozgar Yojana (SGSY)

The SGSY was launched in April 1999, after restructuring of Integrated Rural Development Programme (IRDP) and allied programmes. It is the only self employment Programme currently being implemented for the benefits of rural poor. The objective of SGSY is to bring the assisted Swarozgaris above the poverty line by providing them income generating assets through bank credit and government subsidy. This scheme is being implemented on cost-sharing basis of 75:25 between the centre and states. Upto March 2009, as many as 34 lakh Self–help Groups (SHG) had been formed and 120.89 lakh Swarozgaris have been assisted with a total outlay ₹ 27,183.03 crore.

Mahatma Gandhi National Rural Employment Guarantee Programme (NREGP)

This programme was launched on February 2, 2006. It is a path-breaking initiative, to provide legal guarantee to work and to transform the "geography of poverty". This Act envisages securing the livelihood of people in rural areas by guaranteeing 100 days of employment in a financial year to rural household. The Act provides a social safety net for the vulnerable households and an opportunity to combine growth with equality. In the first phase of implementation during 2006-2007, some 200 districts were covered. Additional 130 districts were brought under NREGP during 2007-2008, in the second phase, making a total of 330 districts covered by the programme. The remaining districts where NREGP will come into force have been notified on 28th September 2007. As such the statutory provision of Act to cover the entire country within five years of its notification is made clear. The main provision of the Act are:

(1) Employment to be given within 15 days of application for work,

(2) If employment is not provided within 15 days, daily unemployment allowance in cash has to be paid,

(3) Employment within 5 km radius, else, enter wages to be paid

(4) At least one-third beneficiaries have to be women

(5) Gramsabha of village panchayat will recommend works.

(6) Gram Panchayat to execute at least 50 per cent works

(7) Panchayat Raj Institutions have a principal role in planning and implementation

(8) Transparency, accountability and social audit would be ensured through institutional mechanism at all levels,

(9) Grievance redressal mechanism to be put in place for ensuring a responsive implementation.

NREGP makes a paradigm shift from all the earlier and existing wage employment programmes, because it is an Act and not just a scheme. It provides a legal guarantee to work. The rural households in the notified districts will have the right to register themselves with the local gram panchayat as persons interested in getting employment under the act. The gram panchayat upon verification will register the household and issue a job card which is a legal document entitling a person to ask for work under the act.

An amount of ₹ 9,10,573.72 lakh has been sent for implementation of NREGP during 2007-2008. The ongoing programmes of Sampoorna Grameena Rozgar Yojana (SGRY) and National Food For Work Programme (NFFWP) would be subsumed with NREGP in the identified districts. The act is a very bold step of the government to eradicate poverty from rural India and to enable the rural poor to survive in the era of liberalization and globalization. The act of aims at creating durable assets and strengthening the livelihood resources base of the rural poor. The works suggested in the Act include restoration of irrigation tanks, deforestation, soil erosion, etc., so that the process of employment generation is on a suitable basis. The experience with NREGP so as far suggests that it is one of the main planks of rapid poverty reduction.[11]

However, this programme is subject to criticism for more than one reason. The NREGP draws away the labour force and thus denies the regular labour supply to the local cultivation. Further, under this act minimum wage fixed is ₹ 100 per day per worker both for men and women. Paying ₹ 100 equally to both men and women labour would amount to increasing the cost of cultivation. Under these circumstances, the farming is found to be unremunerative. Further it is said that all is not well in working of NREGP. Though the act expects to address schemes relating to drought, deforestation, soil erosion etc., the labour is not well utilized for such schemes. It is said that records are only manipulated and funds are misused under this scheme.

Conclusion

Poverty and inequality in income levels continue to be a burning problem in India. The different schemes initiated by the Government need to be implemented effectively. Particularly, the NREGP with equal wages to men and women labour is an appropriate scheme assuring regular employment and income. That apart, Gandhiji's views with regard to poverty and inequality of income have to be honoured. Mahatma Gandhi states that we have resources to meet all our needs, but not our greed. In his words, "Nature produces enough for our wants from day to day, if only everybody took enough for himself and nothing more, there would be no man dying of starvation. If I take anything that I do not need for my immediate use and keep it, it amounts to thieving. He categorically states that this type of behaviour on the part of people is responsible for inequality and poverty. In Gandhiji's words "In India we have got three millions of people having to be satisfied with one meal a day. You and I have no right to anything, that we really have until these three millions are clothed and fed better. You and I, who ought to know better, must adjust our wants, and even undergo voluntary starvation in order that they may be nursed, fed and clothed." The message of Gandhiji is "Enjoy the wealth by renouncing it". If Gandhiji's message is honoured in practice by all in the society the widely prevalent poverty and inequality of income may no longer be a persistent problem in India.

REFERENCES

1. Parthasarathy. G and Rao, Dasarathy D., *"Employment and Unemployment of Rural Labour and the Crash Programme"*, Andhra University, Press Watair, 1974.
2. *Eleventh Plan Report,* Vol. III, p. 80.
3. Working Group on Poverty, *Planning Commission,* 2006.
4. *The Eleventh Five Year Plan,* Vol. III, p. 81.
5. *Ibid.*
6. *Ibid.*
7. *Ibid.*, p. 83.
8. *Ibid.*, p. 84.
9. *Ibid.* p. 5.
10. Venkata Reddy K., *"Agriculture and Rural Development"*, Himalaya Publishing House, p. 69.
11. *Eleventh Plan Report,* p. 133.

22 SOCIAL SECURITY SCHEMES IN INDIA

1. Meaning of Social Security

Social Security simply refers to a planned attempt of providing a benefit package to the needy people such as poor, unemployed, aged, sick aiming at the protection of those who need and deserve outside help. From time to time governments centre and state government have been initiating different social security schemes. The Government is coming forward to assist all those who need help from the state. Thus, the Social Security Benefits in India are need-based. The needy persons are dispersed agriculture labour, landless or assetless poor, old and sick who are helpless without any assets or anyone to take care of them. Social Security Schemes cover a wide range of issues and people.

India has had a Joint Family System, that took care of the social security needs of all members of the family, provided the family had access or ownership of maternal assets like land. In keeping with its cultural traditions, family members and relatives have always discharged a sense of shared responsibility towards one another. To the extent that the family has resources to draw upon, this is often the best relief for the special needs and care required by the aged and those in poor health. However, with urbanization, mobility of occupation, growth of education with increasing and demographic changes, there has been gradual decay of the joint family system. Now a formal system of social security became a necessity.

2. Social Security Schemes in India

Government of India has been implementing several social security schemes[1] from time to time with the aim of extending helping hand to the poor, unemployed, aged, sick and similar such who deserve help. Enumerated below are a few major programmes of social security.

(i) National Social Assistance Programme (NSAP)

This scheme was launched in 1995-96, as a centrally sponsored scheme, administered by the Ministry of Rural Development. But with effect from 2002-03, NSAP was transferred as the State Plan. Funds for the scheme are now provided as Additional Central Assistance (ACA) to the states. The NSA comprises of following schemes:

(a) Indira Gandhi National Old Age Pension Scheme (IGNOAPS)

(b) National Family Benefits Scheme (NFBS)

(c) Annapurna Scheme

(d) Indira Gandhi National Widow Pension Scheme (IGNWPS): Introduced in February 2009, the Scheme is applicable to Below Poverty Line widows in the age group of 40-64 years

(e) Indira Gandhi National Disability Pension Scheme (IGNDPS): Introduced in February 2009, it is applicable to below poverty line persons with severe or multiple disabilities in the age group of 18-64 years.

(ii) Swarna Jayanthi Shahari Rozgar Yojana (SJSRY)

This scheme seeks to provide gainful employment to the urban unemployed or underemployed poor by encouraging them set to up self-employment ventures or provision of wage employment. SJSRY is a centrally sponsored scheme funded on a 75:25 basis between the centre and the states. To avail benefits under this scheme, the concerned Municipal Board of the respective States should be contacted.

(iii) Provident Fund Benefits and Miscellaneous Provisions Act of 1952

This Act applies to specific scheduled factories and establishments employing 20 or more employees and ensures terminal benefits to provident fund, superannuation pension and family pension in case of death of an employee during service. Separate laws exist for similar benefits for the workers in the coal mines and tea plantations.

(iv) The Maternity Benefit Act, 1961

This act provides, for 12 weeks wages for women during maternity as well as paid leave in certain other related contingencies.

(v) The Payment of Gratuity Act, 1972

This Act provides 15 days wages for each year of service to employees who have worked for five years or more in establishments having a minimum of ten workers.

(vi) Schemes for Rural Poor

In recent years, a number of schemes have been introduced benefiting the rural poor, such as Swarnajayanthi Gram Swarozgar Yojana, Sampoorna Grameen Rozgar Yojana, Pradhan Manthri Gram Sadak Yojana, National Rural Employment Guarantee Act of 2005, schemes of rural health, rural housing, rural water supply and sanitation, schemes for elderly and disabled, schemes for rehabilitation of victims of drugs abuse, etc.

(vii) Varishta Pension Bima Yojana (VPBY)

This scheme came into force from 2003-04. The main features of this scheme are: (a) Under VPBY, any citizen above 55 years of age could pay a lump-sum and get a monthly pension, (b) For the lump-sum amount paid under this scheme, 9% per annum is guaranteed by the Life Insurance Corporation of India, (c) The difference between the actual yield earned by the LIC under this scheme and the guaranteed return of 9% per annum will be made up by the Central Government.

(viii) Aam Admi Bima Yojana (AABY)

This scheme was launched on October 2nd 2007. Under this scheme insurance to the head of the family of rural landless households in the country will be provided against natural death as well as accidental death and partial or permanent disability. Upto December 31st 2008, this scheme has covered 60.32 lakh persons.

(ix) Rashtriya Swasthya Bima Yojana (RSBY)

This scheme was launched on October 1st 2007. Under this scheme, all workers in the unorganized sector below poverty line category and their families will be covered for health care. Nagaland is the first state to implement this scheme followed by Arunachal Pradesh, Madhya Pradesh and Andhra Pradesh. Till May 6th, 2009, health insurance cover was extended to more than 2.09 crore persons.

(x) The Unorganized Workers Social Security Act, 2008

This scheme covers workers in the unorganized sector relating to (a) Life and disability cover, (b) Health and Maternity benefits, (c) Old age protection, (d) Any other benefit as may be determined by the Central Government.

(xi) Integrated Child Development Services (ICDS) Scheme

This scheme today is the World's largest programme aimed at enhancing the health, nutrition and learning opportunities of infants. This scheme provides an integrated approach for covering different services such as supplementary nutrition, immunization, health check-up etc. It is centrally sponsored scheme implemented through the state governments with 100% financial assistance from the central government. From 2005-06 onwards, this scheme provides assistance to States for supplementary nutrition also. Because of this scheme, it is noticed that there has been improvement in the quality of life of the children. The different services like nutrition, health and education by this scheme is made available through Anganwadi centres. The government departments like health and rural development and panchayat raj institutions are also involved in ICDS scheme. The Government of India, it is reported, has decided to undertake the construction of Anganwadi centres in different parts of the country, particularly North-eastern states. Below poverty line, is no longer the criteria now for the selection of beneficiaries of supplementary nutrition under the ICDS scheme. Children from all categories of society who are on the rolls are provided supplementary nutrition under the ICDS scheme.

(xii) Bharat Nirman

It is a time-bound plan for action in rural infrastructure. The six areas covered under this plan are Rural Electrification, Rural water supply, Rural housing, Roads, Rural telephone connectivity and Irrigation. Bharat Nirman Programme commenced from the year 2005-06.

3. Rural Electrification

Electricity has be come one of the basic human needs. Rural electrification is a vital programme for socio-economic development of rural areas. Rajiv Gandhi Grameena Vidhyutikaran Yojana (RGGVY)[2] introduced by the Ministry of Power in April 2005, has been brought under the ambit of Bharat Nirman. The goal of this programme is to provide electricity to all the unelectrified villages in the country. As per the census of 2001, as many as 1,25,000 villages remained uncovered, electrification of unelectrified below poverty line households is being financed with 100% capital subsidy of ₹ 1,500 per connection in all rural habitations. Others will be paying for the connections at prescribed charges and no subsidy will be made available, under this scheme, during 2005-06, some 10,000 villages were covered with budget allocation of ₹ 1,100 crores.

4. Rural Water Supply and Sanitation

It is needless to state that water is a basic necessity for survival of all living creatures. The Ministry of Rural Development has been taking steps to provide safe drinking water. Rural water supply is a State subject. States have been taking up projects and schemes from their own resources for providing safe drinking water. However, recognizing the importance of providing safe drinking water in rural habitations, Govt. of India has been providing assistance to State governments. With the goal of providing safe drinking water to all rural habitations, many programmes like Accelerated Rural Water Supply Programme (ARWSP) and Prime Minister's Gramodaya Yojana-Rural Drinking Water (PMGY-RDW) have been implemented to resolve drinking water crisis in rural habitations. These programmes also give importance to rain water harvesting, sustainability of sources and community participation. As per the officially published reports 88.01% of rural habitations have been provided drinking water with an investment of more than ₹ 34,000 crore.

Apart from ARWSP, states are also supported by the Central Government under the rural drinking water component of PMGY. During the Tenth plan period, apart from drinking water, steps were taken for providing sanitation facilities. By the end of the Ninth plan in 2002, it is estimated that 20% of the rural households were covered with sanitary facilities, through the Central Rural Sanitation Programme (CRSP).

The goal of ARWSP is to provide 40 litres per capita per day of drinking water for human beings, 30 litres per capita per day of additional water for cattle in desert and drought prone areas.

5. Desert Development Programme (DDP)

One hand-pump was provided for every 250 persons with water source in the habitation within 1.6 km in the plains and 100 m elevation in hilly areas. Under ARWSP, the funding pattern is 50:50 between the Centre and the States. The funding pattern varies from scheme to scheme. For some schemes or projects taken up by State Governments, the funding pattern is in the ratio of 75:25 between the Central and State Governments.

6. Role of Panchayats

As per the 73rd Amendment to the Constitution of India, the subject of rural water supply is vested with the Panchayat Raj Institutions (PRIs). The Panchayats are expected to play a major role in providing safe drinking water and managing the systems and sources in their respective areas.

The quality of water presents a problem. Quality problems in ground water are of two types, viz., chemical and biological. Groundwater depletion has aggregated water quality problems due to excess fluoride, arsenic and brackishness in certain areas. Steps have been taken for providing water-testing laboratories, particularly mobile water quality testing laboratories. The World Health Organization (WHO) has certified that India has been taking necessary steps in the direction of providing safe drinking water.

Under PMGY-RDW, emphasis has been on schemes for water conservation, rainwater harvesting, recharge and sustainability of drinking water sources in areas under Drought Prone Area Programme (DPAP) and Desert Development Programme (DDP).

7. National Human Resource Development Programme (NHRDP)

The NHRDP was launched by the government in 1994, aiming at building up of a human resource base of trained personnel to serve the needs of the rural water supply and sanitation sector. The programme aims at training beneficiaries, especially women at the grass roots level. It also aims at empowerment of Panchayat Raj Institutions related to water supply.

Under NHRDP, States and Union Territories should set up HRD cells for planning, designing, implementing, monitoring and evaluating appropriate and need-based HRD programmes. The Central Government provides 100 per cent financial assistance for training activities including salary of HRD staff.

There is Research and Development (R&D) wing in HRD. The objective of R&D programme is to provide cost-effective technologies in rural water supply programme.

Rural sanitation is a part of Rural water supply scheme. The concept of sanitation cannotes a comprehensive definition which includes liquid and solid waste disposal, food hygiene, personal, domestic and environmental hygiene. Though the majority of Indian population lives in rural, "their access to a minimum level of sanitation" as Eleventh Plan report observes "is very low". However, in recent years, State governments, have been taking steps towards providing better sanitation facilities in terms on construction of septic tanks, twin-pit pour flush latrines, safe disposal of waste water etc. NGOs are actively involved in creating sanitation facilities fully assisted by the central grants.

8. Central Rural Sanitation Programme (CRSP)

Central Government supplements the efforts of the state in undertaking rural sanitation under the Central Rural Sanitation Programme (CRSP) launched in 1986. CRSP aims at improving the quality of life of the rural people and to provide privacy and dignity to women in particular. The CRSP with effect from April, 1999, has been assisting not only below poverty line category of households but is guided by demand driven approach and assisting may others in rural areas to have better sanitation.

Basing on the Tenth Finance Commission recommendations, the CRSP is not engaged in assisting with financial support for school sanitation facilities. In the eleventh plan it is made clear that "the government is committed to provide 100% coverage of water supply to rural schools.

In order to universalize access to safe drinking water, steps would be initiated to prevent agriculture and other sectors from competiting with drinking water schemes. Also in the Eleventh Plan it is stated that where groundwater quality and availability is unsatisfactory, surface water sources need to be developed. It is pointed out in the plan report. That, involvement of the community in the monitoring of the water supply works should be made a primary condition for release of funds."

The National Rural Employment Guarantee Programme has identified work component relatec to water. The Rural Development Ministry has identified programme[3] such as rainwater harvesting, restoration of water storage tanks, Afforestation programme, Integrated Wasteland Development Programme. etc., for sustainable water supply.

9. Urban Centres and Drinking Water

Urban Drinking water problem is as important as rural drinking water issue. The Eleventh Plan rightly observes "There is a huge gap between the demand and supply of water in urban areas, which is also growing due to population and urbanization. Recycling and release of water, reducing the water demand through harvesting, using water-efficient household equipment, including flushing cisterns would go a long way in counseling water and reducing demand". Proper metering of water use and rational tariff system would certainly reduce water demand and encourage conservation.

10. Jawaharlal Nehru National Urban Renewal Mission (JNNURM)

With a view to provide 100% water supply accessibility to the entire urban population by the end of the Eleventh Five Year Plan,[4] the Government of India launched different schemes. The most important of such schemes is Jawaharlal Nehru National Urban Renewal Mission (JNNURM). The JNNURM is planned development of identified cities. Focus is on efficiency in urban infrastructure and service delivery mechanism, community participation and accountability of urban local bodies towards citizens. The JNNURM was formally launched on 3rd December, 2005. Among others, the scheme focuses attention on urban renewal programme for the old city areas to reduce congestion. Also, provision of basic services to the urban poor including security of tenure at affordable prices, improved housing, water supply and sanitation and ensuring delivery of other existing universal services of the government for education, health and social security.

REFERENCES

1. *Major Programmes of the Govt. of India,* 2006.
2. "Programmes for the People", Govt. of India, 2008.
3. *Economic Survey,* 2008-09.
4. *The Eleventh Five Year Plan Report.*

23 RURAL HOUSING

1. Introduction

Housing is one of basic requirements for human survival and happy life. For a normal citizen owning a house provides significant economic security and dignity in society. According to the 2001 census, rural housing shortage was to the extent of 148 lakhs. Again a sizeable number of people have unserviceable kutcha houses. It has been estimated that with the present population growth rate, houselessness in the country will be of higher magnitude. It is reported that the percentage of houses having grass, or straw roofing is about 33%; houses with walls of mud and unburnt bricks is 6.05% and some 4.22% of households have tents for their residence. Nearly 70% of the rural houses are said to be of a poor quality.

2. National Housing and Habitat Policy

The Central Government announced in 1998, the National Housing and Habitat Policy which aims at providing "Housing for All", with an emphasis on extending benefits to the poor and the deprived. Government had announced that it was committed to the goal of ending all shelterlessness by the end of Ninth Plan period and conversion of all unserviceable kutcha houses to pukka houses by the end of the Tenth Plan period.

3. Different Schemes of Rural Housing

Several housing schemes have been initiated such as Indira Awaas Yojana (IAY), Pradhan Manthri Gramodhaya Yojana, Credit-cum Subsidy Scheme for Rural Housing, setting up of Rural Building Centres, Samagra Awas Yojana, Equity support to Hudco etc.

The Indira Awaas Yojana (IAY) addresses housing shortage as an important component of the poverty alleviation programme in rural India. The Bharat Nirmaan programme has recognized and accorded due priority to the housing problem and envisaged to construct 60 lakh houses by 2005-06. The Ministry of Rural Development through IAY undertakes housing to rural poor as a centrally sponsored scheme. The cost of the house will be shared between Centre and States on a 75:25 basis. Greater emphasis is laid to the states with higher incidence of shelterlessness. 75% weightage is given to housing shortage and 25% to SC and ST component of the population. Grant assistance is provided to the extent of ₹ 25,000 per house for normal areas and ₹ 27,500 for hilly areas. Funds are released in two instalments.[1] By January 2008, it is estimated that some 42 lakh houses were constructed. In order to expedite the programme, Panchayat Raj Institutions are entrusted with the task of implementing the scheme. However, freedom is given to the individuals on the choice of the design of the house.

Under Pradhan Manthri Gramodhaya Yojana, during 2000-2001, ₹ 375 crore has been made available for implementing "Rural Shelter" component of the programme. Under Samagra Awaas

Yojana, the houseless were identified in 25 districts of 24 states. From time to time, steps have been initiated to provide necessary technology for construction of rural houses with locally available resources. From the year 2000-2001, steps were taken for promoting State Specific Technologies, materials, designs etc., for cost-effective rural housing.

4. Central Building Research Institute

The central building research institute (CBRI), Roorkee has been entrusted with a study on "Demonstration and Training on Innovative Materials and Technologies for Rural Housing in seven villages in different regions". The broad objectives of this study are: (1) Cost-effective technologies in rural housing keeping in view local resources. (2) Planning and execution of a time bound action programme. The states identified under this study are Karnataka, Rajasthan, Madhya Pradesh, West Bengal, Sikkim, Andhra Pradesh and hilly regions of Uttar Pradesh.

5. Indira Awaas Yojana

Of all the schemes for rural housing, Indira Awaas Yojana proved to be the best jointly run by the state and the centre. At district level, this scheme is implemented by DRDA and Zilla Parishad. Beneficiaries under this scheme are selected by Gram Sabha out of bonded labour, SC, ST, non-SC/ST and handicapped. The persons whose houses need repairs or having mud houses or without any house are selected under this scheme. Some field studies conducted on the working of Indira Awaas Yojana reveal that the beneficiaries were very happy to get help from the Government for housing. They were proud of having new pukka houses and are living peacefully in them. They do not have any fear of damage to their houses during the rainy season. They are thankful to the government for this.[2] In recent years, Rajiv Swagruha Yojana has been launched to meet the housing requirements of the poor in the country.

REFERENCES

1. *Programmes for the People,* Government of India, 2004-08, p. 13.
2. *Rural Housing,* Ministry of Rural Development, Government of India.

24 HEALTH CARE FOR ALL

1. Introduction

Provision of health for all is an essential component of social security package of Programmes. The national health policy set forth as its objective the goal of "Health for all by 2000 A.D.," through creation of Health Infrastructure in rural areas. In the ultimate analysis, any society would be judged by its ability to meet at least basic minimum needs such as food, clothing, shelter and health care for its people. Numerous programmes and schemes are being implemented under the Minimum Needs Programme to provide primary health care relevant to the needs of the community in the rural areas.

In the Post-independent era, India has built up a vast health infrastructure and health personnel at primary, secondary and tertiary level. In the direction of producing skilled human resources, a number of medical and paramedical institutions including Ayurveda, Yoga and Naturopathy, Unani, Siddhi and Homeopathy (AYUSH) institutions have been set up.

2. Health Care during Planning Era

During the Planning Era, considerable achievements have been made in the direction of health care in terms of life expectancy, child mortality, Infant mortality, and maternal mortality. Small pox and guinea worm have been eradicated to a great extent. There is growing awareness of HIV/AIDS and H_1N_1 or Swine Flu among different sections of society. Nevertheless as XI Plan Report observes, "Problems abound. Malnutrition affects a large proportion of children. An unacceptably high proportion of the population continues to suffer and die from new diseases that are emerging, apart from continuing and new threats posed by the existing ones. Pregnancy and child birth related complications also contribute to the suffering and mortality.[1]

Table 24.1 presents health indicators in selected countries

Table 24.1

Health Indicators in Selected Countries

Country	IMR Per 1000 Live Births	Life Expectancy M/F in Years	MMR per 1,00,000 Live Births	TFR
1	2	3	4	5
India	58	63.9/66.9	301	2.9
China	32	70.6/74.2	56	1.72
Japan	3	78.9/86.1	10	1.35

Republic of Korea	3	74.2/81.5	20	1.19
Indonesia	36	66.2/69.9	230	2.25
Malaysia	9	71.6/76.2	41	2.71
Vietnam	27	69.5/73.5	130	2.19
Bangladesh	52	63.3/65.1	380	3.04
Nepal	58	62.4/63.4	740	3.40
Pakistan	73	64.0/64.3	500	3.87
Sri Lanka	15	72.2/77.5	92	1.89

IMR = Infant Morality Rate
MMR = Maternal Mortality Ratio
TFR = Total Fertility Rate
Source: *XIth Five Year Plan Report,* p. 58.

From Table 24.1, it is clear that IMR in India is still high at 58 per 1000. It is no good of comparing with Pakistan and be satisfied with that. In Japan and Republic of Korea, IMR is low at 3. Life expectancy also needs to be improved in India. TFR for every 100 000 live births of 301 in India is still disturbing as compared to that in Japan where it is only 10. In Republic of Korea, it is only 20. TFR of 2.9% in India is to be cut in the direction of reducing the population growth rate. Life expectancy of both males and females in India is less than that in China, Japan, Malaysia and Sri Lanka.

In the direction of providing health services, particularly meeting the needs of rural poor, primary Health centres were set up. By 1994, there were as many as 21,155 primary health centres in the country. But, it is reported that many primary health centres are ineffective because of lack of doctors or medicines.

We need a comprehensive approach that encompasses individual health care, Public health, sanitation, clean drinking water, access to food and knowledge about hygiene and feeding practice etc. A direct relationship exists between water and health. It is estimated that 80% of diseases are due to the use of contaminated water. Inadequate provision of safe drinking water, improper disposal of human waste and lack of adequate systems for disposal of sewage and solid wastes leads to unhealthy and unhygienic conditions. It is a matter of concern that despite the progress made with water supply, the level of water-related illnesses continue to be high.

Approximately 10 million cases of diarrhea, more than 7.2 lakh typhoid cases and 1.5 lakh viral hepatitis cases occur every year. A majority of them are contributed by unclean water supply and poor sanitation. Micro-level studies reveal that availability of clean water; sanitation and hygiene interventions reduce diarrhoeal diseases on average by between one-quarter and one-third.

As per the 2001 census, only 36.4% of the total population has latrines attached to their houses. In rural areas only 21.9% of population has latrines. The rural poor expressed their inability in meeting 50% share of the cost in rural sanitation programme. Now the central schemes fund to the extent of 75% and share of the state government is 25% under this revised programme, it is hoped that rural poor come forward in utilizing the different schemes in the direction of rural water supply and rural sanitation. Involvement of the community in monitoring of the water supply works should be made a primary condition for release of funds for completed work.

3. Poverty and Ill-health

Poverty creates many problems including ill-health. It is well-known fact that poor are denied of access to reliable health services. High cost incurred on health care drives some into poverty. In 2004-05 the number, below poverty line was estimated at 28.3%[2]. It is rightly observed "The challenge of quality health services in remote rural regions has to be urgently met: Given the magnitude of the problems, we need to transform public health care into an accountable, accessible and affordable system of quality services during the Eleventh Five Year Plan."[3]

Again health is very much linked with sanitation facilities and clean drinking water. For a considerable number of people, particularly rural people, "safe drinking water is a daily problem and will be ready to drink any polluted water."[4] Proper drainage, water supply in urban areas is also far from satisfactory. The inadequacy of quality of rural water and rural sanitation is well pronounced. With regards to rural water supply, the main problems identified in XI Plan are of sustainability of water availability and supply of poor quality water. Sustainability of the rural water supply of programme has emerged as a major issue and the Eleventh Plan aims at arresting the slip backs. The Mid-term appraisal of the Tenth Plan observed that over reliance on groundwater for rural water supply programme has resulted in the twin problem of sustainability and water quality and suggested a shift to surface water sources for tackling this issue. It is estimated that atleast for 2.17 lakh rural habitations, quality of water is poor with problems of excess salinity or fluoride. Desalination plants have also met a similar fate due to lapses at different levels.

The Swajaldhara Programme launched in 2002-03, began tackling the rural water supply. The Bharat Nirmaan Programme aims at addressing water quality in affected habitations. To provide clean drinking water for all, by the end of the Eleventh plan, efforts are on through the Swajaldhara programme and Bharat Nirmaan programme. The government is also committed to provide 100% coverage of water supply to rural schools.

4. National Rural Health Mission

The National Rural Health Mission (NRHM) was launched on 12 April, 2005. The Mission seeks to provide effective health care to rural population throughout the country with special focus on 18 states, which have weak public health indicators and weak infrastructure. These 18 states are Arunachal Pradesh, Assam, Bihar, Chattisgarh, Manipur, Mizoram, Meghalaya, Madhya Pradesh, Nagaland, Odissa, Rajasthan, Sikkim, Tripura, Uttaranchal and Uttar Pradesh.

The National Rural Health mission aims at undertaking architectural correction of the health system to enable it to effectively handle increased allocations as promised under the National Health Mission Programme and promote Policies that strengthen public health management and service delivery in the country. It has a provision of a female health activist in each village giving untied funds to sub health centres; a village health plan prepared through a local team headed by the Health and sanitation Committee of the Panchayat; Strengthening of the rural hospital for effective health care and made measurable and accountable to the community through Indian Public Health Standards (IPHS); and integration of vertical Health and Family Welfare programmes of health like safe water, sanitation, nutrition etc., through an effective District Health Plan. The NRHM also aims to provide help to different programmes including control of Malaria, Blindness, Iodine deficiency, TB, Leprosy and Integrated Diseases Surveillance.

The goals of Mission are (1) Reduction in Infant Mortality Rate and Maternal Mortality Ratio. (2) Universal access to public health services such as Women's health, child health, water, sanitation and hygiene, immunization and Nutrition. (3) Prevention and control of communicable and non-communicable diseases, including locally endemic diseases (4) Access to integrated comprehensive

Primary health care, (5) Population stabilization, gender and demographic balance, (6) Revitalise local health traditions and mainstream AYUSH and (7) Promotion of healthy lifestyles.

Progress of NRHM

Ever since its inception the NRHM is engaged in providing Health Services in different States. As many as 109 District Activity Plans are prepared in different States. Micro Plans were prepared in 255 districts.

5. Janani Suraksha Yojana (JSY)

The JSY is a 100 per cent centrally sponsored scheme and it integrates cash assistance with delivery and post delivery care. The scheme was launched with focus on demand promotion for institutional deliveries in states and regions where these are low. It targeted lowering of MMR by ensuring that deliveries were conducted by the skilled birth attendants at every birth. The yojana has identified ASHA, the accredited social health activist, as an effective link between the government and the poor pregnant women in 10 low performing States[5].

6. Pradhan Mantri Swasthya Suraksha Yojana (PMSSY)

PMSSY began functioning from April, 2006, with the objective of correcting regional imbalances in the availability of affordable and reliable health care services. Also PMSSY intends to augment facilities for quality medical education in the country. PMSSY has two components in its programme (1) Setting up of six AIIMS like institutions, and (2) upgradation of 13 existing government medical college institutions. Steps have been initiated in these directions.

7. National AIDS Control Programme

It is said HIV/AIDS is a very serious ailment. It was estimated that there were 2.31 million persons living with HIV/AIDS in India in 2007. It is estimated that 87% of HIV transmission is through sexual route. Transmission of HIV may be through different routes such as mother to child transmission, transmission through migrants, transmission by the use of certain consumable items like shaving blades, etc.

8. AYUSH Programme

In addition to strengthening the health delivery system under NRHM and related programmes, there are other programmes in the area of health care. These include, Ayurveda, Yoga and Naturopathy, Unani, Sidhha and Homeopathy (AYUSH).

Under AYUSH, there is network of 3,360 hospitals and 21,769 dispensaries across the country. The health services provided by this network largely focused on primary health care. Under AYUSH, specialized therapies like Panchakarma are undertaken in nearly 250 hospitals. The key interventions and strategies in the Eleventh Five Year Plan include training for AYUSH personally, actively involving AYUSH in National Health Care Delivery System, strengthening R&D technology involving accredited laboratories in the government and non-government sector apart from establishing centres of excellence.[6]

9. The Integrated Child Development Scheme (ICDS)

The ICDS has been actively engaged in the health care. Apart from this scheme, two other schemes, namely Kishori Shakti Yojana (KSY) and Nutrition Programme for Adolescent Girls (NPAG) aim at addressing the needs of self-development, nutrition and health status, literacy and numerical skills, vocational skills of adolescent girls in the age group of 11-18 years. Both these schemes are being implemented through the infrastructure of ICDS.

10. Rajiv Gandhi National Creche Scheme

The Rajiv Gandhi National Crèche Scheme for children of working mothers provides services to the children of age group 0-6 years which includes supplementary nutrition, emergency medicines and contingencies. The number of beneficiaries under this scheme by march 2010, is said to be around 8 lakhs. Due to a number of schemes in the direction of child care , the bias against the girl child is gradually reduced. The girl child ratio declined from 945 in 1991 to 927 per 1000 males in 2001. Efforts are being made to ensure the survival of the girl child and her right to be born and grow as a participating member of the society.

11. Dhanalakshmi Scheme

Dhanalakshmi, a conditional cash transfer scheme for girl child with insurance cover was launched as a pilot project in March 2008. This scheme is aimed at providing a set of financial incentives for families to encourage them to retain the girl child and educate her. This scheme provides cash transfer to the family of girl child on fulfilling certain specific conditions such as birth registration, immunization, enrollment and retention if remains unmarried until the age of 18 years. This scheme is being implemented in 11 blocks across seven States.

The Support to Training and Employment Programme (STEP) for women scheme seeks to provide updated skills and new knowledge to poor and assetless women in ten traditional sectors aiming at enhancing their productivity and income generation.

The Self-help Groups (SHGs) are being encouraged in the direction of awareness generation, economic empowerment and social security. Priyadarshini project is rural women's empowerment and livelihood programme of the Ministry of Woman and Child Development. This scheme is funded by International Fund for Agricultural Development, in States where rural women are predominant in agriculture and related programmes. This fund provides assistance for health care of rural women and their children.

Conclusion

Despite all these schemes, even now poor man is denied of the minimum health facilities. In a vast country like India, where population is concentrated in the rural areas, health care for rural poor needs to be addressed effectively. By strengthening all the primary health centres and selected primary health centres with advanced medical facilities in terms of qualified doctors and required drugs, the health care can be effectively addressed.

REFERENCES

1. *The Eleventh Five Year Plan Report,* p. 57.
2. *Agricultural Statistics at a Glance,* Directorate of Economics and Statistics, Ministry of Agriculture, Govt. of India, p. 55.
3. *Eleventh Plan Report,* p. 57
4. *India 2020,* By A.P.J. Abdul Kalam, Penguin Books, p. 251.
5. *Economic Survey 2008-09,* Govt. of India, p. 274.
6. *Ibid.,* p. 275.

25 EDUCATION FOR ALL

1. Introduction

Education of formal type is a process of teaching, training and learning especially in educational institutions like schools or colleges to improve knowledge and develop skills. The goal or purpose of education is not merely to acquire knowledge and skills to make a living, much more is expected from education. Swami Vivekananda said "We want that education by which character is formed, strength of mind is increased, the intellect is expanded and by which one can stand on one's own feet". Gandhiji in his basic education gives equal importance to skill building as well well as character building. He categorically states that education without character building is a great sin.

2. Role and Responsibility of Stakeholders

Education is a crucial input in the process of development and right type of education facilitates the acquisition of required skills as well as right conduct. In this task, responsibility and role of all stakeholders in education becomes equally important. The main stakeholders in education are: the parents, the teachers, the students and the government.

2.1. Parents

Parents are the first and principal educators of their children. More importantly mother is the first teacher who must take the responsibility of moulding well the behaviour of the child. Parents are not relieved of their responsibility as educators of their children with enrolling their children in a school of their choice. They have to take the responsibly of ensuring that the children attend classes regularly and they are regular in doing the homework, in attending physical/Yoga practices if any. Also parents must make sure that their children conduct themselves well and move only with the right type of children. The parents must regularly meet the concerned teacher to find out the progress of the student. Parents-teachers meet becomes necessary and parents must regularly attend such meetings.

2.2. Teachers

The term teacher is taken to mean as one, involved in shaping the child at all levels. Teachers have an indispensable role to play in education system. Quality education is possible only when there are sufficient number of teachers fully qualified, well trained and fully committed to the profession. In a situation of fast growing knowledge, the teachers must necessarily update their knowledge and skills so as to keep themselves abreast of fast growing knowledge. Professionalism is one of the most important characteristics that should identify teachers. The way in which teacher conducts himself in and outside the classroom must be of such nature that teachers become role model. The teacher must be a constant reader himself to deliver goods in the classroom. The teacher must necessarily have a concern for the all-round welfare of the pupils. The teacher must keep himself away from all kinds of narrow loyalties and should be well committed to the profession.

2.3. Students

Students must always have a thirst for knowledge. It is important that their time and energy are chanalised in the right direction. Through well equipped library and appropriate physical education facility, students can be kept fully engaged in the right direction. With vocational education, students can be fully involved in training without diversion of mind. Skill building must become an important component of curriculum and that way students can be well trained. Half the time for classroom training and the rest of the time appropriate practical training facilitating arrangement as in Dayal Bagh Rural University (Autonomous University) in Agra, Uttar Pradesh is worth emulating. "Earn while you learn" arrangement is well organized in Dayal Bagh Institute. This scheme facilitates integration of ethical values into the curriculum.

2.4 Government

The government must necessarily take appropriate steps to provide necessary facilities for imparting right type of education and training. The government must take the responsibility of providing minimum facilities in all schools including schools in remote villages. Even now some twenty per cent of villages in India do not have primary schools, and where schools exist they do not have minimum facilities such as safe drinking water, desks, chair, for teacher, etc. In rural areas one teacher, has to teach five classes and in such situation there is no personal attention towards pupils. In rural areas school-dropout rate continues to be of a high order and all concerned must take necessary steps to ensure regular attendance without discontinuing studies in the course of studies.

3. Literacy Rate during Planning Era

Table 25.1 presents trend of literacy rate between 1951-2001. It is evident from the Table that the percentage of literacy has been on the increase from time to time. We also notice that the literacy of females continues to be less than the literacy of males.

Table 25.1

Literacy Rate: 1951-2001

(In Percentage)

Censor Year	Persons	Males	Females
1951	18.33	27.16	8.86
1961	28.3	40.40	15.35
1971	34.45	45.96	21.97
1981	43.57	56.38	29.76
1991	52.21	61.13	39.29
2001	64.84	75.26	53.67

Source: *INDIA 2009,* Publication Division, Ministry of Information and Broadcasting, Government of India, p. 15.

Before 1976, Education was the exclusive responsibility of the state government. With Constitutional Amendment of 1976, education was brought under the concurrent list. Now with this Amendment, "The Union Government accepts, a greater responsibility of reinforcing the national and integrated character of education. Maintaining quality and standard including those of the teaching profession at all levels and the study and monitoring of the educational requirements of the country."[1] The central government had taken of a number of program such as National Policy on Education (1986), Program of Action (POA) of 1986, establishment of Navodaya schools, expanding Open University system in the states, strengthening of the All India Council of Technical Education (AICTE), strengthening of the central advisory board of education etc.

Despite all these measures, certain gaps were noticed in the education system. Gender gap in literacy was found to be of higher magnitude with low level of rural female literacy.

According to the VII Educational Survey 2002[2], the number of habitations that had a primary schools within a distance of one km was 10.71 lakh (87%), the uncovered habitations numbered 1.61 lakh (13%). Still some 13% of villages do not have primary schools and it is a matter of great concern. The social composition of our school children indicates that 9.97% of Muslim children, 9.54% of Scheduled Tribe (ST) 8.17% of Schedule Caste (SC) and 6.97% of Other Backward Castes (OBC) children were out of school and an overwhelming majority of these school children (68.7%) was concentrated in 5 states namely Bihar (23.6%), Uttar Pradesh (22.2%), West Bengal (9%), Madhya Pradesh (8%) and Rajasthan (5.9%).[3]

4. Expenditure on Education – Plan Wise

Expenditure on different sectors of education is shown in Table. 25.2. It is evident from the Table that relatively higher percentage of amounts has been spent on elementary education in the series of Five Year Plans. Next to elementary education, technical Education received priority, followed by higher education. But yet, some villages are denied primary education facilities and school dropout percentage continues to be higher in rural areas.

Table 25.2

Expenditure on Different Sectors of Education — Plan Wise

Expenditure in Million Rupees
(In Percentages)

Sector	First Plan Expdt. 1951-56	Second Plan Expdt. 1961-66	Third Plan Expdt. 1966-69	Annual Plans Expdt. 1966-69	Fourth Plan Expdt. 1969-74	Fifth Plan Expdt. 1974-79	Sixth Plan Expdt. 1980-85	Seventh Plan Expdt. 1985-90	Annual Plans 1990-92 Expdt.	Eighth Plan Expdt. 1992-97	Ninth Plan Expdt. 1997 2002	Tenth Plan Expdt. (Central Sector)
1	2	3	4	5	6	7	8	9	10	11	12	13
Elementary Education	870 (58%)	950 (35%)	2010 (34%)	750 (24%)	3743 (50%)	5913 52%)	8414 (32%)	28494 (37%)	17290 (37%)	103940 (48%)	145233 (65.7%)	287500 (65.6%)
Secondary Education	83 (5%)	510 (19%)	1030 (18%)	530 (16%)	-	-	5344 (20%)	18315 (24%)	10530 (22%)	52311 (24%)	23227 (10.5%)	43250 (9.9%)
Adult Education	-	-	-	-	126 (2%)	248 (2%)	1533 (6%)	4696 (6%)	4160 (9%)	11421 (5%)	5204 (2.4%)	12500 (2.9%)
Higher Education	117 (8%)	480 (18%)	870 (15%)	770 (24%)	1883 (25%)	3188 (28%)	5604 (21%)	12011 (16%)	5880 (12%)	20944 (10%)	22709 (10.3%)	41765 (9.5%)
Others	227 (15%)	300 (10%)	730 (12%)	370 (11%)	936 (13%)	1071 (9%)	2729 (11%)	1980 (3%)	1180 (2%)	7398 (3%)	3492 (16%)	6235 (1.4%)
Technical Education	215 (14%)	490 (18%)	1250 (21%)	810 (25%)	786 (10%)	1015 (9%)	2563 (10%)	10833 (14%)	8230 (17%)	21987 (10%)	21095 (9.5%)	47000 (10.7%)
Total	**1512 (100%)**	**2730 (100%)**	**5890 (100%)**	**3230 (100%)**	**4747 (100%)**	**11435 (100%)**	**26187 (100%)**	**76329 (100%)**	**47270 (100%)**	**218001 (100%)**	**220960 (100%)**	**438250 (100%)**

Source: *Five Year Plan Reports.*

5. Elementary Education in Tenth Five Year Plan

The Tenth plan laid emphasis on Universalization of Elementary Education (UEE) guided by five parameters – (i) Universal access, (ii) Universal Enrollment, (iii) Universal Retention, (iv) Universal achievement and (v) Equity. The major schemes of elementary education sector during the Tenth Plan includes, among others, Sarva Shiksha Abhiyan (SSA) and National Programme of Nutritional Support to primary education, commonly known as Mid-day Meal Scheme (MDMS).

6. Sarva Shiksha Abiyan (SSA)

This program aims to achieve universal elementary education to all the children in the age group of 6-14 years. Under SSA, the funding pattern in 75:25 between the centre and the states. Under this scheme, a good number of new schools were opened and teachers were appointed. Apart from opening schools with new buildings constructed, free textbooks were also given to students.

7. Mid-day Meal Scheme

This scheme is a nutritional programme covering nearly 12 crore children in more than 9.5 lakh primary schools. This scheme is interested in eliminating hunger to the children. With this scheme, it is noticed that there has been increase in enrollment, more significantly, of girls. It is said that social distance is very much reduced because of this scheme. Under this scheme cooking and serving of mid-day meal is entrusted to self-help groups. Hence, this scheme became instrumental for the growth of SHG's.

8. Plan of Action for the Growth of Elementary Education

The Plan of Action of the Tenth Five Year Plan for the growth of elementary education has been mainly on the following lines:

8.1 Universal Access: In order to make elementary education acceptable some more primary schools were planned and grounded. The number of primary schools in the country increased from 6.64 lakh in 2001-02 to 7.68 lakh in 2004-05. Some additional 55 thousand upper primary schools also began functioning during the same period.

8.2 Universal Enrolment: By setting Village Education Committees, Mother-Teacher Associations and Parent-Teacher Associations and series of campaigns for enrollment, yielded good results. The total enrolment at elementary education level increased from 159 million in 2001-02 to 182 million in 2004-05, an increase of 23 million. Sarva Siksha Abhiyan contributed significantly in bringing down the number of out of school children.

8.3 Universal Retention: It is realized that retaining the children drawn from disadvantaged families is much more difficult than enrolling them into education system. It is estimated that around 22 per cent children dropped out in classes I and II. Dropout rate is found to be more in the case of girl children. It is believed that with the presence of female teachers dropout rate can be curtailed. Therefore, priority was given for women in recruitment of teachers for elementary schools. Some state governments initiated steps to recruit women teachers, particularly, in the states of Goa, Kerala, Puducherry, Tamil Nadu and Delhi. In all these states for every 100 male teachers, there are more than 200 female teachers. No doubt, gradually the dropout rates declined. However, the dropout rates at primary levels for SCs and STs continue to be higher.

8.4. Universal Achievement and Equity: Different surveys conduced by NCERT and other agencies highlight poor quality of learning at elementary school level in the rural areas. Various factors account for poor quality of achievement. The socio-economic background of the family is the major factor. Some parents, out of poverty believe that by sending the children to school they are foregoing some income to the family. Sometimes, the concerned teachers may not evince sufficient interest in imparting right type of education. Some studies reveal that teachers themselves are not regular in attending the schools. One-third of the teachers in Madhya Pradesh, 25 percent in Bihar and 20 per cent in Uttar Pradesh do not attend schools was identified.[4] In order to ensure universal achievement and equity, the parents and teachers have a great responsibility. Gender gap in literacy can be reduced, only if illiterate parents are motivated in right direction to send girl children also to schools.

9. Kasthurba Gandhi Balika Vidyalaya Scheme (KGBVS)

The KGBVS was launched in July, 2004 for setting up of residential schools at upper primary level for girls, predominantly belonging to the SCs, STs, OBCs and minorities. A minimum of 75% per cent of enrollment in KGBVS is reserved for girls and the remaining 25 per cent for girls belonging to the BPL category.

10. District Primary Education Programme (DPEP)

The DPEP is an externally aided project covering classes I to V. The main objective of this scheme is to reduce the dropout rate to less than 10 per cent, reduce disparities among gender and social groups in the enrollment to less than 5 per cent and improve the quality of education. This programme covered as many as 273 districts in 17 states.

11. Mahila Samaikhya (MS)

The MS was started with external assistance but since 2005-06 it is funded by Government of India only. The programme aims at empowerment of women by way of providing facilities to learn at their own homes and according to their choice. The programme is implemented in 9 states covering 83 districts and 20,380 villages.

12. Eleventh Five Year Plan and Elementary Education

At the beginning of the Eleventh Plan, it was noticed that as many as 7.1. million children were out of school and more than 50 per cent dropouts at elementary school level. These being matters of serious concern, SSA programme was reoriented and restructured to meet the challenges of equity, retention, and quality education. It is rightly observed".[5] Unless there is a strong effort to address the issues of regular functioning of schools, teacher competence and attendance, accountability of educational administrators, pragmatic teacher transfer and promotion policies, effective decentralization of school management, and transfer of powers to Panchayat to run institutions, it would be difficult to buildup the gains of SSA".

The Eleventh Plan fixed the following targets for elementary education: (i) Universal enrollment of 6 to 14 age group, (ii) Improvement in quality and standards of education, (iii) All gender, social and regional gaps in enrolments to be eliminated by 2011-12, (iv) to reduce the dropout rate at primary level to 20 percent by 2011-12, (v) All States/UTs to adopt NCERT quality monitoring tools, (vi) Achieve 80 per cent literacy rate with a focus on SCs, STs, minorities and rural women.

13. Higher and Technical Education

Apart from evincing interest in strengthening elementary education, government is committed for the development of higher and technical education. In 2002, there were only 201 universities and by 2007, the number has gone up to 378. The increase in the number of colleges has been highly encouraging. In 2002, there were only 12,342 colleges, and this number has gone up to 18,064 by 2007.

The National Accreditation Assessment Council (NAAC) was set up in 1994, to make quality an essential element through a combination of internal and external quality assessment and accreditation. During the Tenth Plan, NAAC was strengthened with the opening of four regional centres so as to speedup the accreditation process. NAAC as completed accreditation of only some 150 out of 378 universities and some 3500 colleges out of 14000 colleges by the end of 2009. The results of the accreditation process indicate serious quality problems. Only 9% of the colleges and 31%, of the universities are rated as 'A' grade and the rest fall in 'B' and 'C' categories.[6]

14. Autonomous Status

Autonomous Status has been given to selected constitutions to promote new courses and innovative methods in teaching. It is hoped the quality of education would be improved with this arrangement.

15. All-India Council for Technical Education (AICTE)

The AICTE was set up in 1945 and was given statutory status in 1987, for coordinated development of Technical Education, promotion of qualitative improvement and maintenance of norms and standards in Technical Education.

16. Inclusive Growth and Eleventh Five year Plan

In the Eleventh Plan, it is proposed to take following steps in the direction of 'Inclusive Growth': (1) Reduction of Regional Imbalances, (2) Support to Institutions located in remote, hilly and backward areas, (3) Support to institutions with larger student population of SCs, STs, OBCs, minorities and physically challenged students and (4) Support to the SCs, STs, OBC, minorities physically challenged and girl students with special scholarships/fellowships, hostel facilities, remedial coaching and other measures.

17. Quality Improvement

It is proposed to effect quality improvement in higher education through restructuring academic programmes to ensure their relevance to modern market demands; greater emphasis on recruitment of adequate and good quality teachers; broadening the content of science and engineering programmes to strengthen fundamental concepts; improving learning opportunities and conditions by updating text books and learning material. During the Eleventh Five Year Plan period it is proposed to strengthen colleges and universities to enable these institutions to avail the assistance from the UGC.

18. Science and Technology Development through an Integrated Approach

In the Eleventh Five year Plan the proposal is evolving an integrated Science and Technology (S&T) plan involving UGC, Department of Social and Technology (DST), CSIR, Indian Council of Agricultural Research (ICAR), Departments of Atomic Energy and Space to provide the resources needed for substantially stepping up support to basic research, setting up a national level mechanism for evolving policies and providing direction to basic research. Also in the plan it is proposed to promote linkages with other countries in the area of S&T.

19. Polytechnics

In the direction of providing socially relevant education, it is proposed to set up new polytechnics in every district, covering first areas where there are no polytechnics so far, on priority basis. Efforts will also be made to increase intake capacity by using available facilities in the existing polytechnics.

20. Distance Education

Indira Gandhi National Open University (IGNOU) is found to be highly useful to increase access to education. Hence, it is proposed to strengthen the Distance Education arrangement to different states with financial support to different institutions of correspondence courses in the conventional universities.

21. Language Promotion

During the Eleventh Plan it is proposed to undertake New Linguistic Survey (NLSI) of India with the goal of language promotion with focus on 22 languages in the Eighth Schedule and top 15 Non-Scheduled Languages. The survey will be conducted by the Central Institute of Indian Languages (CIIL) Mysore and the Departments in selected universities that have a strong base in Sociology, Anthropology etc.

22. Book Publication Promotion

Apart from language promotion the Eleventh Plan proposes to encourage publication of books. During the Eleventh Plan the National book Trust will strengthen its three regional offices at Bengaluru, Mumbai and Kolkata and also it is proposed to strengthen its activities in other parts of the country. It is stated that the subsidy project for assistance to authors and publishers for producing books of acceptable standards at reasonable prices for students and lecturers will continue.

23. Financing Education in the Eleventh Plan

In the Eleventh Plan, the Central Government envisages an outlay of about 2.70 lakh crore (at 2006-2007 price level) for education. This is four fold increase over the Tenth Plan allocate and thi clearly reflects the priority given to the education by the centre. Education being in the Concurrent list, the State Government shall also accord priority to education in the sectoral plan priorities/ allocation.

24. Right of Children to Free and Compulsory Education Act

This Act has come into force from April 1st, 2010. This Act provides for free and compulsory education to all children of the age of 6 to 14 years. Every child in the age group of 6 to 14 years will be provided 8 years of free elementary education in a school in the vicinity of his/her neighbourhood.

Right to Education Act Rules

(i) The act makes it mandatory for every child between ages of 6 to 14 years to be provided for education by the state. This means that such child does not have to pay a single penny as regards books, uniforms, etc.

(ii) Any time of the academic year, a child can go to a school and demand that his right be respected.

(iii) Private educational institutions have to reserve 25 per cent of their seats starting from class I in 2011 to disadvantaged students.

(iv) Strict criteria for the qualification of Teachers.

(v) The teacher-student ratio of 1:30 in schools ought to be met within a given timeframe.

(vi) The schools need to have certain minimum facilities like adequate teachers, playground and infrastructure. The government will evolve some mechanism to help marginalized schools comply with the provisions of the act.

(vii) The state government and local authorities will establish primary schools within walking distance of 1 km., of the neighborhood. In case of children for class VI to VIII, the school should be within a walking distance of 3 kms. of the neighbourhood.

(viii) Unaided and private schools shall ensure that children from weaker sections shall not be segregated from the other children in the classrooms nor shall their classes be held at places and timings different from the classes held for the other children.

The National Commission for Protection of Child Rights (NCPCR) has been mandated to monitor the implementation of this right. The NCPCR invites all civil society groups, students, teachers, administrators, government personnel, legislators, members of the judiciary and all other stakeholders to join hands and work together to build a movement to ensure that every child of this country is in school and enabled to get 8 years of quality education.

Conclusion

Education is the basis of social progress. It is the backbone of any progress that the country can make. It is the basis for progress in all sectors of economy. Admittedly, there is an imperative need for well structured education at all levels – from the stage of primary education up to higher education level. However, it should be remembered that frequent changes in the system of education is not desirable as education is linked with life of generations of people.

Over a period of time, there has been phenomenal growth of schools, colleges, universities, research institutes, etc., in the country. Education system as a whole, in India, is advancing by leaps and bounds in terms of student enrolment, money spent and the number of educational institutions started every year. Every education system, like any other system needs regular upkeep, maintenance, upgradation and improvement. It has to adjust with times, with development and meet the needs and requirements of the fast changing society. In the functioning of the education system, some gaps are noticed. Among others, gaps and matters of concern relate to quality of education, education of women, Harijan and Girijan still lagging; school dropouts of high magnitude in rural areas, lack of minimum facilities, particularly in schools of rural areas; disproportionate teacher-pupil ratio in both schools and colleges; absence of vocational element in the course structure; student unrest on the campus of general universities and lack of opportunities or uncongenial atmosphere for research within the country, forcing some talented scientists to migrate to other countries.

Education can serve as an instrument of social progress only when there is quality in the education. Further the education of women, Harijan and girijan should receive much more attention. As Mahatma Gandhi says "Mass illiteracy, is India's sin and shame and must be liquidated". The quality of education has to be judged from two angles. First, education received at all levels must equip the student with necessary skills needed either for self-employment or assured paid employment. Second, along with technical skills, character building is important. A German reformer Martin Luther rightly observes "The prosperity of a country depends not on its abundance of revenues or on the strength of its fortifications, but on its men of education, enlightenment and character".

Obviously, the character building of the pupils is necessarily the responsibility of both the parents and the teachers.

In the direction of well designing the structure of education, greater importance must be given to the base structure of the pyramid, namely, primary and secondary level of education. Sound foundation can be laid only with right type of education and training at the school level. For every country, there is a culture which is preserved and passed on from generation to generation through its language. Sanskrit language is the language of India, well endowed with the essence of Indian Culture. Books in Sanskrit have been a treasure of knowledge well presenting Indian culture. Sanskrit should have a special place in the Indian educational system. By offering Sanskrit along with mother tongue, at secondary education level thus, promoting Indianness becomes easier.

Schools in rural areas are inadequate and they also lack minimum facilities. This gap in the education system needs adequate attention. The unemployed educated youth in the rural areas may be encouraged to set up schools in remote villages and run them. The concerned village panchayats may be entrusted with the responsibility of running the school properly. It is absolutely essential that school education is structured on the lines of Basic Education proposed by Gandhiji and on the lines of A.L. Mudaliar's Commission on Secondary Education. In both the systems, it is suggested that at the school level, in the final two years vocational courses of local relevance should be started.

Intermediate (+2) course again is a crucial period in the career making of the student. These are the days of science and technology. Strong foundation in science and Mathematics becomes crucial for pursing higher studies. Compartmentalization of course like Arts and Science course is not desirable. At Intermediate (+2) level, along with two science subjects, there can be one Arts subject and *vice versa.*

The growing unrest in the general education system can be tackled meaningfully by introduction of vocational element in the course structure as recommended by the Kothari Commission. The UGC and the Government should insist on offering of skill-building or socially relevant courses meeting the present day requirements. Students at the undergraduate level, based on their aptitudes, may be advised to choose vocational courses of their choice. Even at the university level, combination of subjects with vocational thrust should be insisted. Such type of curriculum is necessary. In the Dayalbagh model classroom teaching is linked with necessary work skills. Student now acquires necessary skills either for self-employment or assured paid employment. The Dayalbagh model is worth emulating by all educational institutions of higher learning.

In recent years, in some states like Andhra Pradesh, Tamil Nadu, etc., there has been mushroom growth of Engineering Colleges. It is reported that in some colleges there are no minimum facilities in terms of qualified staff, technical equipment etc. It is necessary that the prospectus of employment to the engineering graduates, branch wise need to be well analyzed by the competent authority before granting permission for new Engineering Colleges. It is necessary that fully qualified and competent teachers are recruited at all levels.

The institutes of higher learning cannot be a silent spectators to the poor state of affairs in the rural areas in terms of illiteracy, ill health, poverty and civic inertia. The institutes of higher learning must necessarily extend helping hand in rural reconstruction. The students with a well designed plan of action may regularly visit a chosen village or group of villages and render service as advocated by Gandhiji. "Rural service" must become a part of the curriculum of higher education, particularly in general universities.

REFERENCES

1. *India 2009,* Ministry of Information and Broadcasting, Government of India, p. 208
2. *The Eleventh Five Year Plan,* Vol. II on Education, p. 2.
3. *Ibid.,* p. 3.
4. The Eleventh Five Year Plan, *op. cit.,* p. 4.
5. *Ibid.,* p. 8.
6. *Ibid.,* p. 23.
7. Gandhi M.K., *Towards New Education,* Navajivan Publishing House, Ahmedabad, pp. 77, 79, 81.

26 DR. A.P.J. ABDUL KALAM'S PURA MISSION FOR RURAL RECONSTRUCTION

1. Introduction

From time to time, innovative schemes have been initiated and well grounded by some committed persons for the development of rural areas. A.P.J. Abdul Kalam proposed a scheme known as Providing Urban Amenities in Rural Areas (PURA). Dr. Abdul Kalam is quite aware that the rural people are denied of basic amenities in terms of transport and communication, literacy, health and medical facilities etc. As one well committed to the overall development of India, he proposed a scheme of providing urban amenities in the rural areas. In order to achieve the vision of "Developed India by 2020", several crucial actions are required to ensure speeder growth of infrastructure, energy, quality electric power, roads, telecommunication etc. Dr. Kalam speaks of some short-term measures as well as long-term measures. The long-term action should be aimed at providing world class facilities for all parts of India. He speaks of "rural connectivity" as crucial element even in the short-run for betterment of agriculture and other sectors of economy. Rural connectivity simply means good roads, telecommunications, quality electric power, institutional credit to the needy etc. He speaks of physical connectivity, electronic connectivity, knowledge connectivity and economic connectivity.

2. Outlines of Working of PURA Mission

PURA envisages economic empowerment to a cluster of villages through an integrated approach Chart 26.1 presents outlines of the working of PURA mission in a chosen cluster of villages.

Chart 26.1

Outlines of Working of PURA

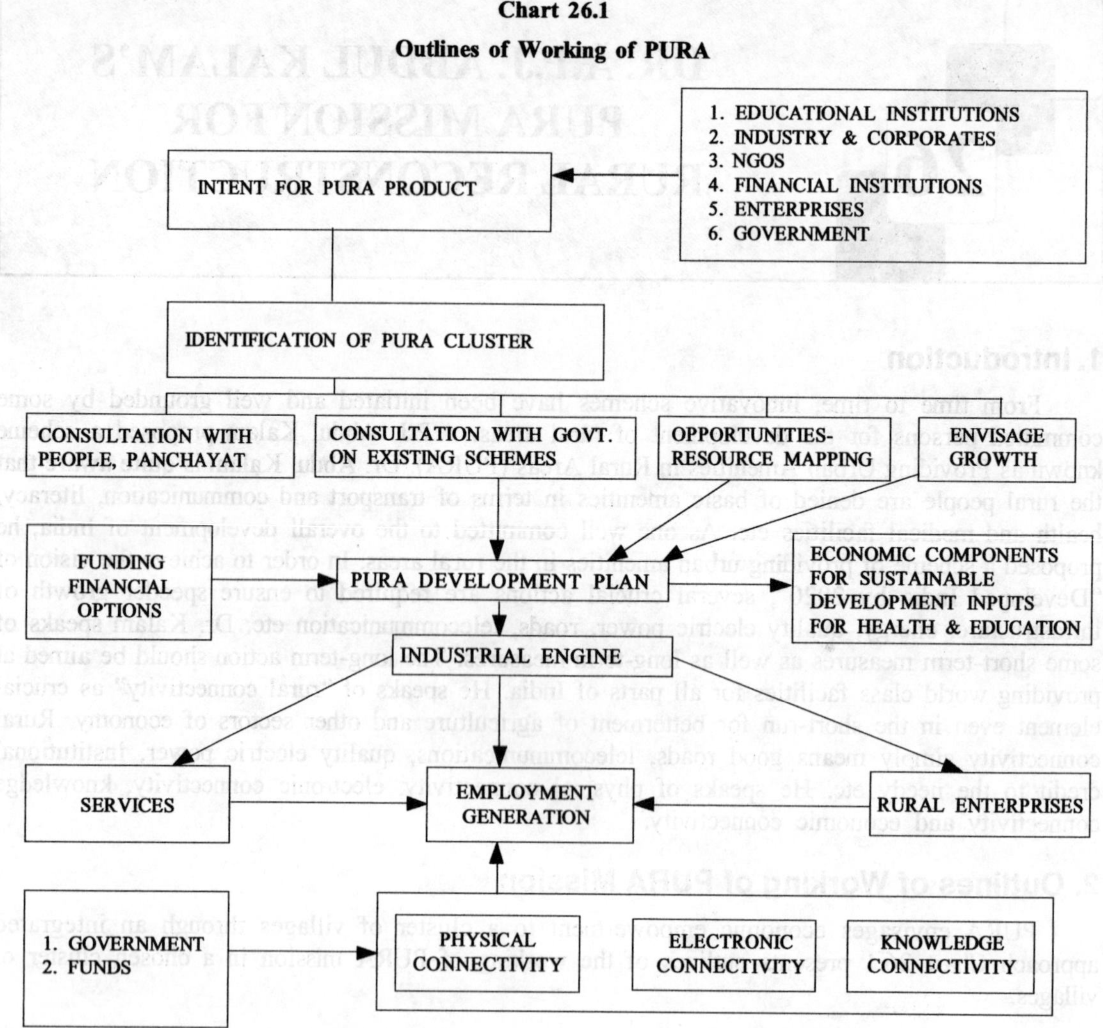

The steps involved in creation and maintenance of PURA are as follows:

1. An institution such as educational/industry/NGO/Financial Institutions/small scale enterprise government intends to create a PURA cluster.
2. Desiring agency identifies the groups of villages in the district which are suitable for creation of an economically empowered PURA cluster.
3. Consultation with the panchayat board members, resource mapping, envisaged growth path and study of planned existing government schemes in the area are carried out parallely.
4. The institution determines the industrial engine for the PURA development based on the core competence and natural resources of the region.
5. Simultaneously, the institution works out the funding requirement including the funds which have been provided for government development schemes envisaged in the area.

6. The industrial engine plans the services and the rural enterprises and determines the total employment generation potential of the PURA complex both during commissioning and subsequently during its operation.
7. The economic empowerment of the PURA is supported by the establishment of ideal physical connectivity, electronic connectivity and knowledge connectivity for the whole complex.
8. The district authorities discuss the whole PURA plan with the intending institution and arrive at an implementation plan which may include the following:
   (a) Government will be responsible for provisioning of the land required for the complex in consultation with local bodies.
   (b) The government may make the funds allotted for the regional development schemes available to the implementing agency to enable implementation of the PURA programme as a turn key project with single point responsibility.
   (c) The implementation agency will create all the connectivities envisaged for the cluster and establish financially visible enterprises leading to provision of planned employment opportunities.
   (d) The implementation agency will also take the responsibility for continuous provisioning of quality health care and education to all people living in the PURA complex. The government will be required to provide the subsidy element to all the eligible categories of people.

3. Thrust Areas of PURA

Thrust areas for PURA would be the following:

1. Creation of employment opportunities for all the employable people, particularly the youth.
2. Capacity building in education – school, value-added employable skills and knowledge.
3. Provision of quality health and timely health care, safe drinking water quality, reliable electric power, energy and water-efficient pucca houses.

He speaks of four types of connectivities, namely, physical, electronic, knowledge and economic.

4. Physical Connectivity

Physical connectivity facilities movement of people and goods, access to schools, health centres and markets. In rural areas, it is well-known fact that there are no adequate and proper roads, rail and public infrastructure. This cluster of villages would be provided physical connectivity by a ringroad. Low-cost buses preferably driven by batteries energized by renewable energy sources, equipped with high efficiency engine would be operated almost throughout the day as shuttle services moving people and goods from village to village and village to school, health centre, fueling stations, farming areas, warehouses, agro industries and other commercial centers.

It is clear that the art of the PURA concept is physical connectivity of the villages by a road. Connectivity thereafter to a rail network and to a nearest city beyond this village cluster would take off from the ring road. All theses roads or links will be of high quality so that the transportation time is minimized.

Providing amenities such as schools, health centers, village markets, warehouses ring roads and commercial centres for servicing a cluster of villages is a cost-effective solution because of economies of scale since it serves a larger population than a conventional village unit. Also, these cluster of villages will become an investment destination for entrepreneurs. It is because of the relatively low transport costs.

5. Electronic Connectivity

The focus here is establishing network, hardware, software and other related equipment, studio and data centre. The typical electronic connectivity components are as follows:

1. Connecting cluster of villages through Wi-MAX wireless link with the min., 2mbps landwidth connectivity in a synchronous mode. Village knowledge centers must be established in this cluster.
2. Interconnecting the PURA Nodal and Domain Service Providers through Broadband or VSAT connectivities with the village knowledge centers using the Wi-MAX.
3. Studies and Data centre has to be established for tele-education, tele-medicine, e-governance, e-business and e-market etc., and internet access. This will serve the backend for the PURA call centre.
4. Collaborative tele-conferencing for expert consultation, training and feedback.
5. Tele-education centre with the necessary hardware, software, multimedia equipment and network.
6. Tele-medicine centre with the necessary hardware, software, multimedia equipment, medical equipment and network.
7. Setting up of PURA portal for web-enabled services and delivery of e-governance services.
8. Community FM radio/Satellite radio services.

6. Knowledge Connectivity

Knowledge connectivity envisages the system oriented approach for the village cluster would require introducing tele-education for farmers and villagers, public call offices, tele-medicines, e-market, etc. It also will provide the opportunity for the villagers to locate call centers, business processing outsourcing and software development centers to use world outsourcing market.

The focus here is value-added services, knowledge empowerment and management, skill development and capacity building.

The typical components of knowledge connectivity are as follows:

1. Establishing PURA Portal at PURA complex, at the nodal centre for Providing Business, marketing, trading-buying, selling of local products, transaction etc., selling of local products, transaction etc., among this cluster.
2. Importing education to the school students of this cluster for the I-XII Stds. in Science and Mathematics subjects, where the guidance is very much required as a supplementary education.
3. Spoken English Courses, IT Courses, Multimedia and DTP course, BPO training for the graduate to prepare themselves and make them employable into call centers and ITES in Chennai, Bengaluru, Hyderabad, Delhi etc.

4. **Agriculture,** Farming guidance following system-oriented approach to the farmers on the following:
   (i) Cultivation of bio-fuel plants such as Neem etc.
   (ii) Cultivation of crops like Rice, Cotton, Chillies, Maize, Groundnut, Coconut, Onion etc.
   (iii) Vegetable cultivation
   (iv) Dairy farming, cattle breeding
   (v) Contract farming and waste land development
   (vi) Organic farming practices, Vermi-compost, pest control and pesticides, drip irrigation and management practices.
5. Weaving and handlooms guidance to the weavers by way of technologies, transforming handlooms into powerlooms and identifying the market potential for the woven cloth.
6. Guidance to setting up of small-scale industries, handicrafts making, dairy and cattle development, paper manufacturing from waste cloths etc.
7. Health care through tele-medicine, expert doctors from the nodal centers will examine the patients at the respective villages in a predetermined timing. Each village knowledge centers will be manned by one nurse equipped with minimal medical equipment such as, stethoscope, BP Monitor, portable ECG etc.
8. Awareness programmes on diseases such as AIDS, TB, POLIO and on hygiene and health.
9. Literacy programmes for adults and women.
10. Youth development programmes career developments, personality development, yoga camp, career guidance, job opportunities etc.
11. Entrance coaching.
12. Computer training, DTP and Multimedia courses training and certificate programmes.
13. Internet, e-mail access, photo printing services, e-mail delivery to home.
14. Empowering SHGs, Micro credit, Entrepreneurship training, etc.

7. Economic Connectivity

Economic connectivity relates to wide range of activities contributing for generation of employment and income. Major economic connectivity areas include small-scale industries, agro industries, warehouses, micro-power plants, renewable energies, village markets etc. all leading to employment generation, women empowerment etc. Economic connectivity is expected to result in decongestion of urban centers. Further, it is hoped that with the economic connectivity purchasing power of the people in PURA cluster area will increase leading to improvement of quality of life.

Employment potential through economic connectivity varies from one activity to another. For example, in a coastal PURA fishing village, the economic activity, will encompass cold storage, fish processing centers, marketing centers, small boat construction facility, boat and net repair facility etc. The focus of economic connectivity is on employment generation, and thereby increase the purchasing power of the people.

The economic connectivity covers the role of banks tied to PURA mission with banking facility, buying and selling between the urban establishments and rural people.

The PURA scheme ultimately aims at providing better facilities in rural areas as in the urban centers. This facilitates better living conditions of the rural people.

Conclusion

The PURA Mission since its inception 2006, during the last four years, tangible results have been achieved in those centers where PURA Mission has been wellgrounded. PURA's Technology and application of scientific methods of working have played a very important role. For example, among others, power through biogas and solar energy is used for household lighting and also for the farmers. Vermi composting; check dams and water purification plants; medicinal and aromatic plants, cultivation, extraction and manufacturing health care products through Self-help Groups; low-cost housing using alternate building blocks, dairy farming, health care and education service connectivity using Wi-MAX technologies and enabling the sustainable development and business processing taking place among the village clutters. Also, it is observed in the villagers where PURA have been functioning some skills useful for livelihood have been well developed. Skill in construction work, tailoring, garment productions, IT and spoken English skills have been well developed. This programme has become instrumental in creating 3000 jobs with the minimum earning of ₹ 3000/- per month which is some three times higher than earlier earning potential.

Keeping in view the present state of affairs in the rural areas and with an eye on the betterment of the rural poor, PURA Mission needs to be popularized in term of its coverage and support from different agencies including government.

REFERENCE

1. PURA Mission to Reality, 2006-07 by Dr. A.P.J. Abdul Kalam, 7th November, 2006, New Delhi.

PART V

CONCERNS IN THE PROCESS OF DEVELOPMENT

27 REGIONAL IMBALANCES AND SOCIAL TENSIONS

1. Introduction

Imbalances of varying nature and magnitude, globally and within each country are well noticed. The consequent social tensions are a matter of serious concern. After Second World War, certain countries are well developed and certain other countries remain less developed. The advancement of Japan has been very significant. During 1970-90, China and Korea progressed tremendously in the manufacturing sector, while India's progress in the service sector has been very significant. For historical and socio-economic reasons, some remain backward and some others emerge significantly. Thus, all over the globe imbalances in the development do exist and consequently social tensions erupt.

2. Disparities among States and Regions within States

Within India, there are disparities among states and regions within the states. The society is divided as urban and rural areas and we notice imbalances between urban and rural areas and imbalances in between different sections of the society have been steadily increasing in the past few years. The Eleventh Plan[1] observes, "The gains of the rapid growth witnessed have not reached all parts of the country and all sections of the people in an equitable manner ..... widening income differentials between more developed and relatively poorer states is a matter of serious concern".

In every state there are regional variations and imbalances in the levels of income of people in different regions. The reasons for regional variations may be different in different states. In regions, where the natural endowment is favourable along with favourable human resource factor, the development of that region is bound to be higher. If the natural resources and human resources are unfavourable and if the human resource is less enterprising then that region is bound to be weak. The regional imbalances in a state is bound to result in social tensions of serious nature, leading to obstruction of the growth of the state. Thus, for historical, socio-economic and natural resource availability, etc., reasons regional imbalances and consequent social tensions do take place.

3. Variations in Growth Rates in Different States

The growth rates in Domestic Product in different states are shown in Table 27.1

Table 27.1

Growth Rates in Sate Domestic Product in Different States

S.No.	State/UT	Eighth Plan	Ninth Plan	Tenth Plan Target	Tenth Plan Actuals #
Non-special Category States					
1.	Andhra Pradesh	5.4	4.6	6.8	6.7
2.	Bihar	2.2	4.0	6.2	4.7
3.	Goa	8.9	5.5	9.2	7.8
4.	Gujarat	12.4	4.0	10.2	10.6
5.	Haryana	5.2	4.1	7.9	7.6
6.	Karnataka	6.2	7.2	10.1	7.0
7.	Kerala	6.5	5.7	6.5	7.2
8.	Madhya Pradesh	6.3	4.0	7.0	4.3
9.	Maharashtra	8.9	4.7	7.4	7.9
10.	Orissa	2.1	5.1	6.2	9.1
11.	Punjab	4.7	4.4	6.4	4.5
12.	Rajasthan	7.5	3.5	8.3	5.0
13.	Tamil Nadu	7.0	6.3	8.0	6.6
14.	Uttar Pradesh	4.9	4.0	7.6	4.6
15.	West Bengal	6.3	6.9	8.8	6.1
16.	Chattisgarh	NA	NA	6.1	9.2
17.	Jharkhand	NA	NA	6.9	11.1
Special Category States					
1.	Arunachal Pradesh	5.1	4.4	8.0	5.8
2.	Assam	2.8	2.1	6.2	6.1
3.	Himachal Pradesh	6.5	5.9	8.9	7.3
4.	Jammu & Kashmir	5.0	5.2	6.3	5.2
5.	Manipur	4.6	6.4	6.5	11.6
6.	Meghalaya	3.8	6.	6.3	5.6
7.	Mizoram	NA	NA	5.3	5.9
8.	Nagaland	8.9	2.6	5.6	8.3
9.	Sikkim	5.3	8.3	7.9	7.7
10.	Tripura	6.6	7.4	7.3	8.7
11.	Uttaranchal (now Uttarakhand)	NA	NA	6.8	8.8

***Source:** The Eleventh Plan Report,* p. 138.

From the Table 27.1 it is evident that there are imbalances in the growth rates in different states. Also we notice that the growth trend from Plan to Plan is not always consistently on higher side. In some states the growth rate in Ninth Plan is lower that what it was during the Eighth Plan. Similar trend can be noticed in some other states too. Table 27.2 presents state-wise growth targets of different sectors of economy. State wise and Sector-wise variations in the targets of the Eleventh Plan clearly reiterate the existence of variations in the levels of development. See Table 27.2.

Table 27.2

State-wise Growth Target for the Eleventh Five Year Plan

S.No.	States/UTs	State Growth Target			
		Agriculture	Industry	Services	GDP Growth
Non-special Category States					
1.	Andhra Pradesh	4.0	12.0	10.4	9.5
2.	Bihar	7.0	8.0	8.0	7.6
3.	Goa	1.7	12.0	8.0	8.6
4.	Gujarat	7.7	15.7	9.0	12.1
5.	Haryana	5.5	14.0	10.5	11.2
6.	Karnataka	5.3	14.0	12.0	11.0
7.	Kerala	6.3	12.0	8.0	9.8
8.	Madhya Pradesh	5.4	12.5	12.0	11.2
9.	Maharashtra	0.3	9.0	11.0	9.5
10.	Orissa	4.4	8.0	7.0	6.7
11.	Punjab	4.4	8.0	10.2	9.1
12.	Rajasthan	3.0	12.0	9.6	8.8
13.	Tamil Nadu	2.4	8.0	7.4	5.9
14.	Uttar Pradesh	3.5	8.0	8.9	7.4
15.	West Bengal	4.7	8.0	7.4	5.9
16.	Chattisgarh	3.0	8.0	7.1	6.1
17.	Jharkhand	4.0	11.0	11.0	9.7
Special Category States					
1.	Arunachal Pradesh	2.8	8.0	7.2	6.4
2.	Assam	2.0	8.0	8.0	6.5
3.	Himachal Pradesh	3.0	14.5	7.5	9.5
4.	Jammu & Kashmir	4.3	9.8	6.4	6.4
5.	Manipur	1.2	8.0	7.0	5.9
6.	Meghalaya	4.7	8.0	7.9	7.3
7.	Mizoram	1.6	8.0	8.0	7.1
8.	Nagaland	8.4	8.0	10.0	9.3
9.	Sikkim	3.3	8.0	7.2	6.7
10.	Tripura	1.4	8.0	8.0	6.9
11.	Uttaranchal (now Uttarakhand)	3.0	12.0	11.0	9.9

Source: *The Eleventh Plan Report,* p.138.

From the Table 27.2, it is evident that the targets of different sectors of economy vary in between sates in the country. Table 27.3 presents variations in the per capita income in between different states.

Table 27.3

Disparity in Per Capita GSDP in between different states

Year	State with Lowest Per Capita GSDP	State with Highest* Per Capita GSDP	Ratio of Minimum to Maximum Per Capita GSDP	Coefficient of Variation	Gini Coefficient
1	2	3	4	5	6
1993-94	Bihar	Punjab	30.537	34.549	0.1917
1996-97	Bihar	Maharashtra	27.586	36.781	0.2071
1999-2000	Bihar	Maharashtra	28.899	37.417	0.2173
2001-02	Bihar	Punjab	21.556	35.610	0.2078
2002-03	Bihar	Punjab	21.608	22.705	0.2771
2003-04	Bihar	Maharashtra	36.686	26.230	0.2290
2004-05	Bihar	Maharashtra	20.105	38.440	0.2409

* Excluding Goa weighted by population
Source: *The Eleventh Five Year Plan Report,* p. 140.

The disparities in the income levels may be due to different factors such as differences in natural resource endowment and quality of human resource. The disparities are found not only in between States but also, in between regions and districts within a State. Even in highly developed states there are regions and districts which remain backward. While some degree of inter-state and intra state disparity is bound to exist, the policy measures must necessarily identify the backward areas wherever they exist and take appropriate steps.

4. Regional Variations in Andhra Pradesh:

Regional variations in development may be analyzed with reference to Andhra Pradesh. Andhra Pradesh is composed of three regions of Telugu speaking people, Coastal Andhra, Rayalaseema and Telangana. The State of Andhra Pradesh came into being in 1956. During these five decades, development of regions has taken place. Yet, regional variations exist. Basing on officially published data relating to Human Development Index, an attempt is made to assess the degree of inter district variations in the development. See Table 27.4.

Table 27.4

Human Development Index — District-wise in Andhra Pradesh

Human Development Index-2007			Poverty Index-2007	
Index Value	**Rank**	**District**	**Index Value**	**Rank**
0.717	1	Hyderabad	0.213	1
0.623	2	Krishna	0.399	3
0.610	3	Ranga Reddy	0.369	2
0.670	4	West Godavari	0.449	5
0.599	5	Guntur	0.428	4
0.586	6	East Godavari	0.465	9
0.573	7	Karim Nagar	0.452	7
0.565	8	Nellore	0.466	10
0.559	9	Khammam	0.500	16
0.558	10	Chittoor	0.461	9
0.553	11	Visakhapatnam	0.504	17
0.550	12	Medak	0.498	15
0.536	13	Cuddapah	0.451	6
0.532	14	Prakasam	0.494	13
0.514	15	Warangal	0.492	12
0.504	16	Nizamabad	0.470	11
0.488	17	Adilabad	0.514	19
0.481	18	Nalgonda	0.513	18
0.473	19	Kurnool	0.494	14
0.458	20	Ananthapur	0.515	20
0.453	21	Srikakulam	0.566	21
0.402	22	Vijayanagaram	0.597	23
0.377	23	Mahaboobnagar	0.592	22

***Source:** Andhra Pradesh Human Development Report*

From Table 27.4, it is clear that out of 16 districts with Human Development Index value of 0.500 points and above, seven districts are in coastal Andhra and again equal number of seven districts are in Telangana region and two are in Rayalaseema region. Out of the four districts of Rayalaseema region, in two districts human development remain backward. In the poverty Index, Hyderabad district of Telangana region is with relatively low level of poverty index of 0.213. Vijayanagaram district of Coastal Andhra region with 0.597 points is with higher degree of poverty, followed by Srikakulam district with 0.566 points and Ananthapur district of Rayalaseema region with 0.515 points and Visakhapatnam of coastal Andhra with 0.504 points. The position of Khammam,

Adilabad and Nalgonda districts of Telangana region is no better. Thus, poverty is not confined to any one region, on the other hand, it is spread over different districts of all the three regions of Andhra Pradesh.

Critical analysis of development of Agriculture Sector, Basic Amenities and Social and Economic sectors are worth examining to locate the backwardness. See Table 27.5.

Table 27.5
District-wise variations in levels of Development

District	Agricultural Sector		Basic Amenities		Economic Sector	
	C.I.	Rank	C.I.	Rank	C.I.	Rank
Srikakulam	0.73	11	0.65	6	0.68	7
Vijayanagaram	0.76	14	0.67	8	0.71	9
Visakhapatnam	0.87	21	0.72	12	0.78	18
East Godavari	0.65	5	0.64	5	0.66	3
West Godavari	0.61	2	0.60	1	0.64	1
Krishna	0.64	4	0.75	16	0.73	11
Guntur	0.60	1	0.68	10	0.68	6
Prakasam	0.71	10	0.75	17	0.75	14
Nellore	0.68	6	0.65	7	0.61	5
Chittoor	0.77	15	0.63	3	0.69	8
Cuddapah	0.80	16	0.73	13	0.77	16
Ananthapur	0.83	17	0.82	20	0.83	20
Kurnool	0.83	18	0.81	19	0.83	19
Hyderabad	—	—	—	—	—	—
Mahaboobnagar	0.84	19	0.86	21	0.87	21
Ranga Reddy	0.87	22	1.00	22	0.99	22
Medak	0.76	13	0.68	9	0.72	10
Nizamabad	0.69	9	0.64	4	0.67	4
Adilabad	0.86	20	0.70	11	0.76	15
Karim Nagar	0.64	3	0.61	2	0.63	2
Warangal	0.73	12	0.77	18	0.77	17
Khammam	0.69	7	0.74	15	0.74	13
Nalgonda	0.69	8	0.77	14	0.74	12

Source: *Indian Society of Agricultural Statistics,* New Delhi,
C.I. = Composite Index

In the agricultural sector, in the list of first 12 ranks, as presented in Table 27.5 seven districts are in coastal Andhra and five are in Telangana region. Rayalaseema districts have no place here. In the basic amenities, in the list of first twelve ranks, seven are from Coastal Andhra region, four districts are from Telangana region, and only one district from Rayalaseema region.

The analysis of inter-district variation reveals that in all the three regions there are backward districts and problem of backwardness in each district need to be well addressed through appropriate planning. Right step would be to identify the backward areas and directly address the problem in all the states. Setting up of Regional Boards, well and budgeted is one way of tackling the problem. Again evolving appropriate industrial policy for the development of backward regions is necessary. Also development of the village panchayats of the backward districts, with adequate powers and resources is an other effective method of development of backward districts.

5. Developmental Key to Fighting Maoism

Maoism problem is not to be viewed as a law and order problem and proceed in that way only. "Lack of development in backward areas such as Jharkhand, Chattisgarh and parts of West Bengal is the main reason for the growth of Maoism and militancy there" says Sri Jayendra Saraswathi, the Sankaracharya of the Kanchi Kamakoti Peetam. Apart from absence of development in the Tribal areas, the Sankaracharya said "the decline in the value systems and the quality of life also had led to many undesirable developments in society. The unity among the people and the feeling of oneness as Indians will go a long way in bringing communal harmony and peace in the country. Unless the government and other agencies took up the developmental issues of the backward regions, militancy and violence cannot be stopped". (*The Hindu,* Tuesday 9, August, 2010)

6. Backward Regions Grant Fund (BRGF)

The development of backward regions has been a major concern of Planners in India. Area Development programmes such as, Drought Prone Area Programme, Tribal Area Development Programme are all intended for development of backward regions in different states. The Mid-term Appraisal of the Ninth Plan observed that these efforts, one of the most serious problems faced in the country was the wide disparity and regional imbalances between States and within a state, between districts.[2]

In the Tenth Plan, a specific programme for the development of backward areas known as Rashtriya Sam Vikas Yojana (RSVY) was initiated in 2003-2004. The RSVY covered 147 districts. Each district was to receive ₹ 45 crore from BRGF. This scheme addressed in the backward areas, the problems of low agricultural productivity, unemployment and fill the critical gaps in infrastructure. Some 33 districts received the full allocation of ₹ 45 crore during the Tenth Plan. The BRGF extends assistance to programmes like Bharat Nirman and NREGP which are explicitly designed to develop infrastructural facilities in the backward areas. The BRGF proposes people's participation in the development of backward areas. In this direction, steps were initiated to associate Panchayat Raj institutions in the implementation of different programmes.

In the direction of addressing inter-state and intra-state the planning commission has chosen to take the assistance of Indian Statistical Institute, Kolkata to identify poverty and its dimensions in different states and different districts within each state. Health problem is critical in the estimation of poverty. During the Eleventh Plan period, steps have been initiated to prepare District Health profiles for all districts in the country, based on an annual health survey undertaken through the Registrar General. Steps are also initiated to prepare State Human Development Reports (SHDR). Already some 18 states have published SHDR. Attempts are also initiated for the preparation of District Human Development Reports (DHDR) in a few selected districts to start with.

7. Redressal Steps

In view of the seriousness of the problem of regional imbalances and consequent social unrest, different State governments are taking action to redress the problem of regional imbalances. In West Bengal,[3] 4,612 revenue villages have been identified for special attention. Female literacy and employment opportunity were used as variables to capture the extent of poverty. Villages with female illiteracy over 70% and villages with 60% of population in the working age group as non-workers or marginal workers have been identified for special attention. The services to be made available in these villages are (1) Food security, (2) Social Assistance, (3) Elementary Education, (4) Nutrition Programmes, (5) Employment Generation, (6) Housing for the Houseless, (7) Formation of Self Help Groups (SHGs) linking with Banks and (8) Sustainable Drinking Water and Public Health Services.

In Andhra Pradesh,[4] identified backward villages have been chosen for eight schemes. These schemes are (1) Pension for all eligible, old persons and widows, (2) Housing for weaker sections, (3) Sustainable drinking water, (4) Power supply to every household, (5) Sanitary latrines in every household, (6) Prevention of school dropouts at the level of elementary education, (7) Anganwadi with building and (8) Improvements in primary health with 100% immunization, 100% institutional delivery and 100% maternity benefit for eligible pregnant women.

Conclusion

'Development' is the right approach to tackle regional imbalances. Each state has its own peculiar problems with regards to imbalance in development at regional and district levels. In states where natural endowment is poor, special care is required. The planning commission must identify backward districts in the country and in each district the districts administration must identify backward areas. The state governments must take steps to set up/revive Regional Development Boards and strengthen them with adequate resources and functions. Appropriate industrial policy giving priority to backward districts is immediately called for. The village panchayats at the village level, are the right bodies to identify the local needs and suggest appropriate steps. The problem of Regional imbalances and consequent social tension that erupt is a serious problem. Local problems must be identified and settled at the local level, regional imbalances and social tensions are linked with poverty, unemployment and illiteracy. An integrated approach in addressing these interconnected problems is required. Decentralization of administration and planning machinery and involvement of panchayat bodies in all development activities and participatory approach involving local people in different development schemes as advocated by Mahatma Gandhi is the right approach to tackle regional imbalances and consequent social tension.

REFERENCES

1. *The Eleventh Plan Report*, p. 137.
2. *Ibid.*, p. 145.
3. *Ibid.*, pp. 150-151.
4. *Ibid.*

28 FOOD SECURITY, AGRICULTURE AND EMPLOYMENT

1. Concept of Food Security Defined

Food is an essential requirement for man's survival along with clothing and housing. Food security, therefore, has become the responsibility of modern states. FAO looks at food security in a comprehensive sense. FAO states "Food Security refers to a situation that exists when all people at all times, have physical, social and economic access to sufficient safe and nutritious food that meets their dietary needs and food preferences for an active and healthy life".[1]

In view of complexities involved in defining the concept of food security, it is necessary to arrive at a generally acceptable definition of food security. Food security refers to all those steps involved in meeting the food and nutritional requirements of all citizens in society for an active and healthy life. Food security has different components: adequate production of food grains of quality; availability of food in the market; access to food grains through adequate purchasing power by the consumers; adequate food stocks in the public godowns and well managed Public Distribution System taking care of the interests mainly of the poor consumers.

At the global level, the South Asian region is found to be centre of more number of chronically food insecure people than any other region in the world. India ranks 94th in the Global Hunger Index of 119 countries.[2] In India most burning problem of great concern today is that of chronic or persistent food insecurity, particularly in terms of nutritional insecurity. The Eleventh Five Year Plan points out that there are serious concerns around food and nutritional security….cereal production has declined in per capita terms. The number of poor was estimated at 300 million in 2005. Low and stagnating incomes among the poor has meant low purchasing power with serious constraint to household food and nutrition security.[3] States like Jharkhand and Chattisgarh are found in the "very high" level of food insecurity, followed by Madhya Pradesh, Bihar and Gujarat where the position is not quite satisfactory. The better performers include Himachal Pradesh, Kerala, Punjab and Jammu and Kashmir. Andhra Pradesh, Karnataka and Maharashtra face the problem of food insecurity and they are in the category of "high food insecurity", a reflection of manifestation of the agrarian crisis. It is not surprising that in States like Andhra Pradesh and Maharashtra rural distress is of high order leading to a high rate of farmers' suicides.

Food security must necessarily be looked as an integrated issue. Nutritional security is an important component of food security. M.S. Swaminathan[3] observes "Nutritional Security involving physical, economic and social access to balanced diet, clean drinking water sanitation and primary health care for every child, women and man is fundamental to giving all our citizens an opportunity for a healthy and productive life". He is very categorical in stating that in our planning, adequate support for agriculture and its vast rural population is very much needed.

2. Food Security and Government Schemes

To ensure food security to weaker sections of the society, Government has been implementing a number of programmes such as Targeted Public Distribution System (TPDS) focused on Below Poverty Line families, Mid-day Meal Scheme for school children, Integrated Child Development Services (ICDS) scheme for children and mothers, Annapurna Scheme, Village Grain Banks Scheme (VGBS) etc. Yet food security is not assured to all for one reason or another.

In the direction of food security, food production needs to be stepped up. It is well-known fact that in recent years, particularly during the Ninth and Tenth Five Year Plan periods, foodgrain productions has been gradually declining. Farmers are driven to the stage of concluding that farming is not a paying proposition. In recent years, we notice a shift of cultivation from food to non-food crops. The per hectare yield in India continues to be much lower than what it is in China, Japan and some other countries. Some farmers are giving up cultivation and migrating to urban centres. The different factors contributing for declining food production in the country need to be well analyzed and appropriate steps must be taken immediately. An integrated approach becomes necessary to ensure higher levels of food production in the country. The crucial areas very much related with food production are irrigation and conservation of water, conservation of common property and biodiversity resources and rehabilitation of waste lands. Swaminathan, M.S., rightly points out "We must explore a horticulture remedy to tide over the nutritional malady."[4]

Several studies reveal that during the planning era there has been significant growth in farm production and related issues. Yet as the Reports of World Food Programme and Swaminathan's M.S. Research Foundation (MSSRF) point out that economic growth so far is not able to improve food security in India, particularly in rural areas.

3. Eleventh Five Year Plan and Food Security

According to the Eleventh Five Year Plan[5] recent trends that have raised concern regarding food security, farmer's income and poverty are as follows: (a) slow down in agricultural growth, (b) widening economic disparities between irrigated and rainfed areas, (c) increased vulnerability to world commodity price, following trade liberalization with all the adverse effect on crops such as cotton and oil seeds, (d) uneven and slow development of technology and inefficient use of available technology, (e) degradation of natural resource base, withdrawal of subsidies on fertilizers, power and irrigation contribute for such degradation, (f) rapid and widespread decline in groundwater table with adverse effects on small and marginal farmers, (g) increased demand for land and water for non-agricultural purpose due to growth of urbanization and industrialization and (h) aggregation in social distress.

As per National Sample Survey of 2003, matters of concern are:

(1) Fragmentation of holdings has been a chronic problem in Indian agriculture. Per capita holding is getting reduced from time to time and it is now less than 1 hectare.

(2) Marginal holdings (of size of 1 hectare or less), in 2002 and 2003 constituted 70% of all operational holdings, small holdings (of 1 to 2 hectares) constituted 16%, semi-medium holdings (or 2 to 4 hectares) constituted 9%, medium holdings (of 4 to 10 hectares) recorded to be 4% and large holdings (of over 10 hectares) constituted less than 1%.

4. Malnutrition and Food Security

The Eleventh Five Year Plan pointed out that 36% of adult population face Protein Energy Malnutrition (PEM).[6] The underweight children in 2005-06, was estimated to be 46%. There is need to look at food security issues not in isolation as being confined to cereal production and consumption, but to examine how nutritional outcomes can be improved for the vast majority of the poor. As the Eleventh Plan Report Points out the challenges that still remain include: High level of adult malnutrition affecting a third of the country's adults; Inappropriate infant feeding and caring practices; High level of undernutrition, particularly in women and children, diet-related diseases and inadequate access to health care etc. These challenges need to be effectively addressed by appropriate steps.

In both rural and urban India, the share of expenditure on food in total expenditure continued to fall throughout the three decades prior to 2004-05. The overall fall was from 73% to 55% in rural areas and from 64.5% to 42% in urban areas. In urban India not only the shares of cereals and pulses have fallen, but also there has been fall in the shares of items like milk and milk products, edible oil and sugar. In rural India, there has been slight increase in the share of non-cereals but the increase in the share of non-cereals is not enough to compensate for the decline in cereal consumption. The magnitude of child malnutrition is well indicated in the Eleventh Five Year Plan report. It is said that nearly half of India's children underage group of 3 are malnourished. India has the largest number of children in the world who are malnourished.[7]

5. National Food Security

It is matter of great appreciation that the government propose to enact a new law-the National Food Security Act, assuring food security to all. Dr. M.S. Swaminathan rightly believes that the goal of food for all can be achieved only through greater and integrated attention to production, procurement, preservation and public distribution.[8] The burning problem is mounting of price level of foodgrains, capacity of poor and even denying purchasing capacity of middle class or even upper middle class. There is a strong case to assess not only those who are really poor but also those whose purchasing is low in the society.[9] That exercise is necessary as a part of provision of the proposed Food Security Act. A scientific resurvey is necessary to arrive at correct picture of BPL families.

6. Integrated Nature of Issues in Food Security

Six major issues of integrated nature become crucial in addressing food security. They are: (1) Assessing the demand for required food, (2) Stability in production of foodgrain, (3) Generation of Employment, income and purchasing power of the poor, (4) Identification of all those below poverty line, (5) Well managing the Public Distribution System and (6) Controlling price level of foodgrain and other consumption items.

6.1. Assessing demand for food

Demand for food is linked with size of population and its growth and economic growth of the country. India's population at present rate of growth is projected at 1.3 billion by 2020. As the economy grows incomes of people also increase and consequently consumption level of people also increases. Further, with increase in income, lifestyle of the people will change. Now with higher incomes, demand for non-cereals like milk, fruits, vegetables, meat etc., tends to increase and the requirement of different items need to be assessed.

Indian agriculture faces pressures from both domestic and international market. India as a member of World Trade Organization (WTO) has obligation to provide market access to other member countries in marketing their products in India. As per WTO Agreement, India has to export 3% of its total production of foodgrains to other countries. These issues will naturally pose a problem in meeting the domestic demand for food.

6.2. Stability in production of foodgrains

Technologies play a critical role in achieving stability in production and ensuring food security in the country. Biotechnology is of crucial importance as it deals with many aspects of basic inputs to agriculture like seeds, plants, soil treatment etc. A.P.J. Abdul Kalam rightly[10] states that apart from Biotechnology, we have to necessarily depend upon conventional technologies even while we target biotechnology. There are a good number of conventional technologies worth honouring. For example, harvesting of rainwater and well preserving the rainwater in tanks or ponds is a crucial conventional technology which needs serious consideration.

As present foodgrain production is mostly confined to irrigated agriculture and introduction of high yielding varieties of crops. The green revolution is confined to irrigated cultivation covering only about 40% of the cropped area. Even now nearly 60% of the cropped area is under rainfed cultivation. The strategy for higher food grains production for rainfed cultivation, in terms of evolving drought resistent seeds etc. need to be developed and popularised. Dry land agriculture must focus attention on production of pulses, oil seeds, fruits and livestock.

There are a good number of technologies which can contribute for stability in agricultural production. Remote sensing or taking electronic pictures of the earth from space is one way of assessing potentialities of natural resources, land degradation, water potentiality as well as to predict crop yield and so on. A.P.J Abdul Kalam, categorically states that our country is and will continue to be a major producer of wheat and rice. Different steps that will ensure stability in production are indicated:[11] (1) To explore possibilities of broadening of the production area of wheat in UP, Bihar, and in the North-east; (2) Increase rice production in traditional area by adopting hybrid variety of seed; (3) Increase production of coarse grains and develop various products which can partly substitute rice and wheat; (4) To take steps for increasing the production of pulses; (5) In view of increase in the demand for vegetables and fruits, appropriate steps need to be taken to step up production of these items. Also cold storage and long distance transport will be essential requirements in this regard; (6) Water being a crucial input for profitable agriculture, every drop of rainwater need to be well harvested and well preserved; (7) Minor irrigation must receive necessary attention. Major irrigation projects need to be so designed as to safeguard the environment as well as the interest of the farm producers and (8) Protected water being an essential requirement for food security, Government must take appropriate steps to ensure supply of safe drinking water.

6.3 Generation of employment, income and purchasing of poor

In rural areas both unemployment and under employment remain a major problem. The NREGP is a one step in the direction of assuring regular employment and income to the rural poor. Further, steps further steps must be initiated to set up small industrial units at village level so as to create non-agricultural employment in the rural areas. This step would prevent the migration of rural people to urban centers. Also SHGs must be promoted and through that pave the way for asset-building so as to ensure assumed income and purchasing capacity of the poor facilitating food security.

6.4 Identification of persons Below Poverty Line (BPL)

The village panchayat at the grass root level are the right type of institutions to identify the really poor, who need help from the government. So the involvement of the village panchayat in identification of those below poverty line is necessary.

6.5 Efficient management of public distribution system

Here again in managing efficiently the public distribution system, the village panchayats need to be well empowered with necessary authority and resources.

6.6 Price control

The burning problem now is one of mounting price level of different foodgrains as well as other consumption items causing hardship to the consumers, particularly low and middle income group. Apart from adequate production of foodgrain and other items, prices levels need to be regulated. It should be made obligatory on the part of every retail shop to display information such as list of all items sold, the wholesale price, the profit margin and the price at which they are sold, the quality of the item and similar such information. This method serves as an effective step to control the price level in the market.

Conclusion

For sustainable food security integrated approach addressing simultaneously, adequate production, procurement, preservation, public distribution and price control become necessary. Food security no doubt requires adequate production of foodgrains and other essential items. But that alone is not enough. The consumer must have necessary purchasing capacity and this depends upon the price level of different goods services required by the consumers and the prices must be well regulated by the government. Ensuring justice to both the consumer and seller is one way of regulating the price level. It should be obligatory on the part of every businessman to display the list of prices of different items sold. If all merchants display such information there is no scope for manipulating the price levels. This is one way of checking the price rise. Also government must procure sufficient stocks and keep the same in godowns to ensure regular supply in the market of different items.

Public Distribution System should be well managed in terms of supply of all essential consumable items to all those poor category of people at subsidized prices. The recommendation of the National Advisory Council (NAC) of the extension of universal food entitlement to either one-fourth of the poorest districts or one-fourth of the poorest blocks in the country merit serious consideration. At the same time, as NAC recommended, in the remaining areas the *status quo* of supply of 35 kgs per house hold per month must continue. Ensuring food security is the responsibility of the Welfare State and if this is not fulfilled there is bound to be unrest in the society.

REFERENCES

1. *FAO, State of Food Insecurity in the World, 2001 Report,* Penguin Books, 2002, pp. 70-72.
2. *A Report by World Food Programme,* February 20, 2009.
3. Swaminathan, M.S., *A Century of Hope, Harmony with Nature and Freedom for Hunger,* East-West Books Pvt. Ltd., Chennai, 1999.
4. *Ibid.*
5. *Eleventh Five Year Plan Report,* Vol. II, p. 128.
6. *Eleventh Five Year Plan Report,* Vol. III.
7. *Ibid.*, pp. 128-129.
8. Swaminathan, *op. cit.*
9. *Ibid.*, p. 130.
10. A.P.J. Abdul Kalam, *India, 2020.*
11. *Ibid.*, p. 131.

29 RURAL DISTRESS AND AGRARIAN CRISIS

1. Introduction

Farmer continues to feed and clothe the world. Despite recent advancements in science and technology, perfect substitutes to foodgrains are not forthcoming. Farmer alone must produce the required quantities for food for the fast growing population and required raw material to the expanding manufacturing industries. But disappointingly, all is not well with agriculture and security of the farmer. In a predominantly agricultural country like India, disappointingly there is rural distress and agrarian crisis.

2. Rural Distress Defined

Rural distress, in the sense of high degree of suffering of rural poor, is the outcome of agricultural and agrarian crises. Subtle differences between agricultural and agrarian crises may be identified. Agricultural crisis may be taken to mean failure of crop production due to adverse climatic conditions; decline in productivity levels of different crops due to degradation of different basic resources, use of inappropriate technology; high degree of input cost, making agriculture unrewarding. On the other hand, the agrarian crisis is structural and institutional in nature as evident from growing marginalization and failure of support systems resulting in serious distress of the farming community. It is evident that agrarian crisis is broader in nature, than the agricultural crisis. Agricultural crisis can as well be attributed to the agrarian crisis to a large extent as the control over resource use, rate of adoption of technology and technological advancements are dependent on structural and institutional changes in agriculture. These two crises together contribute for rural distress and as such they are interlinked and have to be addressed together.

3. Nature of Rural Distress

The most serious manifestation of rural distress is the phenomenon of farmers committing suicide. The nature of rural distress and farmer's suicide need to be well analyzed. Over 65% of population in rural areas have agriculture as the main source of employment and living. Over a period of time, the percentage of rural population in the total population is getting slowly reduced. 82.7% of people were classified under rural population in 1951. In 2001, rural population has come down to 72.2%. The most disturbing factor is the percentage of cultivators in the total agricultural workers is getting reduced. In 1951, cultivators constituted 71.9% and this has come down to 54.4% as per 2001 census. Along with fall in the percentage of cultivators, there has been increase in the percentage of agricultural labourers. In 1951, agricultural labourers constituted only 28.1%. In 2001, agricultural labourers rose up to 45.6%. It is evident that for one reason or another marginal farmers are converting themselves into agricultural labourers. Over the 5 decades, the average size of holdings has been coming down. In 2002-03, the average size of holding was 1.06 hectares. In 2008, it is less than 1 hectare. The farmers are forced to cultivate uneconomical size of holdings and as such, cultivation has become unrewarding. An estimated 27% of farmers did not like farming because it was not profitable. In all, 40% felt that given a choice, they would take up some other career.[1]

4. Factors Accounting for Rural Distress

A research study observes "Rural distress happened due to various factors such as falling employment, declining productivity and profitability, rising indebtedness so as to prevent minimum investment in agriculture;[2] Other factors contributing to rural distress are: shift in cropping pattern towards cash crops, lack of level playing field for farmers in the global markets; increased dependence on high cost inputs which increases the cost of cultivation and indebtedness; enhanced risks, falling profitability and declining public support.[3]

Farmers' suicides could be observed in different states. Many factors account for farmers' suicides. Indebtedness of the farmer seems to be a major factor for farmers resorting to suicides. Table 29.1 presents degree of indebtedness in relation to size of holdings in different states.

Table 29.1

Incidence of Indebtedness Based on Size of Holdings

Sr. No.	State & All-India Farmer Households (upto 1.0 ha of land)	% of Marginal Indebted Households (upto 1.01 to 2.00 ha of land)	% of Small Indebted Farmer Indebted Farmer Households (2.01 to 4.00 ha of land)	% of Semi-medium Farmer Households (4.01 to 10.00 ha of land)	% of Medium Indebted Households (upto > 10.00 ha of land)	% of Large Indebted Farmer
	1	2	3	4	5	6
1.	Uttar Pradesh	71.3	17.4	7.8	3.4	0.3
2.	Maharashtra	36.0	26.2	23.3	12.2	2.4
3.	Madhya Pradesh	33.0	27.1	23.1	13.0	3.9
4.	Rajasthan	43.9	19.8	17.8	14.1	4.5
5.	Karnataka	50.7	22.8	15.9	9.3	1.2
6.	Andhra Pradesh	55.7	21.8	15.1	6.6	0.7
7.	Bihar	86.9	9.2	2.8	0.7	0.6
8.	West Bengal	88.7	8.5	2.4	0.4	0.0
9.	Punjab	53.3	15.8	17.0	11.8	2.2
10.	Orissa	70.3	20.6	7.3	1.7	0.0
	All India	61.0	18.9	12.5	6.4	1.2

Source: Directorate of Economics & Statistics, Ministry of Agriculture, *Agricultural Statistics at a Glance, 2008*, p. 315.

From the Table 29.1 it is clear that all categories of farmers are in indebtedness. We can clearly notice that the percentage of incidence of indebtedness is of higher degree in the case of marginal farmers and small farmers as compared to farmers of higher size of holdings. In the case of marginal and small farmers the land holdings being tiny, agriculture may not be a paying proposition.

5. Extent of Farmers' Suicides

According to the National Crime Records Bureau, between 1997 and 2007, the number of farmers committing suicide stood at 1,82,936 for the country as a whole. The five worst-affected states include Maharashtra, Andhra Pradesh, Karnataka, Madhya Pradesh and Chattisgarh. These states accounted for two-thirds of the farmer suicides in the entire country during 2007, alone. It is surprising to note that the farmer suicides are reported even in prosperous states like Punjab, Tamil Nadu and Kerala.

The geo-physical conditions vary in the country and as such agricultural situation also varies from State to state and also region to region. Over 60% of India's cultivated area is rainfed and is untouched by the fruits of the Green Revolution. Even in the so-called prosperous zones, the farm crisis is reported due to high input costs, lowering of water table, degradation of soil fertility and unrewarding pricing mechanisms.

6. Preventive Steps

M.S. Swaminathan in his report clearly projected that the Minimum Support Price fixed by the government is lower than the cost of production per quintal incurred by the farmer. It is also pointed out in the report that the capital formation in agricultural and allied sectors in relation to GDP started declining. The capital formation in agriculture which was 1.6% of the GDP in 1993-94 had declined to 1.3% of the GDP by 2000-01. It is very disappointing to note that the public investment on agriculture is declining. The increase in the private investment is not able to compensate the declining public investment on agriculture. As noted in Table 29.1, the indebtedness of small and marginal farm families is of higher magnitude.

Here one factor should be noted. Generally the marginal and small farming cultivation is family labour-based. If there is no farm work marginal farmer is quite willing to serve as agricultural labourer and make a living. Therefore, small and marginal farmers though are in indebtedness they may not resort to suicide. In the case of large holdings, agricultural operations are normally mechanized. They may borrow for investment purpose or because of the need for health care. In the case of large holdings, generally suicide is not attempted. It is only medium size holding farmers who resort to suicide for more than one reason. They do not want to work as agricultural labourer. Their cost of cultivation is rising because of input costs. They are not able to sell their produce at remunerative prices. Under these circumstances, for such category of farmers there is debt burden and in a situation of inability to repay, some medium-sized holdings' farmers may resort to suicide (out of mental torture leading to depression.)

The tragic incidents of farmer's suicides in some of the states have been a matter of serious concern. A study[4] with reference to suicide in Andhra Pradesh has identified crop losses, consecutive failure of monsoon, recurrent droughts, mounting debts, monocropping, land tenancy, as some of the main causes which led many distressed farmers to commit suicide. Of the total number of suicide cases reported, 76 per cent of the victims were dependant on rainfed agriculture and 78% of them were small and marginal farmers. An important finding of the study was that 76 to 82 per cent of the victim households had borrowed from non-institutional sources at he interest rates charged on such debts ranged from 24 to 36 percent. The study has recommended several measures to tackle the situation. These include improvement in irrigation coverage, crop diversification, promotion of animal husbandry as an alternative source of income, better accessibility to institutional credit and overall improvement of the marketing infrastructure.

The Indian agriculture is passing through a very critical phase as reflected by declining foodgrain's growth rate. The annual foodgrain growth rate during 1980-81 to 1989-90 was 2.71% and this percentage declined to 1.30% during 1990-91 to 2000-01. It is reported that various factors contribute for agrarian crisis, such as inadequate institutional credit, poor extension services and low coverage of crop insurance.

It is noticed in some studies that farmers report to borrowing and use the amount mainly to give dowry or meeting the extravagant expenses to celebrate the marriage in the family and similar such unproductive expenditure. Such type of expenditure of the borrowed amount becomes a great burden, leading to high degree of indebtedness. A study by the Institute of Development Studies, Madras, states that in the major states where high degree of suicides noticed, it is observed the farmers who have invested huge amounts on commercial crops involving high cost of different inputs, resorted to suicide is a situation of not getting remunerative price for the produce and being in a helpless state in repaying loan.

Suicides in general, among the population as a whole, are also largely concentrated among males. Normally, male member is the head of the family and he takes the responsibility of borrowing and repaying of the loan amount. In a situation of inability to repay, the concerned male member resorts to suicide.

The Government being quite concerned with the problem of farmer's suicides, in the Budget of 2008 announced certain remedial measures. Writing off agricultural loans taken before March 2007 and became overdue as on December 2007 and not repaid as on Feb. 28, 2008 is one such measure. Even after this measure, farmers' suicide did not stop. After this measure, no doubt, there was decline in the suicides to the extent of some 360 only. Writing off loans step applies to amounts borrowed from the Banks, but not money borrowed from the private moneylenders. Even now private money-lenders continue to be one of the main sources of loan. Effective measures must be taken to prevent farmers from borrowing from the private moneylenders. It requires provision of adequate institutional credit to all categories. There has been a gradual decline in the flow of institutional credit to agriculture, for both short and long-term needs which has adversely affected the small farmers. In recent years, in the name of amalgamation of banking, there has been reduction in the number of branches of commercial bank branches in the rural areas. The number of rural bank branches decreased from 34,867 in 1990 to 32,386 in 2003. The share of scheduled commercial banks in the agricultural credit declined from 18% in 1987 to 11% in 2004.

Mahatma Gandhi National Rural Employment Guarantee Programme launched by the Central Government in 2005, aims at assuring employment to the rural poor for 100 days and that way help the rural poor to overcome distress conditions. Apart from this scheme, Government had launched several social and economic security schemes aiming at poverty eradication, generation of employment and livelihood security.

Conclusion

Many factors account for rural distress and agrarian crisis. Among other factors, the macro level policies of the government, gradual decline in the public sector investment in agriculture; reduced public expenditure on health and education; absence of protection to the farmers from market volatility, inadequate institutional credit, particularly to small and marginal farmers; change in the cropping pattern from multiple cropping to monocropping practice; high input costs; mounting indebtedness of the farmers account for rural distress and agrarian crises. An assessment of agriculture credit situation brings out the fact that the credit delivery to the agriculture sector continues to be inadequate, particularly to small and marginal farmers. These categories of farmers are often forced to borrow from moneylenders at a very high rate of interest, apart from many exploitative deeds by the moneylender. The pressure and humiliation from the moneylenders drive some farmers to resort to suicide. Apart from borrowing for agricultural operations, mounting indebtedness is also due to a number of other pressing needs for cash such as expenditure on health, education of the children, marriage expenses etc. The state must take effective steps to ensure remunerative prices, for the produce, provision of timely and adequate institutional credit, marketing facilities including cold storage, and provision of supply of all needed inputs, including irrigation. The farm producers need necessary awareness and technical skills of making farming a paying proposition. Extension services and effective supervision over the end use of credit go a long way, well preparing the ground for making agriculture a paying enterprise.

REFERENCES

1. *World Development Report*: Agriculture for Development, 2008.
2. Nagaraj, K., 2008, Farmer's Suicides in India, Magnitudes, Trends and Spatial Patterns, *Macrascan.*
3. NCEUS (2007), *Report on Conditions of Work and Promotion of Livelihoods in the Unorganized Sector.*
4. Pawar, Sharad (2005), Speech at the Meeting of the National Development Council, New Delhi, June 27.

PART VI

ERA OF NEW ECONOMIC POLICY

30 POLICY OF LIBERALIZATION AND ITS IMPACT ON DEVELOPMENT

1. Genesis of GATT/WTO

After prolonged discussion and consultation with all concerned, India signed the General Agreement on Trade and Tariff (GATT) on April 15, 1994. Arthur Dunkel, the then Director-general played a crucial role in drafting the Agreement and GATT Agreement took the shape of World Trade Organization (WTO) from January 1, 1995. The primary objective of the GATT/WTO is to liberalize the international trade by releasing the tariffs and others so as to enable free flow of goods and associated services across the countries. Presently, through the WTO some 90 per cent of the world trade is carried on.

In addition to being a forum for trade negotiations, WTO is akin to the International court which resolves trade disputes in-between member countries. The WTO has a status similar to the World Bank and International Monetary Fund (IMF). Since the formation of GATT, new issues cropped up and fresh negotiations have been held from time, to time resulting in major departures from the earlier agreement.

2. Major Departures in WTO

(a) While the earlier GATT provisions confined mainly to trade in manufactured goods, now trade in agriculture too received importance.

(b) **Trade Related Intellectual Property Rights (TRIPS):** The TRIPS covers nine types of intellectual property copyrights: (i) trademarks, (ii) trade secrets, (iii) geographical indications and industrial designs, (v) layout designs (topographic) of, (vi) integrated circuits, (vii) patents, (viii) micro-organisms and (ix) plant varieties. The Patent Right has been extended from 7 and 14 years to an uniform period of 20 years for all commodities. Earlier, only process was patented and now product is also patented. The patented product can be exported after the royalties are paid.

(c) Some services like banking, insurance and telecommunications are also included.

(d) **Trade Related Investment Measure (TRIMS):** It means free inflow and outflow of capital of Multinational Corporations (MNCs) without restrictions on local use of materials or export obligation under the WTO.

Opinions are divided on the desirability of different provisions of GATT/WTO. It is argued that some of the provisions are matters of great concern about the very sovereignty of the member countries, particularly a reputed trading country of long standing like India. There are misgivings relating to inclusion, particularly of trade in agriculture, services and intellectual property rights of certain items. The TRIPS Agreement excludes plants and animals from patentability, but requires patenting of "micro-organisms" and "microbiological" process. It is said that the issue of patent is a very tricky one. Patents are given for 'Inventions' and not for 'Discoveries'. In the field of biochemistry there seems to be very thin line dividing the two. Further, there is no clear definition of patentable forms like genetically modified micro-organisms and microbiological processes.

It is to be noted that in the case of plant varieties, the TRIPS Agreement does not lay down the scope of the specifications and standards for protection. Further, the Agreement says that the provisions relating to these may be revised if necessary. At this stage, what is essential is that India must be cautious in properly assessing the impact of WTO Agreement on Indian economy in general and agriculture in particular and take necessary corrective measures.

3. Impact of WTO Agreement on Indian Agriculture

Among others, the important provisions of GATT which influence the agricultural economy and rural development are as follows:

(1) Reduction in agricultural subsidies, (2) Patenting of Seeds, (3) To import 3% of domestic demand for agricultural products and (4) Reduction of Public Distribution System activity, confining subsidized food only to those eligible on well defined criteria of nutritional requirements.

At this stage, it is worth noting the observation made in the Eleventh Five Year Plan report, recent trends that have raised concern regarding food security, farmers' income and poverty are:[1]

(a) Slowdown in growth

(b) Widening economic disparities between rain fed and irrigated areas and.

(c) Increased vulnerability to world commodity price volatility following trade liberalization. This had an adverse effect on agricultural economies of regions growing crops such as cotton and oil seeds.

A research study reveals that due to liberalization of trade, the dumping of China Silk and reduction of import duty on silk from 100 to 30% have adversely affected the Mulberry farmers in Anantapur district of Andhra Pradesh.[2] Globalization has opened up the Indian Economy to international markets. Dumping of certain crop products and reduction of import duties on certain others adversely affected the economic condition of farmers in the country.

The New Economic Policy is expected to promote exports of horticulture, sericulture, aquaculture and other agricultural products and get better prices in international markets. But the export potential of horticultural crops would be realized only if adequate infrastructure in terms of rural transport and communication, cold storage facility etc. is available. Such infrastructure is not available in rural areas. Further, a rise in exports of horticulture crops is not without adverse effects on the poor people. As a consequence of shifts in area from food to horticulture crops, there is bound to be decline in foodgrain production. This is bound to result in price rise of foodgrains and erosion of wages of the poor as revealed by a study of G. Parthasarathy.

A Research study[3] pertaining to drought-prone Anantapur district relating to the impact of globalization reveals that the sericulture farmers in the district were adversely affected by the import of China Silk and reduction of import duty on Silk. A number of industries were closed down due to this, which intensified the unemployment problem in the district. For instance, the closure of industries in the government and private sectors such as the Guntakal Spinning Mills, Andhra Pradesh Lightings, Allwyn, The Bharat Gold Mining Limited, Anantapur Cotton Mills and Pattabhi Forge has thrown thousands of people out of jobs. Again, closing down of a sugar factory led to loss of hundreds of jobs in Hindpur Area. A Study on Agrarian crisis and farmers' suicides[4] pointed out that in Anantapur district more than 70% of oil mills of groundnut, the major crop in the district, were closed down in a situation of inability of the farmers to face the competition from imported products under the policy of globalization. The poor resource base of the district offers very limited opportunities of employment in non-agricultural activities. Many of the workers thrown out of jobs had taken to other occupations such as auto driving and with meagre incomes from these activities their life became miserable.

4. Subsidy Discipline

There is a controversy centered round the provisions of the GATT relating to the subsidies. The subsidies fall broadly into 3 categories: (1) Input subsidization of fertilizer, electricity, canal irrigation and seeds, (2) Export subsidies and (3) Subsidy in government services and programmes such as research programmes, marketing services and infrastructure services.

In India, subsidy on all the 3 items is less than 10%. As per the GATT Agreement developing countries can subsidise their farmers up to 10% of the total value of the product. There is, therefore, no obligation for India under the Treaty to reduce any of the subsidies given to their farmers.

In recent years, there have been attempts of phasing out of input subsidies in India. The fertilizer subsidy which accounted for 3.2% of GDP in 1990-91 was brought down to 2.5% by 1997-98 and its was further reduced to 0.69% of GDP by 2003-04. The subsidy on electricity was also reduced by different State governments in the name of power reforms involving private investments. The water rates for irrigation were also increased to recover the operation and maintenance costs. The trade in seeds and seed industry was thrown open to private trade and foreign investment allowing 100% foreign equity and relaxing the restriction on import of seeds. The reduction in subsidies meant higher cost for the farming community.

5. Institutional Credit during Post-reform Period

There has been a decline in the flow of institutional credit to agriculture during the post-reform period due to the policy of dilution of priority sector lending. The direction to scheduled commercial banks including Regional Rural Banks (RRBs) to adhere to commercial performance acted as a severe constraint on bank credit to agriculture. The number of rural bank branches decreased from 34,867 in 1990 to 32,386 in 2003. The share of Scheduled banks in the agricultural credit declined from 18% in 1987 to 11% in 2004. During the post-reform period, there has been a decline in the flow of institutional credit to agriculture for both short and long-term needs; which has adversely affected mostly small and marginal farmers who have meagre resources at their disposal, mainly depend on credit for agriculture operations. As a result, the small farmers are forced to borrow from private moneylenders at higher rate of interest. That apart the moneylender exploits the poor borrower in different ways such as expecting some service from them and forcing them to sell the produce at a very low price. The proportion of institutional credit to total farmers was about 50% in 2008-09 and the rest was from non-institutional sources. It means private moneylender still plays a dominant role in rural credit.

6. Patenting of Seeds

The Agreement on Patents is in substantial variance from the Indian Patent Act of 1970 and therefore, had given rise to a lot of controversy in India. As per the Indian Law, agricultural products, including seeds and plants, animals and all life forms including microorganism and microbiological process are not patentable There is a lot of ambiguity associated with the concepts of micro-organisms, microbiological processes and sui generic system. Often these concepts are given different interpretation to suit the interests of outsiders and is detrimental to the interests of farmers in the country.

7. Imports to the Extent of Three Per Cent of Domestic Demand

The obligation to import three per cent of the agricultural products required, often, is detrimental to the interests of domestic producers. For example, china rice is allowed to enter into India as per WTO agreement and the rice is sold at ₹ 10/- a Kg. This is understandably detrimental to the interest of Indian farmer who cannot sell at this rate. Under the condition of uneven and slow development of technology inefficient use of available technology and inputs, lack of adequate incentives and appropriate institutions, degradation of natural resource base and rapid and widespread decline in groundwater table, will particularly have adverse impact on small and marginal farmers, the Indian farmers is not in a position to compete with farm producers outside and sell the produce at a competitive price.

Import of agricultural produce, particularly foodgrains under the WTO agreement has turned out to be an important reason for recent farm distress. As stated in the Eleventh Plan Report, "More generally, farmers are now subject to greater risk because variability, farmers are now subjects to greater risk because variability of world prices is much higher than what Indian farmers have been used to in the past."[5]

8. Food Subsidy

The issue of food subsidy has been brought under the purview of WTO. In India Public Distribution System (PDS) has been in operation since many decades. The main objective is to make available selected foodgrains through a network of Fair Price Shops to the vulnerable section of the society at reasonable prices. This would also act as an effective instrument of price stability ensuring food security. The PDS is supposed to insulate poor or weaker sections of population against inflation. In India still some 30 per cent of population is below the poverty line in the ghost of foods society haunting than with the poverty line with the gross of food scarcity haunting them with persistence of nutritional gap. Per capita calorie intake India continues to be lower than what it is at global level and at the Asian level. The average intake of protein in India again is much less and its availability is sharply declaring. The problem in India is one of poverty and lack of purchasing capacity. Under these circumstances the really poor need to be identified and supplied food stuffs at subsidized rates through well managed PDS. Table 30.1 shows the food subsidy available to the farmers.

Table 30.1

Agricultural Subsidies during 2000-01 to 2006-07

(In crores)

	Item	2000-01	2001-02	2002-03	2003-04	2004-05	2005-06	2006-07
1.	Food Subsidy	12060	17499	24176	25181	25798	23077	24014
2.	Fertilizers (Total)	13800	12595	11015	11847	15879	18460	
3.	Electricity	8919	10410	8521	14544	17852	20301	
4.	Irrigation	13259	13009	12794	10921	12508	14625	
5.	Other Subsidies given to Marginal Farmers, Cooperative societies in the form of seeds, development of oil seeds, pulses, cotton, Rice, Maize and Crop Insurance	2733	3234	3173	4132	3598	6504	
	Total	**50771**	**56747**	**59679**	**66625**	**75635**	**82967**	**24013**

From Table 30.1 it is clear that food subsidy has been on the increase from year to year. But in 2005-06 it declined to ₹ 23,077 from ₹ 25,798 crore in 2004-05. In 2006 -07 it again increased to ₹ 24,014, This amount however, is less than the subsidy amount of ₹ 25,798 crore in 2004-05. Subsidies on Food, Subsidy on Fertilizers, Electricity, Irrigation have been on the increase.

Table 30.2 shows agricultural GDP Growth trends from 1991-1992 to 2006-2007.

Table 30.2

Agricultural GDP Growth Trends from 1991-92 to 2006-07

(In Percentages)

	Period	Total Economy	Agriculture and Allied Sectors
Early Reform Period	1991-92 to 1996-97	5.67	3.66
Ninth Plan	1997-98 to 2001-02	5.52	2.50
Tenth Plan Period	2002-03 to 2006-07	7.77	2.47
Of which	2002-03 to 2004-05	6.60	0.89
	2005-06 to 2006-07	9.51	4.84

Source: *Eleventh Five Year Plan Report*, **p. 4**

From Table 30.2 it is evident that the growth of agricultural GDP decelerated from 3.66% (Early reform period) to 2.5% in the Ninth Plan, and subsequent economic reform period. This deceleration of agricultural of GDP during the reform period may be due to various factors, such as unfavorable monsoon, soil degradation and inappropriate technology applied. However, it is evident, that new economic policy of liberalization has not contributed for the betterment of agriculture sector as expected. India's membership in WTO will not come in the way of appropriate steps to step up growth rates of yield of different crops. India has freedom to act for the betterment of agricultural sector. Through appropriate policy of subsidy, input supply must be ensured at reasonable price paving the way for higher yields of the crops, diversification of production and higher level of agricultural output. Removing the input subsidies and adjusting it on producer prices may not be a feasible method to encourage crop diversification or even to retain the suitable cropping pattern that has already been achieved in certain regions.[6]

9. Expectations from the New Economic Policy

The change in the economic policies was found necessary and expectations from the New Economic Policy are many. The crises in the balance of payments, huge external beds, inflation of high magnitude by 1991, necessitated the government of India to Yield to the advice of IMF and World Bank to adopt a new economic policy of Liberalization. The performance of public sector was not functioning in a manner expected of it. The New economic policy of liberalization is expected to (i) increase competition and give the consumers higher equality goods and services at lower prices,

(ii) debureaucratise economic activity and promote entrepreneurship and individual initiative, (iii) increase productivity, eliminate social waste and create additional wealth with the utilization of advances in science and technology and (iv) enable utilization of the vast potential of the global market in terms of export promotion and development of competitive spirit of high order.

Ever since India signed the GATT, there has been controversy mainly centered round the patents right and intellectuals property rights relating to seeds, subsidy, public distribution system, role of multinationals in India and corporatization of agriculture. Those favoured India joining WTO, believe that liberalized trading system under WTO is "Silver lining for exports" of Indian agriculture, However, those opposed to it, view the whole exercise as an unholy alliance between the unequals and a planned design of subjugation of Indian interest. With advancement of information technology, different nations are brought closer towards "one world". It is but appropriate, that Indian economy is integrated with the global economy. Dr. Abid Hussain, who has authored a report on small-scale industries, reforms, describes, "globalization as a process driven by the forces of science and technology and no country can remain isolated from them."[7] However, India should make sure that her interests are not jeopardized in any way as a member of the WTO.

India has certain basic advantages. It has her own well proven native technology and plant breeders who have evolved new strains of high quality. Further, the very geographical size of India, the size of her population wide domestic market for different goods and services provide strength to her trade and bargaining capacity in international markets. In recent years a new brand of entrepreneur class from the rural sector is emerging and this is attributed to liberalized economic policies. Of late, there has been a trend towards non agricultural occupations within the rural sector. Also, trend towards dairying, poultry, cultivation of grapes and other horticultural crops is a clear indication of emergence of entrepreneur class in the rural areas. Liberalization and globalization policy provide ample opportunity for the growth of entrepreneur class in rural areas. Emergence of such promising entrepreneurs put an end to the investment flight from rural to urban and migration of people from rural to urban centers.

10. Limitations of Policy of Globalization and Privatization

The expected benefits out of the policy of liberalization of WTO are based on many conditions and assumptions linked with individuals enterprise qualities and investment capacity. The agriculture sector continues to suffer for various reasons including lack of adequate investment, particularly, on infrastructure development at village levels. Creation of infrastructure at village level such as roads and communication, electricity, godowns, processing centres cold storage facilities and market facilities. In recent years, for various reasons, there has been gradual decline in public investment on agriculture. The increase in private investment is not adequate enough to compensate the extent of fall in the public investment. G. Parthasarathy rightly identifies the problem and states that export potential of horticulture crops would be realized only if adequate infrastructure is built up. Further, a rise in protection of horticulture crops is not without adverse effects. As a consequence of shift in area to horticulture crops there is bound to be decline in foodgrain production and increase in further market dependence. This may result in price rise of foodgrains and erosion of wages of the poor.[8]

Special Economic Zone, an outcome of New Economic Policy of Liberalization is subject to criticism. Some research studies noted that SEZs are detrimental to the interests of the farming community and food security in the country. Under liberalization policy, there is a tendency to shift the cropped area from the foodgrains to high value export crops and this is bound to create food shortage. Food security to people is more important and pressing than earning through export of food grains and other agriculture products. India has not reached the take-off stage in agricultural production so as to concentrate on exports leaving behind millions who suffer from chronic hunger. Under these circumstances, understandably, there is bound to be resistance for export of foodgrains.

11. Withdrawal of Subsidies and Small Farmers

In India nearly 70% of the cultivators are of small and marginal holdings category with less than 2 hectares. The benefits of globalization, if any would only go to small number of farmers of large size holdings. The farmers of small holdings being subsistence cultivators, have little marketable surplus. Any attempt of withdrawal of subsidy even partly on fertilizer or electricity, would cause a great hardship to the small and marginal farmers.

It is feared that liberalization policy benefiting the rich farmers would further accentuate the existing disparity in the income levels between small and large farmers.

It is to be noted that some of the developed countries continue to provide subsides to fertilizer and resort to raising high tariff walls to prevent imports from the developing countries. India must be well guarded against such attempts. The Indian farmers should not be placed at disadvantage in matters of enjoying subsidy and tariff benefits.

12. Protecting Indian Products

Each country has its own Patent Laws including India. India is entitled for TRIPS in certain items. In India, Scientists and Plant Breeders have evolved new strains such as Basmati Rice, Darjeeling Tea, Neem and Turmeric-based products. It is rightly pointed out that Indian products which have acquired a commercial significance by usages over time as that of Basmati Rice, etc., may be vulnerable to encroachments by others. As an example, it is stated that recently a company has filed an application in Britain for "American Type of Basmati Rice". It is necessary that we enact a suitable legislation for protecting products with geographical indication and products having specific characters in order to prevent such encroachments in future.[9]

Two decades after the new economic policy has been in vogue, the goals set remain far from met, and on the other hand, there are indications of unbalanced growth and widening of income disparities. The conclusions of some studies[10] of an autonomous organization on the outcome of the new economic policy of liberalization and globalization are worth noting. It is pointed out "Poverty and unemployment situation in the country has worsened in the nineties and income disparities have widened than never before..... These polices lead to greater inequality in the society. In 1999-2000, only 0.05 per cent of the population earned in the stock market ₹ 4,00,000 crores, while this was the income in the entire agricultural sector, on which 67 per cent of the population depends," notes Dr. Arun Kumar. It is also pointed out that in the years of liberalization, over 63 per cent of the workforce engaged in the services sector obtains a little over half of the GDP. In this context, says Dr. Kabra, "No one policy will trigger development; human needs have to be addressed directly; a comprehensive approach is needed .. sustainable development should be rooted in processes that are socially inclusive."[11] (*The Hindu,* The other face of liberalization, August 31, 2000).

13. Eco-friendly Agricultural Development and New Economic Policy

The need of the hour is sustainable or eco-friendly development, particularly sustainable agricultural development in India. Overexploitation of groundwater using pressure irrigation systems, excessive dosage of chemical fertilizer in areas of assumed irrigation, prawn farming in the coastal belt, encroachment of irrigation tanks, large-scale grazing, tree felling, hunting, clandestine timber trading and consequent damage to forest wealth, etc., endanger environment and bio-diversity. There must be a check on the investment pattern of the multinationals in India so as to ensure that there is no exploitation of nature and the poor people. With the role of many multinationals, and functioning of SEZs, there is bound to be environmental degradation in the country.

14. R&D Facility

Strengthening Research and Development (R&D) in the private sector, with adequate state support is necessary to address, among others, seed multiplication and distribution with a focus on dry-farming technology. Appropriate steps need to be taken to protect the indigenous seed industry against competition from the multinational corporations.

We cannot expect miracles to happen always through privatization. Due to privatization of Dairy sector, for example, the cooperative sector which has sustained the livelihood of landless poor and small and marginal cultivators is on the verge of collapse in some parts of India. A healthy competition between private, cooperative and public sectors needs to be encouraged and that would ensure better quality products which can enjoy international market.

15. Steps to Ensure Quality of Farm Products

To meet the interests of farming community and requirements of the international markets, the farm production needs to be diversified. Fruits, vegetables, better quality rice, prawns and other aqua-cultural products have to be produced on a priority basis. The quality of Indian farm produce is under cloud in international markets. Effective steps to ensure quality of farm produce are immediately called for to enjoy the benefits of export promotion under liberalization policy. The grade specifications and standards of agricultural products in India should be harmonized with international standards and specifications. As Gokhul Patnaik indicated, the package of practices recommended for application of chemicals and pesticides in agriculture should be suitably modified to ensure the level chemical residences in agricultural foodstuffs remained within the tolerable limits prescribed by importing countries". At the same times, export should be allowed only after meeting fully the domestic requirements.

16. Inflow of Multinationals and Impact on Economic and Social Life

It is feared that the inflow of multinationals and foreign capital under globalization and liberalization may result in capital intensive "Mass Production" with exploitative element and not labour intensive "Production by Masses" as pleaded by Mahatma Gandhi. Further, there is a fear that free flow-in of foreign capital into India, apart from adversely affecting, small producers – both farm and non-farm categories, "may create a class that becomes a part of global consumerist class with all its adverse effects on social life and well cherished moral values of India[11].

Today, India, as a member of WTO and with a policy of economic liberalization, has to face a great technological challenge. The international competition should be met through technology advancement, emergence of entrepreneurship and quality up gradation of varying product profile.

17. Export-Import Policy (EXIM Policy) and Its Impact

The EXIM policy announced on March 31, 1992, aims to further carry forward the process of liberalization. With specific reference to agriculture, the objectives of the EXIM policy (1997-2002) are stated as "To enhance the technological strength and efficiency of Indian agriculture, industry and services, thereby improving their competitive strength while generating new employment opportunities and encourage the attainment of internationally accepted standards of quality and to provide consumers with good quality products at reasonable prices".

Under the EXIM policy, it is said that now there is freedom and licensing, quantitative restrictions and other regulatory and discretionary controls have been substantially eliminated or relaxed. All goods, except those coming under the negative list, may be freely imported and exported.

The negative list consists of three categories: (1) Prohibited, (2) Restricted by Licensing or otherwise and (3) Canalized. As per the EXIM policy, effective from April 1, 1997, there are three prohibited items of import, namely, animal, fallow – fat and/or oils; animal rennet and wild animals including their parts and products and ivory. The prohibited items of export cover all forms of wild animals including their parts and products with certain exemptions; exotic birds; animal fallow – fat and/or oils excluding fish oil; beef; human skeletons; sandal wood excluding fully finished handicrafts and machine finished sandalwood products; certain wood and wood products and certain chemicals.

Through licensing the export or import transaction are restricted with certain terms and conditions stipulated by the licensing authority. The terms and conditions of licensing include (1) quality description and value of the goods, (2) actual user condition, (3) export objective, if any (4) value addition to be achieved, (5) minimum export price and, (6) the country of origin and destination of the goods.

The restricted items in the negative list of imports include consumers goods, animals, birds and reptiles; aircraft and helicopters and certain drugs. The restricted items of exports cover cattle, camel; chemical fertilizer; hides and skins; industrial leathers and paddy.

The third category comes under canalization. Canalization means that the import or export of the goods concerned may be done only through the designated public sector agencies like State Trading Corporation (STC), Minerals and Metals Trading Corporation (MMTC) of India, Indian Oil Corporation (IOC), National Dairy Development Board (BDDB) etc. There are some canalized items of imports, namely, petroleum products; vegetable oils like coconut oil, in seeds like copra and groundnut, and fertilizers. There are some canalized items of export like petroleum products; certain mineral ores and concentrates; wiger seeds; and onion.

The new EXIM policy is quite in line with the economic reforms introduced in India. The expectation is that the technological strength and efficiency of Indian agriculture along with industry and services development would enhance leading to employment generation in the rural sector. If once quantitative restrictions on import of agricultural commodities go, after India's position in Balance of Payments becomes quite comfortable, as part of EXIM policy, the apprehensions of adverse effects of such policy are genuine. India's strategy in the ongoing and coming negotiations on the agreement on agriculture in WTO should be one of protecting agriculture, rural development and livelihood of rural masses.

18. Corporatization of Agriculture

As a part of globalization and liberalization, corporatization of agriculture is advocated and promoted in certain parts of country. The case for such policy is based on: (1) economies of scale in production, storage and marketing, (2) potential for higher investment in agriculture, drawing the surpluses from industry and service sectors, (3) acceleration exports to earn much needed foreign exchange for imports, and (4) upgradation of farm technology to ensure cost-effectiveness and competitiveness in international markets.

The critics argue that the strategy if fraught with certain adverse consequences like: (1) Opening up of opportunities for individual farmers, (2) Accentuation of regional inequalities in the levels of agricultural development, (3) Curtailment of employment opportunities in the farm front, (4) Small and marginal farmers will be driven to the ranks of agricultural labourers, (5) Market dependence for food articles will increase leading to reduction in food security, and (6) Prospects for reduction in rural poverty become bleak, rendering the policy of land reforms ineffective. In short, it is felt that in the corporate agreement, equity issues become secondary.

Conclusion

A.P.J. Abdul Kalam warns that "In coming years we cannot address our agricultural problems in isolation. The WTO agreements have implications for the future course of agricultural research and development. These relate to giving market access to other countries in selling their products in India. This will place a demand on quality and efficiency in our own agricultural operations. Restrictions in terms of sanitary and phyto-sanitary measures both for import and export of agricultural commodities will be imposed. This means there will be demand that residues of pesticides and chemicals be reduced to the internationally acceptable standards.... Thus, use of agro-chemicals and fertilizers has to often conform to international specifications."[12]

In view of the adverse consequences, hasty pursuit in the matter is not warranted. The liberalization policy may be continued and at the same time the Government must be well guarded at all stages to ensure and protect the interests of all people in the country.

REFERENCES

1. *Eleventh Plan Report,* Vol. III, p. 65.
2. Sainath, "Farm Crisis" in Series on Farmers' Suicides in Andhra Pradesh, 2004.
3. G. Sridhar, "Agrarian Crisis and Farmers Suicides": A Study in the Drought-prone Anantapur District of Andhra Pradesh, A Paper presented at a National Workshop organized by Centre for Rural Studies, Lal Bahadur Shastri National Academy of Administration, Mussorie, Nov-23-24, 2007.
4. Sainath, *op. cit.*
5. *Eleventh Five Year Plant Report,* p. 6.
6. Dantwala M.L., "Prices and Cropping Patterns", *Economic and Political Weekly,* April 19, 1986.
7. *The Hindu,* "The Other Face of Liberalization", August 31, 2000.
8. G. Parthasarathy, "Effect of Central Government Trade Policy and Andhra Pradesh Agricultural Exports – An Assessment" in R.P. Singh (Ed.), *Implication for GATT/WTO on Agriculture and Rural Development,* NIRD, Hyderabad, 1998, p. 164.
9. Gokul Patnaik, "Implication of GATT on Indian Agricultural Exports", in R.P. Singh (Ed.), *"Implication of GATT on Agriculture and Rural Development",* NIRD, Hyderabad, 1998, p. 19.
10. *The Hindu,* "The Other Face of Liberalization", August 31, 2000.
11. Gokul Patnaik, *op. cit.,* p. 22.
12. A.P.J. Abdul Kalam, *India 2020,* Penguin Books, 2002, pp. 65, 66.

31 SPECIAL ECONOMC ZONE (SEZ)

1. Introduction

Special Economic Zone is a geographical region, specially identified by the government for the purpose of setting up of industries. Such zone has economic laws that are more literal than the countries typical economic laws. Usually the goal of SEZ is one of attracting investments including foreign investment and that way promote industrialization.

As early as 1980, the government of the People's Republic of China under Dung Xenoping designed Special Economic Zones and well planned to promote industrialization. This experiment of Special Economic Zone in China proved to be very successful and SEZ in a small village gradually within 20 years emerged as a city with a population over 10 million. Following the Chinese example, Special Economic Zones have been established in different countries including India, Pakistan, Russia, Iran, Jordon, Poland, Kazakhstan, the Philippines and Ukraine. In the United States SEZs are referred to as "Urban Enterprise Zones".

2. Special Economic Zones in India

India was one of the first in Asia to give thought to production of Export-related products and plan for setting up of Export Processing Zone (EPZ) model in Kandla in 1965. A number of hurdles were experienced in the functioning EPZ. With a view to overcome series of hurdles and attract foreign investments, the Government announced Special Economic Zones (SEZs) policy in April 2000. The SEZ in India began functioning from 01.11.2000, under the provision of the Foreign Trade Policy and fiscal incentives through the provisions of relevant statutes. To install confidence in investors and signal the governments commitment to a stable SEZ, a comprehensive draft bill was prepared after extensive discussions with the stakeholders. The draft SEZ Rules were given wide publicity. It is reported that as many as 800 suggestions were received on the draft rules. After extensive consultations, the SEZ Act 2005, came into effect on 10th February, 2006. The main objectives of the SEZ Act are

(a) Generation of additional economic activity;

(b) Promotion of exports of goods and services;

(c) Promotion of investment from domestic and foreign sources;

(d) Creation of employment opportunities, and

(e) Development of infrastructure facilities.

It is expected that this will trigger a large flow of foreign and domestic investment in SEZs infrastructure and productive capacity, leading to generation of additional economic activity and creation of employment opportunities. The SEZ Act 2005, envisages key role for the State Governments in Export Promotion and creation of related infrastructure.

The policy provides for setting up of SEZs in the public, private, joint sector or by state governments. For this a policy was introduced on 1.4.2006. Some of the existing Export Processing Zones as listed below were converted into Special Economic Zones.

Kandla and Surat (Gujarat), Cochin (Kerala), Santacruz (Maharashtra), Visakhapatnam (Andhra Pradesh), Noida (Uttar Pradesh), Nanganeri and Tirunalveli (Tamil Nadu). Thus, these eight export processing zones (EPZs) have been converted into SEZ. Currently India has 811 units in operation in 8 functional SEZs each an average size of 200 acres. All these SEZs are in different parts of the country in the private and joint sectors or by the State Government.

The SEZ Rules provide for different minimum land requirement for different class of SEZ. Every SEZ is divided into a processing area where alone SEZ would come up and the non processing area where the supporting infrastructure is to be created.

The SEZ Rules provide for single window clearance on matters relating to approval from Central as well as State Governments for setting up of a unit under Special Economic Zone Scheme.

Approval Mechanism: The developer submits the proposal for establishment of SEZ to the concerned State Government. The State Government has to forward the proposal with its recommendations within 45 days from the date of receipt of such proposal to the Board of Approval. The applicant also has the option to submit the proposal directly to the Board of Approval. The Board of Approval has been constituted by the Central Government in exercise of the power conferred under the SEZ Act. All the decisions are taken in the Board of Approval by consensus. The Board of Approval has 19 Members. Its constitution is as follows:

1.	Secretary, Department of Commerce	Chairman
2.	Member CBEC	Member
3.	Member IT CBDT	Member
4.	Joint Secretary, Department of Economic Affairs, Ministry of Finance, Govt. of India	Member
5.	Joint Secretary (SEZ), Department of Commerce	Member
6.	Joint Secretary, DIPP	Member
7.	Joint Secretary, Ministry of Science & Technology	Member
8.	Joint Secretary, Ministry of Small-scale Industries and Agro and Rural Industries	Member
9.	Joint Secretary, Ministry of Home Affairs	Member
10.	Joint Secretary, Ministry of Defence	Member
11.	Joint Secretary, Ministry of Environment and Forests	Member
12.	Joint Secretary, Ministry of Law and Justice	Member
13.	Joint Secretary, Ministry of Overseas of Indian Affairs	Member
14.	Joint Secretary, Ministry of Urban Development	Member
15.	A nominee of the State Government concerned	Member
16.	Director General of Foreign Trade or His nominee	Member
17.	Development Commissioner Concerned	Member

18. A Professor in the Indian Institute of Management or the Indian Institute of Foreign Trade — Member

19. Director or Deputy Secretary, Ministry of Commerce and Industry, Department of Commerce — Member

3. Functioning of SEZ

The functioning of the SEZs, is governed by a three-tier administrative set-up. The board of approval is the apex body and is headed by the Secretary, Department of Commerce. The Approval Committee at the zone level deals with approval of units in the SEZs and other related issues. Each Zone is headed by a Development Commissioner, who is ex-officio chairperson of the Approved Committee. Once the Report at Central Government has notified the area of the SEZ, units are allowed to be set up in the SEZ. The performance of the SEZ units are periodically monitored by the Approval Committee and units are liable for penal action under the Foreign Trade (Development and Regulation) Act, in case of violation of the condition of the approval.

4. Facilities and Incentives to the SEZs

The incentives and facilities offered to the units in SEZs for attracting investments into the SEZs including foreign investments include:

1. Duty free import/domestic procurement of goods for development, operation and maintenance of SEZ units.
2. 100% Income Tax exemption on exports income for SEZ units under 10 AA of the Income Tax Act for first 5 years, 50% exemption for next 5 years, thereafter and 50% of the ploughed back export profit for next five years.
3. Exemption from minimum alternate tax under section 115JB of the Income Tax Act.
4. External Commercial borrowings by SEZ units upto U.S $500 million in a year without any maturity restriction through recognized banking channels.
5. Exemption from Central Sales Tax
6. Exemption from Service Tax
7. Single window clearance for central and state level approvals
8. Exemption from state sales tax and other levies as extended by the respective state governments.

The major incentives and facilities available to SEZ developers include:

1. Exemption from customs/excise duties for development of SEZs for authorized operations approved by the GoI.
2. Income Tax exemption on income derived from the business of development of the SEZ in a block of 10 years in 15 years under section 80-1AB of the Income Tax Act.
3. Exemption from minimum alternate tax under section 115JB of the income tax act.
4. Exemption from dividend distribution tax under section 115O of the income tax act.
5. Exemption from Central Sales Tax (CST)
6. Exemption from Service Tax (Section 7, 26 and Second Schedule of the SEZ Act)
7. Exemption from Export performances

Opinion is divided on the rational and merits of SEZ. The SEZs in India are set up in different Sates cutting across political ideology. It is claimed that through SEZ industrialization can be well grounded and employment opportunities can be increased. It is agreed that agriculture is not a paying proposition and large scale migration of rural population to urban centres is an indication that agriculture can no longer accommodate any more. Industrialization, it is agreed, is the only answer and SEZs would very well pave the way for industrialization and creation of employment opportunities.

However, the States in which SEZs have been approved are facing intense protests from the farming community accusing the government of forcibly snatching fertile land from them, that too at a heavily discounted prices as against the prevailing prices, particularly in the commercial real estate industry. As a part of the economic reforms SEZ, are being set up in different states. It is reported in the press since SEZ deprives the farmers of their fertile land, it is harmful from the point of livelihood of farming community and overall agricultural production in the country. It is estimated that government acquired 5 million hectares of land for SEZ between 1991 and 2003, with all its adverse consequences. In view of mounting protests against government acquiring private lands well under cultivation some reputed companies like Bajaj and others have suggested that barren and wastelands be used for setting up of SEZs. In a situation of shortage of food grains and mounting cry for food security, converting fertile cropped land for SEZs has become highly questionable.

5. SEZs and Its Impact on Agriculture

In response to the cry of farming community, the Ministry of Commerce asked the state governments not to acquire agricultural land and also ensure prevailing market value to farmers. But yet, in Punjab, where almost the entire land is irrigated, SEZs are being set up on prime agriculture land. In Himachal Pradesh where the average size of holding is only about 0.4 hectares, the government as reported in the press is keen to convert 35,000 hectares in the Kangara valley into SEZ. Near Mumbai one of the biggest SEZ is coming up spread over 14,000 hectares predominantly on double cropped area. In Orissa despite massive protests by farmers, there is a proposal to set up steel unit under SEZ with 1,600 hectares of land well under cultivation. The CPM government in West Bengal, as reported to the press has acquired some 400 hectares of fertile land for the Tatas to set up an automobile factory at Singur, near Kolkata. In Kerala too the Communist Government is going ahead with the plan of acquiring private land for SEZ. The protest against SEZ is answered by stating that the whole scheme would ultimately benefit the poor creating employment opportunities.

Provision of assured employment and food security should be the responsibility of the State and hence, this should be the national priority. Due to gradual reduction in the area under cultivation of foodgrains, trend of the farm producers moving towards commercial crops, and migration of labour from rural areas to urban centres, the State should necessarily be alive to these problems and take appropriate steps. Industrial growth, no doubt, facilitates employment generation. But at the same time care must be taken to ensure that food production is not adversely affected, with a threat to food security. Also economic growth with social justice is the need of the hour. Keeping issues of food security and equity principle in view, SEZ must be designed. Confining SEZ to waste and barren lands preventing the domination of capitalists with role of public sector or joint enterprise of public and small investors is the right approach in the proper functioning of Special Economic Zones.

Conclusion

While the government asserts that SEZ policy had worked wonders in attracting investments, boosting exports, creating infrastructure and generating employment, civil society groups continue to be skeptical. "The governments claims are highly exaggerated on all counts", none of the milestones it has set have been achieved. Moreover, these claims will have to be weighed against the large scale displacement of farmers" says Usha Seethalakshmi who did extensive research and conducted peoples's audits on SEZs. "What is the point in claiming to have generated employment for 55,000 after five years while displacing of poor and vulnerable section ten times over" she asks! When a two-day national people's audit of SEZs was held in New Delhi, the call from the audit was a thorough review of approval of SEZs performance through a participatory public process.

REFERENCE

1. K. Venkateswarlu, SEZ: "Tall on promise, Short on Delivery", *The Hindu,* Monday, May 17th, 2010.

32 CORPORATE SOCIAL RESPONSIBILITY (CSR)

1. Introduction

Corporate Social Responsibility (CSR) is also known as corporate responsibility, corporate citizenship, sustainable responsible business or corporate social performance.[1] In so describing, the nature and scope of CSR is made clear. It is a form of corporate self-regulation integrated into a business model. Ideally, CSR policy would function as a built-in, self regulating mechanism whereby business would monitor ensuring ethical standards and international norms, keeping primarily in view the interests of environment, consumers, employees, communities, stakeholders and all other members of the public sphere. CSR is expected to promote the public interest by encouraging community growth and development and voluntarily eliminating practices that harm the public sphere, regardless of legality. Essentially, CSR is the deliberate inclusion of public interest into corporate decision making.

2. Functioning of CSR

The term CSR came into common use in the early 1970s, after many multinational corporations formed and began functioning with a spirit of understanding, keeping in view the realities. CSR has been on the agenda in India for a considerable period. Most major Indian corporations are engaged in some CSR activities. As is the case in many countries, the private sector is generally more active in this area than the public sector. Several major CSR initiatives have been launched in India since the mid-1990s. With initiative by the Confederation of Indian Industry (CII) India's largest business association, "Desirable Corporate Governance: A Code", a voluntary code of corporate governance came into existence in April 1998. A National Foundation for Corporate Governance (NFCG) has been established by the Ministry of Corporate Affairs. This is a partnership with the confederation of Indian Industry, the Institute of Company Secretaries of India (ICSI) and the Institute of Chartered Accountants of India (ICAI). The purpose of the NFCG is to promote better corporate governance practices and raise the standard of corporate governance in India towards achieving stability and growth. It is hoped that the existing gaps in labour houses, Right to Organize and Collective Bargaining, Minimum Age Convention (1973), Elimination of the Worst Forms of Child Labour (1999) etc., would be taken into consideration and necessary steps initiated by CSR.

No doubt, the business sector generates wealth and value for the shareholders, but simultaneously we see the problems of poverty, unemployment, illiteracy, malnutrition etc., facing the nation. The corporate sector is seen as widening gap between rural and urban sectors. This gap between the rural and urban needs to be bridged. The schemes of the government alone are not sufficient. The business sector also needs to take the responsibility of exhibiting socially responsible business practices that ensures the proper distribution of wealth without scope for environmental degradation and widening of inequalities in the incomes of the people in the society.

No doubt the Indian business has traditionally been socially responsible. It is not enough if selected rich business remain philanthropic. Many more must come forward to extend their helping hand in the task of eradication of poverty and social tensions in the country. R. Bandyopadhyay, Secretary, Ministry of Corporate Affairs, very rightly points out, "By exhibiting socially, environmentally and ethically responsible behaviour in governance of its operations, the business can generate value and long-term sustainability for itself, while making positive contribution in the betterment of the society. We still continue to face major challenges on the human side in India. The problems like poverty, illiteracy, malnutrition, etc., have resulted in a large section of the population remaining an "unincluded" from the mainstream. We need to address these challenges through suitable efforts and interventions in which all the State and non-State actors need to partner together to find and implement innovative solutions, the culture of social responsibility needs to go deeper in the governance of the business taking into the challenges faced in our country as well as the expectations of the society".[2]

3. Policy of CSR

Each business entity should formulate a CSR policy to guide its strategic planning and provide a roadmap for its CSR initiations, which should be an integral part of overall business policy. The Ministry of Corporate Affairs listed out core elements of CSR policy.

(i) Care for all stakeholders

The business units should respect the interests of all stakeholders including share holders, employees, customers, affected persons, if any, society at large, etc. They should take stakeholders into confidence and inform them the direction in which the company is moving and respect their views at all stages.

(ii) Ethical functioning

The companies should necessarily be guided by ethics, transparency and accountability. They should not engage in practices which are unfair, corrupt and antisocial.

(iii) Respect for workers rights and welfare

The corporate bodies should provide a safe, hygienic and humane approach and conducive environment with which workers work with dignity and honour. They should provide all employees with access to training and necessary work skills for career advancement. The workers should be given freedom of forming an association and respect the view of association and redress the grievances of the workers from time to time. They should not employ child labour. They should maintain cordial relationship with workers and provide equality of opportunities to all sections of the society without any discrimination on any grounds in recruitment, promotion and provision of service welfare measures.

(iv) Respect for human rights

Workers in any establishment like any section of the society want certain rights. Companies should respect human rights for all and avoid friction and conflict with human rights.

(v) Respect for Environment

Protection of environment is the responsibility of every citizen. All business establishments should necessarily prevent pollution of any type and ensure safety of natural resources in a sustainable manner. The land, water, atmosphere around should be well protected, responding to the challenges of climate change. Apart from these steps, at every stage, environment-friendly technologies should be followed.

(vi) Social and Inclusive Development

The corporate bodies have social responsibility. In this direction, they should undertake activities for economic and social development of communities and geographical areas all around. These include, among others, education, skill building for livelihood of people, health, cultural and social welfare, mainly targeting disadvantaged sections of society.

The Ministry of Corporate Affairs, Government of India has specifically indicated guidance for the implementation of the different core elements. The CSR policy should specify implementation strategy. The strategy should include identification of projects and activities, setting physical targets with time frame. Companies may associate local authorities, civil society and non-governmental organization. From time to time, independent evaluation need to be undertaken in the working of projects and in the light of evaluation work, necessary changes may be incorporated.

Companies should allocate specific amount in their budget for CSR activities. This amount may be related to profits after tax, cost of planned CSR activities or any other suitable parameter.

To share experiences and network with other organizations, the company should engage responsible bodies ensuring business practices and CSR activities. This step would help companies in effectively projecting the image of the company.

The companies should disseminate information on CSR policy, activities and progress to all their stakeholders and the public at large through their annual reports, website and other communication media.[3]

4. Criticism Against CSR

Inspite of all the guidelines given by the Ministry of Corporate Affairs and watching the functioning of the corporate business units, the practice of CSR is subject to much debate and criticism. (1) It is argued that in this system, there is not expected social responsibility. The company is a business unit and as such, it is guided mostly by profit motive. (2) It is a major business unit and there is bound to be environmental problem. (3) The practice of CSR is likely to lead to concentration of wealth in the selected few rich capitalists. This system is capital-intensive and as such employment generation may not be that rosy as is expected. (4) It is argued that the guidelines are simply superficial window-dressing, as an attempt to pre-empt the role of governments as a watchdog over powerful multinational corporations. (5) It is argued that the guidelines are only descriptive and enforcement is not that easy. (6) In business, they have their own business ethics which is not always transparent. (7) Under this system, there is scope for exploitation of labour in some form or the other. (8) The practice of CSR is most likely to generate inequalities of income in the society resulting in social tensions.

Conclusion

No doubt the practice of CSR helps industrialization in the country and that way generate employment. From this angle, the practice of CSR is justified. The Government must take effective steps to safeguard the interests of the local entrepreneurs and business establishments. Both Government and the public at large must act as watchdogs to ensure the CSR to function in the interests of the community at large.

REFERENCES

1. *Report from RNE* — Royal Norwegian Embassy, New Delhi.
2. Corporate Social Responsibility — Voluntary Guidelines, 2009, *India Corporate Week*, Dec. 14-21, 2009, Ministry of Corporate Affairs, Govt. of India.
3. Wood D., (1991), Corporate Social Performance Revisited, *The Academy of Management Review*, Vol. 16, No. 4, October 1991.

33 NATIONAL AGRICULTURAL INSURANCE SCHEME (NAIS)

1. Origin and Growth of NAIS

The NAIS was introduced in the country with effect from Rabi season of 1999-2000, replacing the Comprehensive Crop Insurance Scheme (CCIS) which was in operation in the country since 1985. This scheme is being implemented by the General Insurance Corporation (GIC) on behalf of the Ministry of Agriculture. The main objective of NAIS is to protect the farmers against losses suffered by them due to crop failure on account of natural calamities such as flood, drought, cyclone, hailstorm, fire, pest, diseases etc., so as to restore their creditworthiness for the ensuring season.

This new scheme is available to all farmers regardless of their size of landholding or indebtedness. It envisages coverage of all food crops like cereals, millets and pulses, oil seeds, horticultural/ commercial crops. Among the commercial crops, eleven crops namely sugar cane, potato, cotton, ginger, onion, turmeric, chillies, jute, tapioca, annual banana and pineapple are presently covered.

This new scheme operated on the basis of an area approach, i.e., defined areas for each notified crop for widespread calamities and on an individual basis for localized calamities, such as, hailstorm, landslide, cyclone and flood. Individual-based assessment in case of localized calamities would be implemented in limited areas, on an experimental basis initially, and shall be extended in the light of operational experience gained. Under the new scheme, each participating State/UT is required to reach the level of Gram Panchayat as the unit of insurance in a maximum period of three years.

The premium rates are 3.5% (of sum insured or actuarial rates whichever is less) for bajra or oilseeds, 2.5%, for other kharif crops; 1.5% for wheat and 2% for other rabi crops. In the case of annual commercial/horticultural crops, actuarial rates are being charged. A 50% subsidy in the premium is given to small and marginal farmers. The subsidy is shared equally by State/UT and Central Government. The subsidy in premium will be phased out on a sunset basis over a period of five years.

At present the scheme is being implemented by 21 States/UTs. These States/UTs are Andhra Pradesh, Assam, Bihar, Chattisgarh, Goa, Gujarat, Himachal Pradesh, Jharkhand, Karnataka, Kerala, Madhya Pradesh, Maharashtra, Meghalaya, Orissa, Sikkim, Tamil Nadu, Tripura, Uttar Pradesh, West Bengal, Andaman & Nicobar Islands and Pondicherry.

2. Pilot Scheme on Seed Crop Insurance

In order to strengthen confidence in the existing seed breeders/growers and to provide financial security to them in the event of failure of seed crops, the Government of India introduced a Pilot Seed Crop Insurance Scheme from the Rabi 1999-2000 season covering. 'Breeder', 'Foundation' and 'Certified' seeds of paddy, wheat, maize, jowar, bajra, gram, red gram, groundnut, soyabean, sunflower and cotton. The states of Andhra Pradesh, Gujarat, Haryana, Karnataka, Madhya Pradesh, Maharashtra, Orissa, Punjab, Rajasthan and Uttar Pradesh are covered under this scheme. The scheme covers all natural risks at the following stages:

(i) Failure of seed crop either in full or in part due to natural risk;

(ii) Loss in expected raw seed yield;

(iii) Loss of seed crop after harvest;

(iv) Loss at seed certification stage.

The sum insured is equivalent to the average of preceding three/five years foundation and certified seed yield of the identified unit area multiplied by 'procurement price' of the seed crop variety prevailing in the previous season by National Seed Corporation/State Seed Corporations. The Premium rates for the seeds of wheat and groundnut are 2% of the sum insured, 2.5% for sunflower, 3% for paddy, 3.5% for jowar and 5% for gram, red gram, cotton, bajra, soyabean and maize.

3. Livestock Insurance

Livestock insurance, consisting mainly of cattle insurance, is being implemented by the four public sector general insurance companies, *viz.*, (i) National Insurance Co. Ltd., (ii) Oriental Insurance Co. Ltd., (iii) New India Assurance Co. Ltd., and (iv) United India Insurance Co. Ltd. Under the various Livestock Insurance Policies, cover is provided for the sum insured or the market value of the animal at the time of death, whichever is less. Animals are insured upto 100% of their market value normally. The fall in number of animals insured in the past few years is due to reduction in the number of low value animals, such as, sheep, calves and goats which at present constitute about 25% of animals covered.

REFERENCES

1. *Indian Budget,* 2001-02.
2. *Economic Survey,* 2008-09.

NATIONAL COMMISSION ON FARMERS

1. Introduction

The Government of India had set up a National Commission on Farmers headed by a reputed agricultural scientist Dr. M.S. Swaminathan. The Commission has submitted report in 2006, with clear analysis of different issues concerning the agricultural sector and the farmer.

2. Findings of the Commission[1]

The main findings of the National Commission on Farmers are worth considering for policy making purpose.

The cost of production of different crops is invariably higher than the Minimum Support Price (MSP) guaranteed by the government. An examination of the projection of cost of cultivation for 12 foodgrain crops given by the Commission for Agricultural Costs and Prices (CACP) for the crop season 2005-06 with the minimum support price prevailing in 2004-05, clearly shows that cost of production per quintal is not covered by MSP in most states. The data for paddy and wheat is presented in Tables 34.1 and 34.2.

Table 34.1

Projected Cost of Production for Paddy (₹/Qtl)

2005-2006

Sl.No.	States	Cost of Production/Quantity	MSP/Qtl (2004-05)	Return/Qtl Over Cost
1	Andhra Pradesh	578	560	-18
2	Assam	564	560	-4
3	Bihar	512	560	48
4	Haryana	685	560	-125
5	Karnataka	602	560	-42
6	Kerala	765	560	-205
7	Madhya Pradesh	682	560	-122
8	Orissa	558	560	2
9	Punjab	481	560	79
10	Tamil Nadu	620	560	-60

11	Uttar Pradesh	511	560	49
12	West Bengal	573	560	-13

Source:

1. Department of Agriculture and Cooperation, Ministry of Agriculture.
2. Reports of the Commission for Agricultural Costs and Prices (2005-06).

Table 34.2

Projected Cost of Production for Wheat Crop (₹/Qtl)

2005-2006

S. No.	States	Cost of Production/Quantity	MSP/Qtl (2004-05)	Return/Qtl Over Cost
1	Bihar	612	640	28
2	Gujarat	617	640	23
3	Haryana	516	640	124
4	Madhya Pradesh	656	640	–16
5	Punjab	516	640	124
6	Rajasthan	525	640	115
7	Uttar Pradesh	528	640	112

Source:

1. Department of Agriculture and Cooperation, Ministry of Agriculture.
2. Reports of the Commission for Agricultural Costs and Prices (2005-06).

From Table 34.1, it is clear that except in Bihar, Orissa, Punjab and Uttar Pradesh, in all other States, the cost of production per quintal of paddy is higher than MSP per quintal. In the case of wheat, as indicated by Table 34.2, only in Madhya Pradesh cost of production per quintal of wheat is higher than MSP per quintal. In other wheat growing States position is relatively better. MSP should be regarded as the bottom line for procurement both by Govt. and private traders. Purchase by Government should be MSP plus cost escalation since the announcement of MSP. This will be reflected in the prevailing market price. The commission recommends that the government should procure the staple grains needed for PDS, at the same price private traders are willing to pay to farmers. Thus, the procurement price must necessarily be higher than the MSP and then only justice will be done to the farmer.

Aim of government purchase is to feed the PDS, while the aim of private trader is for making a profit. Thus, government purchases foodgrains for public good, while private traders purchase for commercial profit. By purchasing at prevailing market rate, government ensures that both farm families and consumers get a fair deal. The government by acting promptly must be able to check the private trade to buy and store large quantities of staple grains and sell them when the prices go up substantially.

3. Capital Formation in Agriculture

Capital formation in agriculture and allied sectors in relation to GDP started declining in 1980s. The capital formation in agriculture which was 1.6% of the GDP on 1993-94, had declaimed to 1.3% of the GDP by 2000-01. Similarly, the share of agriculture and allied sector in the total gross capital formation declined from 14.3% in 1970-71 to 7.1% in 2000-01. During the nineties, the profitability in agriculture declined by 14.2% mainly due to stagnancy of yield growth and increase in prices of purchased inputs outpacing increase in output prices. The declining profitability in agriculture led to stagnation of private sector investment in agriculture.

4. Trend in Public and Private Investment in Agriculture

Table 34.3 shows the trend in public and private investment in agriculture and capital formation in agriculture.

Table 34.3

Trend in Public and Private Investment in Agriculture

As 1999-00 Prices		Investment in Agriculture (₹ in crores)		Share in Gross Capital Formation		Investment in Agriculture as a Percentage of GDP at Constant Prices
Year	Total	Public	Private	Public	Private	
1999-00	43473	7754	35719	17.8	82.2	2.2
2000-01	38176	7018	31158	18.4	81.6	1.9
2001-02	46744	8529	38215	18.2	81.8	2.2
2002-03	45867	7849	38018	17.1	82.9	2.1
2003-04	47833	12809	35024	26.8	73.2	2.0
2004-05	43123	12591	30532	29.2	70.8	1.7

Source: Report of National Commission on Farmers.

From Table 34.3, it is clear that private sector investment in agricultures during 1999-00 was ₹ 35,719 crores at 1999-00, prices and this investment increased to ₹ 38,215 crore in 2001-02, consistently fell thereafter and reached a low of ₹ 30,532 crore in 2004-05. On the other hand, public investment which was ₹ 7,754 crore in 1999-2000, though declined in the next year 2000-01 to ₹ 7,018 crore, thereafter continued to increase and reached ₹ 12,591 crore in 2004-05. Both Public and Private investment put together, it was ₹ 43,473 crore in 1999.00 and it has come down to ₹ 43,123 crore in 2004-05. It is evident that overall investment in agriculture is not encouraging. While the investment in agriculture as a percentage of GDP at constant prices stood at 2.2% in 1990-00, it has come down to 1.7%, in 2004-05. The Commission very rightly points out "Public Sector investments are specially needed in the poorer, low rainfall areas of the country, which must now play a larger role in achieving rapid agricultural growth. These areas do not attract much private investment, which generally prefer irrigated and developed areas. It is expected that the increased public sector investment in such areas would also attract private on farm investments and also in due course increased investments in agro-based industries". The Commission further states "Investment in Watershed Development and the water saving technologies in the rain fed areas could help in improving the incomes of the rural households considerably. Access to even limited irrigation (small water ponds filled up by rain water) could overcome drought conditions during critical growth periods, which would substantially increase production and incomes". The Commission further observes, "Agricultural research which could help the farmers to diversify into higher value products and developing technologies which could reduce the impact of long dry period on crops and enable them to have a diversified income flow by mix of crops, horticulture, tree crops and animal husbandry could help in stabilizing their incomes. The need is to increase investment in local research and make the research institutions accountable".

5. Meagre Resources

Resource flow to agriculture sector is declining and indebtedness of small and marginal farm families is rising. Input costs are increasing while productivity is declining. The cost-risk-return structure of farming is becoming adverse, to over 80 million farming families operating small holdings.

6. Non-profitability of Agriculture

Non-profitability situation of agriculture as is evident from the fact that the cost of production of different crops is not covered by MSP in different states is shown in Table 34.4.

Table 34.4

Non-Profitability of Agriculture: State-wise in India

Name of the Crop	States where the cost of production is not covered by MSP, 2004-05
Paddy	AP, Assam, Haryana, Karnataka, Kerala, Madhya Pradesh, Tamil Nadu & West Bengal
Jowar	AP, Karnataka, Madhya Pradesh, Tamil Nadu and Maharashtra
Bajra	Gujarat, Haryana, Uttar Pradesh and Maharashtra
Maize	AP, Himachal Pradesh, Karnataka, Madhya Pradesh, Rajasthan and Uttar Pradesh
Ragi	Karnataka, Tamil Nadu
Tur (Arhar)	AP, Gujarat, Karnataka and Orissa
Moong	AP, Maharashtra, Orissa and Rajasthan
Urad	Madhya Pradesh, Maharashtra, Orissa, Rajasthan, and Tamil Nadu
Gram	Haryana, Rajasthan
Barley	Rajasthan

Source: From Commission Report

7. Non-viability of Farming

Fragmentation of holdings in India makes farming an utterly unprofitable activity and this has been a burning problem all though. Fragmentation of holdings has been a chronic problem in Indian agriculture. Marginal holdings of size 1 hectare or less in 2002-03 constituted 70% of all operational holdings, small holdings of sizes 1 to 2 hectares constituted 16%, semi-medium holdings of 2 to 4 hectares 4% and large holdings over 10 hectares constituted less than 1%.

The Agriculture Commission based on NSC Report shows that on average the total monthly income of farmers households for landholding upto 2 hectare was lower than the total consumption expenditure indicating the non-viable status of these farmer households.

The Commission states that "Indebtedness of farmers is rising not only because of farming related expenditure, but also because of the need for health care. "The public health care system in villages is in a state of collapse".

It is rightly pointed out by the commission: If we do not attend to the problem of small farm and landless agricultural labour families with a sense of urgency and commitment, the Indian enigma of the coexistence of enormous technological capability and entrepreneurship on the one hand, and extensive undernutrition, poverty and deprivation on the other, will not only persist, but will lead to social disruption, violence and increasing human insecurity".

8. Distress Sales

The Commission rightly highlights the problem of distress sales. "Distress sales by small/ marginal farmers to square off their debts or for immediate consumption purposes soon after harvest are quite common. It is normal for a farmer to get 10-15 per cent discounted price for spot payment for his produce. According to reliable sources, about 50 per cent of the marketable surplus of small/ marginal farmer is disposed of in this manner".

9. Inadequate Credit and Insurance

Natural calamities like drought, flood and pest infestation are serious and crippling risks. Rescheduling, restructuring of their loans are not enough. "Waiver of loans is also needed." An Agriculture Risk Fund, set up with contributions from the Central and State Governments and banks in a predetermined fashion, could provide relief to farmers in the form of Waivers in full/part of loans and interest. "Also, interest should be waived on loans in areas hit by droughts and floods and for crops under heavy pest infestation". Quite a good number of farmers are forced to drive to the state of lapsing their insurance policies after initial payments. There are provisions in the insurance laws that allow LIC to revive the lapsed policies. If revived, this would help to get large sum back in the farmers[1] accounts and give them a sense of confidence. Crop Insurance now comes only about 14 per cent of the farmers and this has to be extended covering more crops as well as cattle.

10. Poor Marketing

The inadequacies of rural marketing are to be addressed with a sense of urgency. About 70% of marketing costs are estimated to be avoidable losses during handling, storage and transport. Cold storage facilities need to be provided and rural transport must be improved.

11. Suicides of Farmers

The hardships faced by small farm families need to be well identified and rural distress and agrarian crisis well addressed as a social and moral responsibility. The solutions for the problem must be identified and effective steps need to be taken.

12. Extension Services and Input Supply

The inadequacy, if not, almost absent situation of extension services, is well known. Linkage between the research units and field is very much required. "Extension services have often little to extend by way of specific information and advice on the basis of locations". Also, good quality seeds at affordable prices are in short supply and spurious pesticides and bio-fertilizers are being sold in the absence of effective quality control systems. Needed input supply particularly in dry farming areas is very much lacking. Micronutrient deficiencies in the soil as well as problems, relating to soil physics are crying for attention.

13. Growth Rates of Foodgrain and Non-foodgrains

The productivity of different crops in India continues to be low and something is missing in our technology is evident if we compare growth rates with those in other countries. Further, gradual decline in growth rate of yields of both food grains and non food grains is a matter of serious concern. Table 34.5 presents annual yield growth rate of foodgrains, non-foodgrains and all crops.

Table 34.5

Growth Rate of Food and Non-foodgrains 1970-71 – 2000-01

Period	Yield Growth Rate of Food-grains (Annual)	Yield Growth Rate of Non-Foodgrains (Annual)	Yield Growth Rate of All Crops (Annual)
1	2	3	4
1970-71 to 2000-01	2.13	1.66	1.93
1970-71 to 1979-80	1.06	1.00	1.03
1980-81 to 1989-90	2.71	2.28	2.52
1990-91 to 2000-01	1.30	1.08	1.19

Source: National Commission on Farmers.

Table 34.6 presents position of yield levels of selected crops in India in comparison with the yield level in the World as a whole and yield level in selected exporting countries.

Table 34.6

Yield Levels in India in Comparison with Yield Positions in the World and Exporting Countries (1999-2001)

Crop	Indian Yield as a Percentage of Average World Yield	Indian Yield as a Percentage Average Yield in Top Five Exporting Countries
1	2	3
Rice	0.76	0.64
Wheat	0.98	0.81
Pulses	0.77	0.29
Seeds	0.45	0.41
Seed Cotton	0.41	0.28
Onion	0.57	0.29
Tomato	0.63	0.18
Potato	1.16	0.47

Source: National Commission on Farmers.

14. Declining Growth of Food Production

From Table 34.5 it is clear that the growth rate in foodgrains production was 1.3% and it is below the population growth rate of 1.8%. It implies human numbers are increasing faster than our capacity to realize the goal of "Food for All". At the same time, consumption per capita is not going up, due to inadequate purchasing power at the household level of power". A famine of jobs/livelihoods as a result of poor growth of opportunities for employment in the rural non-farm and off-farm sectors is leading to a "food famine: at the household level". The decline in per capita net availability of cereals and pulses over the last 15 years from 510 grams per capita a day in 1991 to 463 grams in 2004, has been unprecedented. Estimate of requirements of cereals in 2020, range from 224 million tonnes to 296 million tones. The target of 230 million tonnes for 2010, is yet to be realized and reaching the figure of 296 million tonnes in another two decades appears to be not that easy with the present state of affairs of farming community.

To double annual food gram production from the present 210 million tonnes to 420 million tonnes within next 10 years, i.e., by 2015 will call for producing of least 160 million tonnes of rice from 40 million hectares and 100 million tonnes of wheat from 25 million hectares. Pulses, oil seeds, maize and millets will have to contribute substantially to reach the target of production of foodgrains.

15. No Risk Cover

There is no Agricultural Risk Fund Policy support for risk mitigation and price stabilization is also quite inadequate. Investment in agriculture has been declining. As such cost-risk-return is becoming adverse.

16. Poverty and Agricultural Production

As per the Eleventh Plan report, the proportion of households below the poverty line was as high as 28% in 2004-05. At this rate 300 million persons come under the poor category. Also, there are regional variations in the incidence of poverty. Across the nation, the states identified as the poorest are Orissa, followed by Bihar, Madhya Pradesh and Assam. In the direction of tackling this problem, Agricultural Commission observes. "Development of agro processing is important to increase farmer's income and also to create employment. It would, however, be necessary to introduce reforms in the agriculture sector to facilitate greater private corporate sector investments in agro-processing not only in new units but also in modernizing the established units". The food processing sector is dominated by small-scale producers including traditional village industries. However, the link up of small units with the large units including MNCs has not developed adequately. A system where initial processing could be decentralized in the rural areas in small units, (creating employment and reducing transport costs) and the final processing, quality control, packaging and marketing under brand name could be done in a decentralized manner.

17. Rural Unemployment

The employment growth in the rural areas is continuously decelerating as shown in Table 34.7.

Table 34.7

All India Employment Growth Rates in Rural Areas

1972-73 to 1983	1980-81 to 1993-94	1993-94 to 1999-2000	1990-2000
Agriculture	1.59	1.38	0.20
Non-farm	4.54	3.37	2.34
Total	2.12	1.77	0.68

Source: Agricultural Commission on Farmers.

It is evident from Table 34.7 that employment growth rate has been on the decline. Non-farm sector fared well with 4.54% growth and this growth also has come down to 3.37%. In view of poor performance of agriculture in employment, the total employment growth rate is reduced to 2.12% (1980-94) and it is further reduced to 1.77% (1993-2000).

The Commission observes, "the widening disparity in per capita income between farm and other than farm sector, the very slow growth in agriculture, the declining profitability, extremely weak social security arrangements, weakening family and community based mechanism of social protection, lack of employment opportunities etc., and the rising aspirations are building up social unrest, which if not arrested, could lead to threats to internal peace and security". So much so, as per the 59th Round of NSS, 40 per cent of the farmers wish to quit farming. Commission states "We cannot be silent onlookers to a situation where 30% of India is shining and 70% is weeping".

18. Recommendations of the Commission

1. To improve the economic viability of farming by ensuring that farmers earn a "minimum net income" and ensure that agricultural progress is measured, by the advance made in improving that income. Focus more on the economic well-being of the women and focus must be on men feeding the nation rather than just on production.
2. To ensure Nations Food security and self sufficiency. In this direction, the commission recommended that "MSP to be at least 50% more than the cost of production". The "net take home income" of farmers should be comparable to those of civil servants. The Commission on Agricultural Costs and Prices (CACP) should be an autonomous statutory organization with its primary mandate being the recommendation of remunerative prices for the principal agricultural commodities of both dry-farming and irrigated areas. The scope of the MSP programme should be expanded to cover all crops of importance to food and income security for small farmers. Arrangements should be made to ensure MSP at the right time at the right place, particularly in the areas coming within the scope of the National Rain-fed Area Authority (NRAA).

In the direction of Assured and Remunerative Marketing Opportunities, 'Price Stabilization Fund' should be established jointly by central and state governments and financial institutions to protect farmers during periods of violent fluctuations in prices as in the case of perishable commodities like onion, potato, tomato.

Commodity-based farmers' organization like Small Cotton Farmers' Estates, Small Farmers' Horticultural Estates, Small Farmers' Poultry Estates and Small Farmers' Medicinal Plant Estates shall be promoted to combine decentralized production with centralized services such as post-harvest management, value addition and marketing facilitating direct farmer-consumer-linkage.

The Commission believes that "An efficient marketing system with farmers' organizations as important players could significantly add to farmers' income."

3. The banking system needs to meet the large unmet credit potential needed to raise agriculture to higher thresholds and for the growth of rural and agri-business enterprises and employment at 4% interest rate.
4. In the direction of promoting sustainable livelihoods for the poor certain steps need to be taken as follows:

   (i) Financial Services (insurance for risk, health, crop and livestock)

   (ii) Infrastructure (finance for roads, power, market and telecommunication)

(iii) Investments in human development, agriculture and business development services (including productivity enhancement, local value addition, and alternate market linkage and

(iv) Institutional development services (farming and strengthening various producers organization such as self-help groups, water user associations, forest protection committees, credit and commodity cooperatives, empowering panchayats through capacity building and knowledge centres).

The commission stated that credit without insurance is an added risk factor. Farmers need user-friendly insurance instruments covering production, right from sowing to post-harvest operations and also to cover the market risks for all crops throughout the country. Drought-prone areas should have a 4-5 years repayment cycle for crop loans, taking into account the management of risk. The scope of Agricultural Insurance Policies should become wider and should also cover health insurance, as envisaged under the "Parivar Bhima Policy". Seed companies should provide insurance in the case of GM crops. The commission recommends compensation for damage caused to livestocks or crops by wild animals.

5. Rural Energy: "A comprehensive integrated Rural Energy Programme during the Health Plan Period is needed for meeting the needs of rural families in their totality." Particular attention should be paid to renewable energy technologies like Biogas plants, Biomass gasification, Mini Hydro Power and Biofuel Technologies. The availability of energy is also essential for non-farm enterprises including agro processing.
6. Commission recommends, "Coverage of farmers, particularly small and marginal farmers and landless agricultural workers, under a comprehensive National Security Scheme is essential for ensuring livelihood security.
7. Indian Trade Organization (ITO) should be established to help the Government to operate a livelihood security box and link global policies with local action in a manner beneficial to farmers. The livelihood security box should have provision to impose quantitative restriction on imports and or/increases in import tariffs, under conditions where imports of certain commodities will be detrimental to the work and income security of large numbers of farming families.
8. Agricultural progress should be measured by the growth in the net income of farm families. Along with production growth rates, income growth rates should also be measured and published by the Economics and Statistics, Directorate of Ministry of Agriculture.
9. The Commission recommended that the Ministry and Departments of Agriculture both in the centre and the states may be restructured to become Ministry/Department of Agriculture and Farmers' Welfare in order to highlight their critical role in ensuring the income and work security of the farmers. Leading farmers should be inducted at the senior level in the Ministry for specific tasks and specific periods. It is rightly stated "Agriculture should not just be a food producing machine for the urban population, but as the major source of skilled and remunerative employment and global outsourcing hub".
10. The commission made another major recommendation, namely to include Agriculture in the concurrent list so that important policy decisions like those relating to prices, credit and trade are taken by the Government of India.
11. The 73rd Amendment Act, 1992 entrusts Panchayats "with responsibility for agriculture including agricultural extension". The Panchayats to perform this particular function,

should be empowered with needed resources, information, training and tools for discharging this responsibility.

12. The commission recommended that an "Agriculture Risk Fund" should be set up to insulate farmers from risks arising from recurrent droughts and other weather adversities.
13. It is also recommended that multi-stakeholder National Food Security and Sovereignty Board chaired by the Prime Minister be set up.
14. Again, it is recommended that state government should set up a State Farmers Commission with an eminent farmer as chairperson. The membership of the commission should include all the principal stakeholders in the farming enterprise.
15. "Increase in small farm productivity and creating multiple livelihood opportunities through crop-livestock integrated farming systems as well as agro-processing and value addition to biomass have become urgent tasks for increasing farmers income".
16. The commission recommended that appropriate steps be taken for rural non-farm livelihood opportunities by way of strengthening non-farm employment programme.
17. The Commission specially stated that prime farm land must be well conserved for agriculture and the land should not be diverted for non-agricultural purposes and for programmes like the Special Economic Zone.

The Commission apart from the above stated recommendations, also made certain other recommendations. Wherever possible landless labour households should be provided with at least one acre per household, which will give them space for home gardens and animal rearing; 10% increase in the present level of water use efficiency in irrigated project may help to provide crop life saving irrigation in large areas; establishing Livestock Feed and Fodder Corporations at the state level for ensuring availability of quality fodder and feed; Women SHGs can produce hybrid seeds on contract for seed companies, with proper technical guidance and training in seed technology, In the case of new varieties, foundation seeds could be provided to SHGs; Every family should be issued with a Soil Health Pass Book, which contains integrated information on the physics, chemistry and microbiology of the soils on their farm; A Gram Panchayat Mahila Fund should be established to enable SHGs and other Women's groups to undertake community activities that help to meet essential gender-specific needs. Joint Pattas for both houses and agricultural land are essential for women to get access to credit with alternate collateral till the Pattas are issued for women.

The Commission gives due importance to Extension and training services, retraining and retooling of existing extension personnel, promote farmer to farmer learning.

Conclusion

The recommendations of the Commission should necessarily be respected *in toto* and implement them with all seriousness. The farm producer must be assured of remunerative price reasonably higher than the cost of production.

REFERENCE

1. *"National Commission on Farmers"*, Prof. Swaminathan Report.

35 INCLUSIVE GROWTH

1. Introduction

From time to time, policy measures are being announced with catchy phrases, keeping in view the aspirations of the people. In the Nehru Era, the concept of "Socialistic Pattern of Society" became popular. Indira Gandhi introduced very popular theme, "Garibi Hatao". In the subsequent years, the concept of "Growth with Social Justice" came into vogue. The Planning Commission, had declared that the goal of the Eleventh Five Year Plan is not merely enhanced growth, but "Inclusive Growth". Now this concept is very much talked about.

During the planned development, spread over 60 years so far, there has been significant growth in different sectors of economy. India became self-sufficient in foodgrain production. Industrial production has gone up by more than ten times. In the field of transport and communication, progress has been significant. Literacy percentage has been stepped up very significantly. In the field of health and sanitation too, the progress has been quite satisfactory.

2. A Case for Inclusive Growth

In the words of Planning Commission "a major weakness in the economy is that the growth is not perceived as being sufficiently inclusive for major groups, especially, Scheduled Castes (SCs), Scheduled Tribes (STs) and Minorities. Gender inequality remains a pervasive problem and some of the structural changes taking place have adverse effect on women".[1]

Under the given socio-economic-polity structure in India, the obstacles in the path of 'inclusiveness' are bound to be many. Society is composed of diverse castes, sub-castes, religious groups with varying interests. During the days of Independence Movement, Gandhiji tried to bring people within the umbrella of nationalism. The aim of nationalism was not limited to the attainment of freedom but Gandhiji envisaged creation of a qualitatively different society to avoid caste and religious antagonisms. Gandhiji focused attention on two major issues of equal importance, namely, social emancipation and secondly struggle for political movement or political swaraj. In his Plan 'Rural Development' is an integral part of Independence Movement or political swaraj. It means social reordering and independence movement must go hand in hand. Gandhiji was well aware of the need for incorporating the marginalized sections and thus creating an inclusive society. In Gandhian Programme, abolition of untouchability occupied a central position. Dr. B.R. Ambedkar, architect of the Indian Constitution very rightly observed "we must make our political democracy a social democracy as well. Political democracy cannot last unless there lies at the base of it social democracy. Social democracy simply means a way of life which recognizes liberty, equality and fraternity as the principle of life". Dr. Ambedkar was categorical in stating that a system of equality in political life and inequality in social and economic life will never be acceptable.

3. Concept of Inclusive Growth Questioned

K.N. Panikar[2] questions "was Indian nationalism inclusive?" In his words, "In a highly differentiated society, inclusiveness is indeed a process which takes place, in three ways: Politically through common struggles, socially by overcoming of internal social barriers and culturally by identifying a common past by involving indigenous cultural consciousness. He very rightly states that appropriate steps need to be taken to see that marginalized sections identify themselves with the nation. He lists out series of emergence of movements if the marginalized section are neglected. In this connection, he makes mention of Satyasodak Samaj in Maharashtra in the 19th century, the Dravida Kazhakam in Tamil Nadu, the Sadhu Jana Paripalana Sabha in Kerala and indeed the movement led by Dr. Ambedkar. The upper caste people must necessarily be conscious of appropriate steps needed to avoid giving room for social tension and that they ensure social democracy and pave the way for 'inclusiveness'. However, still caste prejudices persist, modern economy and enhanced urbanization only ruined the joint-family system, but not the caste system as such. Caste consciousness of higher degree continues and some people continue to be bypassed and remain outside the development process.

Indian economy is a market economy. In this system, the trend is concentration of capital in the hands of selected few. Here there is only 'mass production' with environmental problems. In such system, despite labour laws, there is exploitation of labour in many ways. There is no proper relationship between labour and management. In India, agriculture is the major employer and agricultural operations are a 'gamble in the monsoon' and employment in the sector is fluctuating. Further, labour laws are inadequate and ineffective. Agricultural labour continue to be a marginal segments of the society. Thus, both in the manufacturing and agricultural sectors, labour participation is marginal. The economy operates by excluding the labour as the labour has no meaningful participation in the management of factory or land nor modern avenues are available to labour here. There is scope for gulf in the inequalities of income in this type of system. The agricultural labour and tenant cultivators are marginalized segments of the society. Thus, the labour in the manufacturing sector and agriculture continue to be marginalized and they are yet to be brought under the umbrella of 'inclusiveness' in the growth process. In India, growth in service sector is found to be significant. V. Anil Kumar observes "Corporate work cultures in service sectors do not encourage or equip the workers with the consciousness of being part of a common labour force. The faster the Indian economy grows, the more exclusionary it becomes. Modern economy and enhanced urbanization are supposed to unpack caste and religious identities and create a more secure society; it is wishful thinking that by opening up economy and increasing its economic pace we increase the pace of social change towards some variety of 'modernity'.[3]

In the field of polity, enfranchisement is taking place in India, without sufficient economic enfranchisement. With voting right to all citizens of 18 years and above and 30 per cent reservation for women in the local bodies, there is certainly participation in the polity. Very soon, reservation to women in the Parliament to the extent of 50 per cent is most likely to materialize. There is a demand for reservation for BC, OBC and minorities from the quota of women. Now what is a reality? It is only those who are financially strong are able to get into law making bodies. Even if reservation is made for BC, OBC and minorities, here again, those who can afford to spend alone can get into the position and not the poor. Economic enfranchisement is meaningful only when the poor can get elected. Will it happen?

4. Major Hurdle in the Path of Inclusive Growth

The major hurdle in the path of "inclusiveness" is poverty. The problem of poverty needs to be tackled effectively and that way ensure social democracy. Poverty is responsible for illiteracy, ill health lack of motivation for development and civic inertia. Women illiteracy is of higher degree. Hence, poverty must be addressed and that way ensure a smooth path for 'inclusive growth'. The Eleventh Plan observes 'The persistence of poverty on the scale at which it still exists is not acceptable. A decisive reduction in poverty and an expansion in economic opportunity for all sections of the population should therefore be a crucial element of the vision for the Eleventh Plan."[4]

While the literacy rate has gone up from 18.3% in 1951 to 64.8% in 2001, the number of illiterate persons still exceeds 304 million, making India the country with the highest number of illiterate persons in the world. In the direction of 'inclusive' growth illiteracy must be addressed with all seriousness. Very recently, Parliament enacted a bill ensuring free and compulsory education to all children in the age group of 8 and 14 years. It is a welcome step to eradicate illiteracy. All Government run schools would admit children without insisting on any payment. In the case of convents or corporate schools it is reported in the press that government will reimburse to the extent of 25% of admissions. It means, the rest have to pay as usual and seek admission. Understandably, poverty comes in the way and poor are denied quality education. Life expectancy has increased from approximately 32 years for both males and females in 1951 to 63.9 years for male and 66.9 years for females in 2001-06. Yet this is well below the life expectancy of around 80 years in industrialized countries and 72 years in China. Although, Indian women have now higher life expectancy than Indian men, India has an adverse sex ratio with only 933 women per 1000 men. India's maternal and infant mortality rates are much higher than those of countries in East Asia, showing poor access to essential health care services.[5]

The growth trend in agricultural production is disappointing. The Eleventh Five Year Plan observes "Agriculture has grown very slowly from the 9th Plan onwards, and this has widened the rural-urban divide and also contributed to the severe distress in certain rural areas. Some of the backward areas are yet to experience any significant growth.[6] It means agricultural sector must be brought under "inclusive growth".

5. Inclusive Growth and Agriculture

There is a strong case for the inclusion of agriculture under the umbrella of inclusive growth. The Eleventh Five year Plan indicates the following steps for strengthening the agricultural sector.

1. Growth of agriculture at the rate of 4% is critical for achieving greater inclusiveness.
2. It is important to focus research on raising the yield potential in rain fed areas and for the higher public expenditure on agricultural research.
3. Evolving suitable strategies for each agro-climate zone.
4. More effective extension services meeting the technical needs of different categories of farmers.
5. Preparation of district specific agricultural plans to identify crop productivity constraints and suitably guide the farmers.
6. More public investment to develop irrigation including modernization of the existing irrigation systems.
7. Encourage private investment by farmers in the areas of land development, lift irrigation, agricultural machinery etc.

8. To strengthen the cooperative banking system as recommended by the Vaidhyanathan Committee.
9. In the direction of restoring soil health to restructure the fertilizer subsidy to make it nutrient-based.
10. To take appropriate steps to supply quality seeds.
11. To strengthen the marketing machinery, ensuring proper linkage to the markets with transparent pricing, cold storage facility, all aiming at fair price to different crops.
12. Encouraging contract farming ensuring small and marginal farmer adequate bargaining strength.
13. To encourage formation of farmer's groups or associations to safeguard their interests.
14. To take appropriate steps to control price fluctuations so as to ensure fair price to the producer providing price signals and management of risks.
15. Improving rural infrastructure including rural roads and rural electrification.

These different initiatives, no doubt, are relevant. But in rural areas what is a reality? Poverty, illiteracy, ill health and civic inertia are all concentrated here, particularly in the agricultural sector. Agricultural labour, tenant cultivators and women are much exploited segments of the rural society. There is an unending migration of rural poor to the urban centres seeking better employment and safer life. The ability to generate adequate productive employment opportunities with a fair remuneration to all categories will be a major factor on which the inclusiveness of growth will be judged. India is currently at a stage of 'demographic transition'; where population growth is slowing down, but the population of young people entering the labour force continues to expand with a trend of migration to urban centres. It is essential to evolve a strategy of creating employment to the new entrants to the labour force on one hand, and on the other to plan well for employment in the non-agricultural sector for workers leaving agriculture. Migration needs to be prevented by creating non agricultural occupations within the rural areas.

The right approach could be the Gandhian Plan of designing supplementary occupation to every agricultural household such as dairying, poultry etc. Self-help Groups need to be popularized involving all categories of women and that way pave the way for asset-building by way of supplementing occupation with supplementary income. Also Gandhian approach of production by masses by way of adequate number of work places in the rural areas and not mass production need to be given serious thought. It simply means labour-intensive small-scale industries must be well planned. Inequalities of income of higher degree in the society is bound to create social tensions. Hence the Gandhian proposal of Trusteeship, i.e., voluntarily handing over the surplus land or any other property to a Trust for common good over and above need of the family need to be popularized. The certainty of social unrest due to gulf in the levels of incomes in-between people must be made clear. That way in a non-violent and peaceful way economic justice is to be ensured in the society. In rural areas cottage and micro enterprises must be popularized. It should be the responsibility of the government to provide infrastructure and through self-help groups, promote asset building and cottage industries. Schumacher in his publication has all the appreciation for the Gandhian ideology of rural development. Harsh Singh[7] attempted a study of ground level realities in local governance through the panchayatraj addressing the problems of agricultural productivity through meaningful extension services and with a provision of value added service. The role that the business sector could play in rural transformation is well indicated with public-private-panchayat partnership for inclusive growth.

6. Panchayat Raj Institutions and Inclusive Growth

Basing on empirical study in sixteen poorest districts in India, it is noted that despite teething problems, Panchayatraj Institutions (PRIs) play important role in the betterment of lives of the poorest. In many districts, their importance was rated next only to the public distribution system, but higher than the institutions such as schools and hospitals. It is also observed that panchayat leaders often see their role as confined to resolving local disputes and implementing small works sanctioned by the governments and place the main responsibility for agriculture and other development activities to agencies such as cooperatives and government departments. The study concludes that "PRIs must transform the narrow and lop-sided vision of their mandate and accept a direct leadership role in agriculture which is the mainstay of the rural economy". It is rightly pointed out that extension services in the agricultural sector have not kept pace with the new challenges and opportunities. Insofar as the delegation of power to local government institutions is concerned, measures suggested are partial and ineffective. A research study in Dungarpur district of Rajasthan pertaining to the tribal families concluded that enhancement of incomes to the poor tribal families is possible through investment in milch animals and horticulture. This model envisages public investments in augmenting water resources and creating suitable access for farmers to different inputs. Specifically it is pointed that public investments are required for infrastructure creation aimed at water harvesting and recharge. Private investments could provide water delivery and irrigation services, farm mechanization services, agri-inputs, insurance and risk management, credit access and market linkages. The institutional mechanism at community level includes water user groups and community-based organizations for governance of water and other natural resource. In this model, panchayatraj institutions have to be fully involved. Their resources and powers need to the suitably strengthened so as to enable them to function effectively. In this model, panchayatraj institutions would have a dual role. First, is a two-way interaction with the extension agency on one hand and stakeholders on the other in the context of local planning to promote convergence of development activities at the local level. Second, is monitoring of services delivery by the extension agency so as to ensure suitable participatory mechanism without any foul play. The study expects public-private panchayat partnership for inclusive growth.

7. Towards Right Approach to Inclusive Growth

'Inclusive Growth' becomes meaningful and operational if only the gaps in the process of development are well identified by the competent authorities. Major areas and associated issues that come under the purview of 'Inclusive Growth' may be classified under four broad areas: (1) Regional disparities and related issues, (2) Rural-Urban Divide and related issues, (3) Agricultural and related issues, and (4) Poverty and inequalities of income and related issues. Multi-pronged approach is required for each of these four streams.

Dr. C. Hanumanth Rao[8] rightly points out that the Indian Economy continues to be constrained by the insufficiency of physical and social infrastructures where the government plays a major role. At a little over 4% of GDP in the base year of the Eleventh Five Year Plan, Public Investment in infrastructure was seriously deficient. This is explained by years of underinvestment relative to the requirements: (World Bank, 2008). Inadequacy of infrastructure in the less developed regions and rural areas in general in the country is responsible for the inability of such areas to fully benefit from the opportunities opened up by economic reforms including globalization leading to growing regional and rural-urban disparities (Rao, 2009). There is a strong case to set up adequate number of small-scale industrial units, in the rural areas. Also basic amenities must be provided in the rural areas benefiting the rural people. Adequate investment in physical and social infrastructure in the less developed regions and rural areas in terms of primary health, rural roads, power and communication, market information and marketing network etc., become necessary to reduce the regional disparities, as well as reduce the gulf between the rural and urban.

Another crucial area to be covered by "inclusive growth" relates to agriculture. In the area of farm production, issues to be addressed, among others, are (1) Improvement in technology of rainfed cultivation, (2) Water management including rain water conservation, (3) Remunerative price to farm production, (4) Marketing facilities including cold storage, rural roads and market information, (5) Adequate institutional credit, (6) Extension Services, (7) Insurance scheme covering both crop production and livestock and (8) Provision of supplementary occupation assuring income to all farm households.

The strategy of "Inclusive growth" must addressed the president poverty and growing inequalities of income ensuring equity to all sections. From the socially and economically disadvantaged sections like scheduled castes, scheduled tribes, Backward Classes and Minorities who are benefited by the programmes of the state, and reasons for such situation have to be identified and appropriate steps need to be taken. It is well-known fact that women and children have not been adequately benefited from different schemes in operation. Growth by itself, far from reducing the risk of poverty for the socially excluded groups has in fact increased such risk due to market failures arising from discrimination as well as public interventions that are both inadequate and discriminatory in operation (Rao, 2009). There is pressing need for empowerment of women. Social and economic empowerment of women and poor category of people should receive priority. Apart from social security, economic security is very much needed for the category in terms of ownership of capital assets, like land. Land reforms should receive necessary attention, both by way of ensuring land ownership rights as well as safeguarding the interests of tenant cultivators. Authentic land records with details of land ownership land tenure, tenancy cultivation and related issues become necessary.

Raising farm productivity should necessarily be given priority and therefore, the schemes of SPECIAL ECONOMIC ZONES, INDUSTRIALIZATION AND URBANIZATION should not be allowed to encroach upon fertile cropped area adversely affecting the interests of the farm community. Special care of tribal population is needed. Land alienation should not under any circumstances cause a problem to tribal people. Appropriate legal protection to the Tribal people in matters of exploiting natural resources and undertaking major irrigation projects is needed. In addressing issues relating to poverty and inequalities of income, the scheme of linking micro finance with Bank credit must be popularised involving all sections of the poor category. People at large must become partners in the development process and that is possible through strengthening of Village Panchayats with adequate funds and functions. Path can be well set for 'inclusive growth' through decentralization of administration and planning mechanism with village as growth centre. This is the right approach in the strategy of 'Inclusive growth'.

REFERENCES

1. *Eleventh Five Year Plan Report,* "Inclusive Growth Vision and Strategy", p. 1.
2. K.N. Panikar, "Was Indian? Nationalism Inclusive?, The Hindu, Tuesday, February 23, 2010.
3. V. Anil Kumar, "A Universal Paradox: Can Market Economy become Inclusive?, The Hindu, Tuesday February 9, 2010.
4. *Eleventh Five Year Plan Report,* p. 2.
5. *Ibid.*
6. *Ibid.*
7. Harsh Singh, "Public-Private Panchayat Partnership for Inclusive Growth", *The Hindu*, Wednesday, April 28, 2010.
8. C.H. Hanumanth Rao, "Inclusive Growth — An Overview of Performance and the Challenges Ahead", Presidential Address at 92nd Annual Conference of the Indian Economic Association (27-29 December, 2009), KIIT University, Bhuvaneswar.

36 INDIA TO BE A DEVELOPED NATION

1. Introduction

Rural India has been in the process of development and it should emerge as a developed nation. Our experience with six decades of planned development, clearly indicates that along with development there has been persistent poverty, inequality, regional variations and consequent social tensions. These problems of serious nature must be addressed. Also, we have to be very clear about the different aspects of the development and the yardstick with which we assess the development and the ways and means of tackling persistent burning problems of poverty, inequality and social tensions.

2. Gandhian Approach to Development

Gandhiji opined that development should not be assessed in terms of gross national product, per capita income and similar such monetary terms only. The non-monetary aspects, such as family ties, cordial relationship among the different sections of the society, development of different Arts, etc., become relevant in assessing development. Gandhiji's approach to the problems of poverty, inequality and social tensions was very realistic and solutions indicated are basically humanitarian. Gandhiji desired reduction of wants to a minimum level, bearing in mind the poverty of India. To quote Gandhiji "It is the fundamental law of Nature, without exception, that Nature produces enough for our wants from day to day; and if only everybody took enough for himself and nothing more, there would be no pauperism in this world, there would be no man dying of starvation in this world". Gandhiji speaks of the need for economic equality. In his words" "economic equality means, the levelling down of the few rich in whose hands is concentrated the bulk of the national wealth on the one hand and the levelling up of the semi-starved naked millions on the other. A non-violent system of government is clearly an impossibility, so long as the wide gulf between the rich and the hungry million persists. He proceeds further and speaks of a method of reducing the gulf between the rich and the poor. In his words "As for the present owners of wealth they would have to make a choice between class war and voluntarily converting themselves into Trustees of their wealth". These views of Gandhiji clearly indicate that he was more revolutionary in his thinking. As one who was committed to non-violence, he appealed to the wealthy class to voluntarily handover their surplus wealth to a Trust intended for social good without giving room for social conflict. He pleaded for "Production by Masses" and not "Mass production". Production by masses simply means establishment of eco-friendly micro enterprises in large numbers spread over the rural areas. Mass production is capital-intensive, highly mechanized system which gives room for amassing wealth by selected few capitalists.

Dr. Schumacher in his thought provoking work, *"Small is Beautiful"*, in line with the Gandhian thinking states, "Man is pulling the earth and himself out of equilibrium by applying only one test to everything he does: Money, profits and therefore, giant operations. We have got to ask, instead, what about the cost in human terms of happiness, health, beauty, and conserving the planet"

Schumacher very rightly speaks of "Buddhist Economics" while the materialist is mainly interested in goods, Buddhism is mainly interested in liberation from the attachment to wealth. He observes "It is not the wealth that stands in the way of liberation, but the attachment to wealth: not the enjoyment of pleasurable things, but craving for them. The keynote of Buddhist economics, therefore, is simplicity and non-violence." He further points out that it is not a question of choosing "Modern Growth" and "traditional stagnation". It is a question of finding the right path of development, the middle way between materiliastic heedlessness and traditionalist immobility — in short, of finding right livelihood. It is this type of middle way that has been advocated by Gandhiji and Pandit Nehru.

3. Pandit Nehru and Development of Science and Technology

Pandit Nehru rightly thought of development of science and technology in the country and apply the same in the process of development, meeting the needs of modern society. Pandit Nehru advocated 'socialistic pattern of society' or mixed economy which is very relevant and worth pursuing even now. The scheme of public-private partnership which is in force in selected parts of India is a corollary of mixed economy advocated by Pandit Nehru.

During the planning era of six decades so far, considerable progress has been achieved in all sectors of economy. India is now self-sufficient in the foodstuffs production, of course, at the existing level of consumption. India is able to export foodstuffs as well as industrial goods and earn foreign exchange. Now our balance of payments position is safe. After the introduction of structural reforms, Indian economy got integrated into global economy. Globalization process integrated the Indian economy with the economies of the west, which enabled India to move on to a higher growth path and reduce poverty, withstand keen competition from foreign markets. However, Indian financial system could not escape completely from the impact of the US financial crisis. The Reserve Bank of India had well managed and supervised the Indian banking system effectively and ensured adequate capital base for the banks. However, as Dr. Thimmaiah, an Economist observes, "notwithstanding prudent management of the Indian financial system, the Indian economy could not escape from the adverse effects of the meltdown of the real economy in the west. The efforts of the successive governments, after the introduction economic reforms to reduce poverty by achieving higher growth rate of GDP were made ineffective by the decline in the growth rate of GDP and large-scale job losses on account of the meltdown of the real economy of India". The growth rates of both exports and imports went on declining until the middle of 2009-10, as is evident from the economic survey 2009-10. But encouragingly enough, the Indian economy started recovering from the slow-down, towards the end of the fiscal year 2009-10. It is evident from the 7.2 % growth rate of GDP. Except agriculture, with negative growth rate of 0.2% all other sectors have shown recovery. It is therefore of utmost importance to give special attention to agriculture sector and ensure higher growth of agriculture.

4. Nature of Environmental Problems in India

In the process of development, environmental problems crop up. In India too there are environment problems. Environmental problems in India are different in nature and magnitude as compared to that in the western countries. Again environmental problems of urban centres are different from those in rural areas within India. The increase in population densities in rural areas fortunately has not yet led to problems of air, soil and water pollution of the magnitude that we have in urban India. The size of the average village is such that most of the household effluents are ground-absorbed whereas, garbage is composed for manure. Industrial waste is still not an appreciable hazard in a rural environment, the treatment of chemical fertilizers with the water ways is a problem, but not as yet an insuperable one. Similarly, the danger of the use of chemical pesticides has still not reached the alarming proportion of the developed world. Rural population pressures, as it is, have an impact on natural forest resources, the water regime and wildlife. When we see the urban environment the situation is different and alarming. Exploitation of groundwater is not limited to recharge potential or natural potential. With the adoption of modern technologies like high dams for large storages and deep tube wells the potential for utilizing available quantities of water has considerably increased with growing pressure on cropped land, there is overexploitation of both groundwater and surface water. As a part of development process, we notice gradual increase in the use of water for domestic and industrial purposes and this will give rise to the problem of deteriorating quality of water. Good quality of water is essential for better environment and for the health of the people as well as cattle population. In the past, groundwater was lifted from the dug-wells by manual devices or by use of animal power. The draft was limited and could hardly exceed the natural recharge potential. Under the conditions of use of diesel or electricity for pumping water, the capacity to draw water from greater depths has increased. The exploitation of groundwater is beyond tolerable limits. Under the situation of over exploitation of groundwater and excessive mining activities, great damage is done to the environment.

Again, the forest resources are equally over exploited in terms of firewood, timber for house construction and growing mining activities as a part of development. As the rural population increases, and landholdings become uneconomical in size, the capacity to invest in land improvement also declines. Now many farmers become marginalized in terms of the size of holdings with the adverse effects on land productivity. Under these circumstances, there is bound to be migration of labour to the urban centres. A stable social structure is a vital component of the total environment and when this is disturbed, violence follows and both the physical and social environment are ravaged.

5. Development Key to Fighting Naxalism

Persistent poverty and inequality of income is a burning problem. Lack of development and inequality of higher magnitude being the main reason for the growth of Maoism or Naxalism and militancy, the government and other agencies must act quickly in the direction of appropriate development schemes as well as reducing the inequality both economic and social. Gandhiji's thoughts and his life were truly an expression of the philosophy of our ancient civilization in which peace and harmony, non-violence, and truth, human dignity and compassion are being given greater prominence. Every citizen of this sacred land has the responsibility of honouring the ideology of Gandhiji in terms of limiting one's own earnings to the extent of meeting the needs only; keeping in view the welfare of the many in the society, extending helping hand, treating all sections of the society as equals without any discrimination; designing adequate number of eco-friendly, micro enterprises in the rural areas so as to create employment and assured income to those who need it most. People must be involved as partners in the development process at all levels. The rural poor must develop work culture and they should try to lead a life within their resources. Formation of Self-help Groups and that way the poor should try to mobilise their own resources and build up assets as an effective step in tackling poverty. The rural poor must develop work culture and they should try to lead a life within their own resources. Every individual should try to be self-reliant and that way pave the way for the village also to be self reliant. Gandhiji opined that in the international trade too, as for as possible, a country should try to be self-reliant and avoid depending upon foreign goods.

6. Steps in the Direction of Development

Most appropriate step would be to retain rural population in the rural areas only. This is possible when work places are created by setting up of micro enterprises on a large scale in rural areas. Also, each rural household may plan for a supplementary occupation through mobilization of finance by forming self-help groups.

Soil erosion and land degradation coupled with declining per capita availability of land and fresh water are posing a serious threat to the environment. As indicated earlier, Joint-forest management and participatory watershed management need to be well designed and implemented as an effective step of environmental protection paving the way for higher productivity.

Agriculture plays a crucial role in terms of generation of employment and supply of basic necessities of life. But, Indian agriculture still suffers from: (i) Low productivity, (ii) Declining water-table levels, (iii) Inadequate institutional credit, (iv) Distorted market, (v) Increased role of intermediaries who only increase the cost but do not add much value (vi) Poor private investment (vii) Lack of remunerative price to the produce, (viii) Poor infrastructure and ix) Inadequate research and technology.

Foodgrain productions must necessarily be stepped up through increase of yield per hectare of land of different crops. The reasons for low level of productivity of different crops in India in comparison with that of other countries must be identified and accordingly appropriate steps be taken immediately. Still, some 30 per cent of the cropped area is under the dry land agriculture. Advanced technology in farming must be extended to the rain fed agriculture also. The problems in areas with rain fed agriculture need to be identified and appropriate steps have to be initiated. The research should focus on evolving drought resistant strains and popularize them. India, being a member of W.T.O. has to face severe competition from the member countries. The use of agro-chemicals and fertilizers has to conform to the international specifications. Food security in the country demands adequate foodgrains production. Technologies play a crucial role in achieving higher levels of production and that way ensure food security. Apart from depending on conventional agricultural technologies, attention should also be focused on the biotechnology. It is crucial because biotechnology deals with many aspects of basic inputs to agriculture, namely, seeds, plants, soil treatment etc. More than any other input, irrigation is crucial in agriculture and hence there is an urgent need for integrated approach in the management of water resources. Also there is a case for linking of all major rivers ensuring water supply to water scarce areas. Through appropriate legislation, all major rivers have to be declare as a national property. The rain water harvesting and protecting forest resources should become the social responsibility. All the villagers unitedly work for well conserving the rain water and utilize the water. Apart from water harvesting, drip irrigation method should also be followed for better management of water.

Apart from stepping up of rice and wheat production, production of pulses, vegetables and different horticulture crops should also receive attention. Cold storage facilities in rural areas become necessary to promote dairying. Post-harvest technologies and agri-food processing should receive adequate attention. Modernization of storage and processing facilities will not only reduce losses but also help a more efficient use of the by-products. The simple indigenous technologies should necessarily be adopted at all levels to attain food security in a sustainable manner.

Food security requires adequate purchasing capacity on the part of the consumers. The NREGP scheme should be effectively implemented to ensure regular income to the poor and that way, ensure food security. By setting up of adequate number of small scale and cottage industries in the rural areas, migration of rural labour to the urban centres can be prevented.

India is a vast country with varying geo-physical conditions. Some regions are well endowed with resources and potentialities for development, where as some other regions are denied of such resources. The backward and drought-prone areas need special care of the government. The programmes initiated in this connection must be strengthened and implemented effectively. Some research studies indicate that only those who are relatively better placed in the society are able to get benefit from these schemes and not the poor. There is a case for identifying the really poor and meet the needs of such people. Strengthening the village panchayats and making the village a growth centre would be an effective step in addressing the problem of regional imbalances and consequent social tensions. Timely and adequate institutional credit facilities must be strengthened and role of private money-lenders in rural credit must be completely eliminated.

7. Role of Citizen in the Task of Development

A.P.J. Abdul Kalam, strongly believes that India should emerge as a developed nation. He is categorical and out spoken in his observation "India should become a developed nation by 2020. A billion people are our resource for this transformation. All of us have a role to play, actions are many but the goal is one. The Japanese are proud of their country's capability. They want to excel in their work. If each of us attempts to do so in our spheres of work, the status of developed India will arrive sooner than we expect because our country has many natural core strengths and competitive advantages" He very rightly points out the importance of human resource in the process of development. In all sectors of economy and in all establishments there is one common element, namely, the people who work. He opines that it should not be difficult to achieve the goal of "Developed India" if only all of us, without any exception work more and work better. To quote, Dr. Kalam again, "If you are a clerk, say in a government department, you can decide to work slightly more efficiently in clearing a public demand or a new project. If you can be an instrument in creating a feeling that the government (central, state municipality) works speedily and justly, you have created necessary conditions for a developed India. Don't think what can one person do. Many drops make a flood. A worker in a factory can decide to increase his or her productivity a little more and give attention to quality. The Japanese have an organized system to obtain and act upon the suggestions at the grass roots level. We don't have such a system. You can be the imitator of such an effort. At every level, a feeling of contributing concretely towards a developed India is a must. The larger the number of persons who act, the better it is". Specifically for teachers, his appeal is, "If you are a teacher, in whatever capacity, you have a very special role to play because more than anybody else, you are shaping generations". He indicated two functions for a teacher. "First, let them think about a developed India in their own way and enthuse the students. Secondly, they should update their own knowledge because the student is only as good as the teacher. Let them constantly try to upgrade their own skills so that they can enthuse the children to think big. Let us not transmit our frustration to them. If you can convey a message about bright future and encourage them, that will be a great service you will be doing to then and also the country". He repeatedly reminds that we have to work and work and a great nation is made of contributions from a large number of ordinary persons.

Conclusion

'Development' of right type and in the right direction is the answer to the burning problems of poverty, regional variations in growth and inequality of high magnitude with consequent social tensions. Development ultimately means development of human resource. The need of the hour is to identify such areas and such people by-passed by development programmes so far. Development of only those who need help is of utmost importance. All citizens in the country have a responsibility to contribute for the development. Those who are well placed in the society must act with high degree of responsibility with social consciousness aiming at with general welfare of the society.

The Eleventh Five Year Plan period will soon come to an end. The planning commission has already started the exercise of drafting 12th Five Year Plan. At the end of the Eleventh Five Year Plan matters of concern are: Slowdown in agricultural growth rates, widening income disparities between irrigated and rain fed areas, adverse effects of trade liberalization policy on farm production, neglect of indigenous technology on one hand and inefficient use of modern farm technology on the other, lack of required incentives to the farming community leading to farmers resorting to suicides and growing environmental degradation in terms of decline in ground water table and over exploitation of natural resources, particularly mineral resources leading to environmental degradation adversely affecting the livelihood of the poor.

It is most likely that even 4% growth rate in agriculture by the end of Eleventh Plan may not materialize. Farm production depends primarily on the decision by individual farm producers. They must be willing to invest on farm operations and they must also have an ability to invest. Keeping this reality in view necessary incentives have to be provided for the farm producers to act. Dr. M.S. Swaminathan recommendations need to be honoured and implemented immediately, particularly by way of the following steps:

1. Provision of institutional credit at interest rate not exceeding 4% to different farm operations from the purchase of seeds till the harvesting and marketing operations.
2. Provision of remunerative price, namely "Minimum support price to the at least 50% more than the cost of production".
3. An efficient marketing system with farmers' organizations as important players.
4. Promoting sustainable livelihoods for the poor in terms of insurance for risk, health, crop and livestock.
5. Strengthening of rural energy in terms of biogas plants, many hydro power and bio-fuel technologies.
6. Empowerment of village panchayats with needed resources, information and training to discharge the responsibility of "Agricultural Extension" besides other developmental programmes.

Apart from implementing these recommendations, the policy of 'inclusive growth' must identify all those sections in the society not benefited by the developmental schemes so far. The village panchayats, particularly the gramsabhas are right bodies to identify who really need help and extend help only to those who need help. The public at large must be alert without any bias and act with a sense of social consciousness and responsibility.

Environmental degradation is an issue affecting the interests of all people of both the present generation as well as future generation. Since all are responsible, directly or indirectly for environmental degradation, all must consider the environmental protection as a social responsibility and act suitably. We have to protect common property resources such as grazing lands, waste and barren lands. By means of a scientific micro-water shed management programme, it is possible to conserve every drop of water wherever and when ever it rains and that way prevent soil erosion and bring most of barren land under grasses. The scheme of NREGP should be utilized for planting various species such as teak wood, sandalwood, bamboo etc. on tracks of barren and wastelands. Such plantation will prevent indiscriminate cutting of trees from forest land for domestic use. Protection of forest resources should be taken as a social responsibility. The scheme of joint forest management with formation of Vanasamrakshana Samithies should be popularized and strengthened. Self-help Group formation by rural poor, harvesting and preserving of rainwater, safe guarding of common property resources, protecting forest resources become crucial in the process of development. The non governmental organizations have to play a constructive role in creating necessary awareness and involving all sections of society in the process of development. Instead of looking towards government for help in all matters, people at different levels-village, town, city should form associations and act as social capital or collective force in addressing their own problems. However, government must necessarily take the responsibility of providing minimum basic facilities such as primary education, primary health, infrastructure, protected water supply, etc. Gandhiji's life of simplicity needs to be emulated by one and all.

Work culture on the part of every citizen at every level with a sense of social responsibility is the need of the hour. If work culture becomes the motto of all the citizens, coupled with value-based native culture of this sacred land, path is well set and then India is bound to emerge as a Developed Nation.

REFERENCES

1. M.K. Gandhi, *Thus spoke Gandhi,* Compiled by Dr. Ramesh Bharadwaj, New Delhi – 110002.
2. M.K. Gandhi, *Panchayiti Raj,* Navajeevan Publishing Home Ahmedabad-14.
3. M.K. Gandhi, *Trusteeship,* Navajeevan Publishing Home Ahmedabad-14.
4. A.P.J. Abdul Kalam, *India 2020 – A Vision for the New Millennium,* Penguin Books, 2002.
5. Ramesh Golait, *"Reserve Bank of India Occasional Papers",* Vol. 28, No. 1, 2007.
6. Vasant Gowariker (Ed.), *Science, Population and Development,* Publications & Information Directorate CSIR New Delhi - 110 012.

BIBLIOGRAPHY

Abdul Kalam, A.P.J., *India 2020,* Penguin Books, Chennai, 2002.

Ahluwalia Montesi, S., "Rural Poverty and Agricultural Performance in India", *Journal of Development Studies,* Vol. 14, April, 1978.

Anirudh Krishna, *Active Social Capital,* Oxford University Press, New Delhi, 2003.

Amritananda Das, *Foundations of Gandhian Economics,* St. Martin's Press, New York, 1979.

Anil Khandelwal, K., "A Micro Finance Development Strategy for India" in *Economic and Political Weekly,* March 2007.

Anker Desmond, L.W., "Rural Development : Problems and Strategies", *International Labour Review,* 1973.

Aurora, R.C., *Integrated Rural Development,* New Delhi, 2nd Revised Edition, S. Chand and Co. Ltd., 1986.

Arthur W. Lewis, "The Economic Development with Unlimited Supplies of Labour" in *Economics of Underdevelopment,* (Eds.) A.N. Agarwala and S.P. Singh, Oxford University Press, 1953.

Aurora, D., "Drought Prone Areas: Some Development Issues", New Delhi, Institute of Public Administration, 1977.

Belshaw, H., "Agricultural Credit in Economically Underdeveloped Countries", *FAO Agriculture Studies,* Rome, No. 46, 1959.

Brown, Lester, R.G. "The Agricultural Revolution in Asia", *Foreign Affairs, An American Quarterly Review,* July, Vol. 46, No. 4, 1968.

Budumuru, Y., and Jesfa, G.B., "Participatory Watershed Management for Sustainable Rural Livelihoods in India", a paper presented at southern Agricultural Economics Association Annual Meeting, Florida, Feb, 2008.

Carle, C.Z. and Pritirim Sorokin, *Principles of Rural-Urban Society,* New York, H. Hott & Co., 1929.

Centre for Environmental Concerns, Report on Joint Forest Management in Andhra Pradesh, Hyderabad.

Climate Action Network South East Asia (CANSEA), Report on "Towards Climate Justice, Social Equity and a Sustainable Future", 2008.

Coale, Ansely, J. and Hoover Edger, M., *Population Growth and Economic Development in Low Income Countries,* New Jersey, Princeton University Press, 1958.

Dandekar, V.M., "Planning in Indian Agriculture, Indian *Journal of Agricultural Economics,* Vol. XXII, No. 1, January-March, 1967.

"Agriculture, Employment and Poverty", *Economic and Political Weekly,* Vol. XXI, No. 38 and 39, Sep. 20-27, 1986.

Dandekar, V.M. and Neelakanta Rath, "Poverty in India" in *Economics and Political Weekly,* Vol. VI, No.1, Jan. 1971 and Vol. VI, No. 2, Jan, 1977.

Dantwala, M.L., "Preface to Comparative Experience of Agricultural Development in Developing Countries of Asia and South-East since World War II", Paper, Proceedings of International Seminar held at New Delhi on Oct, 25-28, 1971, *Indian Society of Agricultural Economics,* Mumbai, 1972.

______ "Agrarian Structure and Economic Development", Presidential Address, All-India Agricultural Economics Conference, Chandigarh, 1960.

______ "Agrarian Structure and Agrarian Relations in India", in *Indian Agricultural Development Since Independence,* Indian Society of Agricultural Economics, Mumbai, Oxford and IBH Publishing Co. Pvt. Ltd., New Delhi, 1986.

______ "Prices and Cropping Patterns", Economic and Political Weekly, 1986.

Dakshina Murthy, K.S., "Politics of Environment", *Economic and Political Weekly,* Vol. XXI, No. 18, May 3, 1986.

David, L.Sills (Ed.), *Encyclopedia of Social Sciences.* The Macmillan Company and the Free Press, 1968.

Desai, S.S.M., *Rural Banking in India,* Himalaya Publishing House, Mumbai, 1986.

Dey, S.K., "Radical Theory is the Right Alternative", *Kurukshetra,* Vol. XXIX, October 1, 1980.

Dube, S.C., *Rural Society in India,* Popular Publications, 1978.

Edmund Des. Brunner *et. al.,* *Farmers of the World: The Development of Agriculture Extension,* New York, Columbia University Press, 1954.

Eicher and Witt (Eds.) *Agriculture in Economic Development,* New York, Columbia University Press, 1954.

______ *Encyclopedia of Britannia,* Chicago, Helen Hemingway Beutar, 1981.

FAO Report on Conditions of Work and Promotion of Livelihoods in the Unorganised Sector, NCEUS Report on "State of Food Insecurity in the World, Penguin Books, 2002.

Fei, C.H. and Gustar Ranis, *Development of Labour Surplus Economy – Theory and Policy,* Homewood, Yale University, Irwin, 1964.

Gadgil, D.R., Economic Effects of Irrigation, Pune, Gokhale Institute of Politics and Economics, 1948.

Gandhi, Mahatma, *Basic Education,* Ahmedabad, Navajeevan Publishing House (NPH), 1951.

______ *Hindu Dharma,* NPH, 1950.

______ *Towards Non-violent Socialism,* NPH, 1951.

______ *Trusteeship,* NPH, 1960.

______ *Sarvodaya: Its Principles and Programmes,* NPH, 1951.

Ganguli, B.N., "The Economics of Technology, Output and Employment" in *Reflections on Economic Development and Social Change,* (Eds.) C.H. Hanumantha Rao and P.C. Joshi, New Delhi, Allied Publishers, Pvt., Ltd., 1979.

Gopalaswamy, T.P., *Rural Marketing Environment: Problems and Strategies,* Wheeler Publishing, New Delhi, 1998.

Gupta, D.B. *et. al.,* *Development Planning and Policy: Some Essays* in Honour of Professor V.K.R.V. Rao, Wiley Eastern Ltd., New Delhi, 1982.

Hanumantha Rao, C.H., *Agricultural Production and Returns in India,* Institute of Economics Growth, Delhi, Asia Publishing House, 1965.

______ "Agricultural Growth and Stagnation in India", *Readings in Agricultural Development,* (Ed.) A.M. Khusro, Mumbai, Allied Publishers Pvt. Ltd., 1st Ed., 1968.

Hanumantha Rao, C.H. "Inclusive Growth — An Overview of Performances and Challenges Ahead" — Presidential Address of 92nd Annual Conference of Indian Economic Association (24-29 December 2009), KIIT University, Bhuvaneswar.

Hanumantha Rao, C.H and Joshi, P.C (Eds.), *Reflection on Economic Development and Social Change*, New Delhi, Allied Publishers Pvt., 1980.

India, Government of Ministry of Agriculture and Irrigation, *Reports of the National Commission on Agriculture,* 1976.

______ Different Five Year Plan Reports.

______ Ministry of Education, *Challenge of Education,* New Delhi, 1985.

______ *India 2009.*

______ *Royal Commission on Agriculture in India Report,* Calcutta, Central Publications Bureau, 1928.

______ *Programmes for the People,* 2008.

______ Ministry of Rural Development, *Annual Report 2000-2001.*

______ *Working Group of Poverty,* Planning Commission, 2006.

______ Ministry of Rural Development, Report on *Rural Housing,* 2008.

______ Economic Survey, 2008-2009.

______ Intergovernmental Panel on Climate Change (IPCC), *Climate Resilient Development* – Synthesis Report, Swiss Agency for Development and Cooperation.

______ *National Sample Survey*, Different Round.

Jagadish Bhagwati, *The Economics of Underdeveloped Countries,* World University Library, 1966.

Jain, S.C, *Problems of Agricultural Development in India,* Allahabad, Kitab Mahal, 1985.

——— *Rural Development – Institutions and Strategies,* Jaipur, Rawat Publications, 1985.

Jain, S.P., "The Social Structure in Rural India" in *Rural Development in India – Some Facets,* NIRD, Hyderabad, 1987.

Forgenson Dale, W., "The Development of a Dual Economy", in *Economic Journal,* Vol. LXXI, June, 1961.

Kapar Singh, *Principles, Policies and Management,* Sage Publication, 3rd Edn. 2009, New Delhi.

Karl Brandt, *The Reconstruction of World Agriculture,* London, Allen and Unwin, 1945.

Khusro, A.M. (Ed.), *Readings in Agricultural Development,* Mumbai, Allied Publications, 1968.

Knight, Frank H., *Journal of Political Economy,* March, 1944.

Kumudini Dandekar, "Prospects of Employment Guarantee Scheme" in *Lokrajya,* October 6, 1983.

Kuznets, Simon, *Economic Growth and Structure – Selected Essays,* London, Heinemann Educational Book Ltd., 1966.

Lakdwala, D.T., "Direct Taxation of Agriculture", in *Indian Journal of Agriculture Economics,* Vol. XXX, No. 4, Oct-Dec, 1975.

Lele, Uma, *Design of Rural Development – Lessons from Africa,* London, John Hopkins University Press, 1975.

Mahbub-ul-Hag, *Reflections on Human Development,* Oxford University Press, 1997.

Marshall, Alfred, *Principles of Economics,* London, Macmillan Co., 1930.

McNamara, S.Robert, "Striving for a Decent Human Life", Presidential Address to the Board of Governors, Kenya, *Eastern Economist,* October 5, 1973.

Memoria, C.B., *Agricultural Problems of India,* Allahabad, Kitab Mahal, 9th 1979.

Mellor, John W., *The Economics of Agricultural Development,* Mumbai, Vora & Co., Publishers, 1964.

——— "The Role of Agriculture in Economic Development", in *American Economic Review,* Vol. 51, Sept. 1961.

Misra and Puri, *Indian Economy,* Himalaya Publishing House, Mumbai, 2009.

Mollison, *"Permaculture",* The Deccan Development Society and Perm Culture – India, October 1990.

Mosher, A.T., *Getting Agriculture Moving,* New York, Frederic A. Praeger Inc. Publisher, 1966.

Myrdal, Gunnar, *Asian Drama – An Enquiry into the Poverty of Nations,* Vol. II, London, The Penguin Press, 1968.

——— *The Challenge of World Poverty — A World Anti-poverty Programme in Outline,* Penguin Books, 1970.

Nagaraj. K, Farmers' Suicides in India, Magnitudes, Trends and Spatial Patterns, Macroscan, 2008.

Nanavati, B. and Anjaria, J.J., *The Indian Rural Problems,* Mumbai, Indian Society of Agricultural Economics, Revised Edition, 1970.

Narayanan, K.P., "Technology at the Doorstep" in *The Hindu,* 11th October, 1986.

Nirodbaran, *Sri Aurobindo for all ages – A biography,* Sri Aurobindo Ashram, Pondicherry, 2004.

Nurkse, Ragnar, "Trade Fluctuations and Buffer policies of Low Income Countries" in *Agriculture in Economic Development,* (Ed.) Eicher and Witt, Mumbai, Vora & Co. Publishing Pvt. Ltd., Fourth Indian Report, 1970.

Palkivala, N.A., *India's Priceless Heritage,* Bharatiya Vidya Bhawan, Mumbai, 2003.

Parthasarthy, G. and Prasada Rao, B., *Implementation of Land Reforms in Andhra Pradesh,* Calcutta, Scientific Book Agency, 1969.

______ Dasarathy D., "Employment and Unemployment of Rural Labour and Cash Programme", Andhra University Press, Waltair, 1974.

Parthasarathy, G., "Land Reforms and Changing Agrarian Structure" in *Agricultural Development of India – Policy and Problems,* (Ed.) Shah, C.H., Mumbai, Orient Longman, 1979.

______ "Integrated Rural Development: Concepts, Theoretical Base and Contradictions, in *Rural Development in India* by Mathew (Ed.), New Delhi, Agricole Publishing Academy, 1981.

______ *The Green Revolution and the Weaker Sections,* Mumbai, Thacker & Co., Ltd., 1972.

______ "Changes in the incidence of Rural Poverty and Recent Trends in some Aspects of Agrarian Economy", Presidential Address delivered at the 46th Annual Conference of the Indian Society of Agricultural Economics held at Udaipur (Rajasthan), published in *Indian Journal of Agricultural Economics,* Vol. XLII, Jan-Mar, 1987.

Patel, Bihari L., "Biogas Technology and Rural India" in *Kurukshetra,* Vol. XXXIV, No. 6, March, 1986.

Pawar, Sharad, Speech at the meeting of National Development Council, New Delhi, June 2, 2005.

Prabhakar Singh, *Community Development Programme in India,* New Delhi, Deep Publications, 1982.

Raghavan, B.S., "Eying Global Markets", in *The Hindu,* Survey of Indian Agriculture, 1994.

Rajagopalachari, C. and Kumarappa, J.C. (Eds.), *The Nation's Voice,* Ahmedabad, Navjeevan Publishing House, 1947.

Ramachandran and T.K. Mahadevan (Eds.), *Guest for Gandhi,* New Delhi, Gandhi Peace Foundation, 1970.

Ramabhai, B., *Silent Revolution,* Delhi, Tiwan Prakashan, 1959.

Ramaswamy, N.S., "How Valuable is Animal Energy?" in *Kurukshetra,* Sept, 1978.

Ramesh Golait, Reserve Bank of India: Occasional Papers, 2007..

Ranga Reddy A. (Ed.) *The State of Rayalaseema,* Mittal Publications, New Delhi, 2003.

______ *Ethics Environment,* Mittal Publications, New Delhi, 2009.

Rao, V.K.R.V., *Education and Human Resource Development;* New Delhi, Allied Publishers, 1966.

______ "Alleviating Rural Poverty, But How?" in *Kurukshetra,* Vol. XXXII, Oct. 1983.

Reddy, Raghava G. and Subramanyam P., *Dynamics of Sustainable Rural Development,* Serials Publications, New Delhi, 2003.

Reddy, V.R. *et. al.,* "Participatory Watershed Development in India: Can it Sustain Rural Livelihoods?" in Development and Change, 2004.

Reserve Bank of India *General Report of the Committee of Direction,* Abridged edition, 1955.

______ The All-India Debt and Investment Survey, 2001.

______ Regional Rural Banks – Report of the Review Committee (Dantwala Committee).

Sainath, "Farm Crisis" in Series of *Farmers' Suicides in Andhra Pradesh,* 2004.

Sathis Chandra Nair and N.D. Jayal, *Forest Policy,* The INTACH Environmental Series.

Schultz, T.W., *Economic Crises in World Agriculture,* Ann Arbor, University of Michigan Press, 1965.

______ *Transforming Traditional Agriculture,* Ludhiana, Lyall Book Depot, 1970.

Schumacher, *Small is Beautiful,* Penguin Books Ltd., London, 1973.

Schumpeter Joseph, A., *The Theory of Economic Development,* Harvard, Cambridge University Press, 1951.

Sen, S.R., *Strategy for Agricultural Development,* Mumbai, Asia Publishing House, 2nd Edn., 1966.

______ "Growth and Instability in Indian Agriculture", in Agricultural Situation in Indian *Agriculture", in Agricultural Situation in India,* Jan. 1967.

Shridevi, S., *Gandhi and the Emancipation of Women in India,* Hyderabad, Gandhi Sahiteeya Pracharalaya, 1969.

Singh, R.P., *Implications of GATT/WTO on Agriculture and Rural Development,* NIRD, Rajendra Nagar, Hyderabad, 1998.

Singh, R.L. and Singh Rana, P.B., *Rural Habitat Transformation in World Frontiers,* Tokyo, 24, IGC Publication, 1980.

Sivaraman, B., "The Alternative?" in *Kurukshetra,* Vol. XXXIX, October 1, 1980.

Sonachalam, K.S., *Agricultural Development of India,* Annamalai Nagar, Annamalai University, 1983.

——— *Land Reforms in Tamil Nadu – Evaluation of Implementation,* Delhi, Oxford and IBH Publishers, 1970.

Southworth, Herman, M. and Johnson Bruce, F. (eds.), *Agricultural Development and Economic Growth,* Ithaca, London, Cornell University Press, 1967.

Srinivas, M.N., *India's Villages,* Mumbai, Media Promoters and Publishers Pvt. Ltd., 2nd Edition, 1978.

——— *India: Social Structure,* Delhi, Hindustan Publishing Corporation, 1982.

Subramaniam, C., "Strategy for Integrated Rural Development", *Community Development and Panchayat Raj Digest,* April 8, 1977.

Sushila Mehata, *A Study of Rural Sociology,* New Delhi, S. Chand & Co. Ltd., 1st Edn., 1980.

Swaminathan, M.S., Report on *National Commission on Farmers.*

——— *A Century of Hope, Harmony with Nature and Freedom from Hunger,* East-West Books Pvt. Ltd., Chennai, 1999.

Tharner Daniel, *Agrarian Prospect in India,* Delhi, Delhi University Press, 1956.

Todaro, Micheal P., *Economics of Developing World,* London, Longman Group Ltd., 1977.

Udai Pareek, *Education and Rural Development in Asia,* New Delhi, Oxford and IBH Publications, 1982.

United Nation "Economic Development and Planning in Asia and the Far East: The Agricultural sector", *Economic Bulletin for Asia and the Far East,* Vol. III, No. 3, 1957.

——— *Human Development Report,* Oxford University Press, 1997.

Venkata Reddy, K., *Agricultural Productivity in Andhra Pradesh,* Tirupati, S.V. University, 1977.

——— "Education, Extension and Agricultural Productivity", in *Khadigramodyog,* March 1970.

——— "Population Growth Trends and Food Requirements in India", in *Man and Life,* Jan-June, 1980.

——— "Rural Banking and Agricultural Finance – A Case Study of Rayalaseema Region," *Financing Agriculture,* July-Sept., 1980.

——— "Rural Development: Retrospective and Prospective", in *Bharatiya Vikas,* Jan-Mar., 1981.

——— "Supervisal Credit for Rural Development" in *Kurukshetra,* Nov.-Dec. 1981.

——— *"Rural Development in India (Poverty and Development)"*, Himalaya Publishing House, Mumbai, 1988.

——— *Agriculture and Rural Development,* Himalaya Publishing House, Mumbai, 2001.

——— "Ethical Empowerment and Sustainable Development" in *Ethics and Environment* (Ed) A. Ranga Reddy, Mittal Publications, New Delhi.

Venugopal Reddy, Y., "Regional Planning in India", in *Planning and Development of Backward Regions – A Case Study of Rayalaseema,* Vol. I.

Warren C. Bawm and Stokes M. Jobbert, *Investing in Development,* Oxford University Press, 1985.

Whiden Hague *et., al.,* *Towards a Theory of Rural Development,* United Nations Asian Development Institute, 1975.

Wood, D., *Corporate Social Performance Revisited,* The Academy of Management Review, Vol. 16, No. 4, October, 1991.

World Bank *World Development Report 2008,* Oxford University Press.

——— "Development and Climate Change", *World Development Report 2010,* World Bank – Washington D.C.

INDEX

ACRONYMS

AA	Automobile Association; Alcoholics Anonymous
AAA	American Automobile Association; Amateur Athletic Association
AABY	Aam Admi Bima Yojana
AAY	Antyodaya Anna Yojana
ABC	American Broadcasting Company
ABTA	Association of British Travel Agents
AC	Air Conditioning
ACA	Additional Central Assistance
ADB	Asian Development Board
AGDP	Gross Domestic Product from Agriculture
AGM	Annual General Meeting
AgMark	Agricultural Marketing
AIBP	Accelerated Irrigation Benefit Programme
AICTE	All India Council of Technical Education
AIDS	Acquired Immuno Deficiency Syndrome
AMUL	Anand Milk Producers Union Limited
ARWSP	Accelerated Rural Water Supply Programme
ASBO	Anti Social Behaviour Order
ASHA	Accredited Social Health Activists
ATC	Agreement in Textiles and Clothing
ATM	Automated Teller Machine
ATMAs	Agriculture Technology Management Agencies
AYUSH	Ayurveda, Yoga & Naturopathy, Unani, Siddha and Homeopathy
B.A.	Bachelor of Arts
BE	Budget Estimates
BBC	British Broadcasting Corporation
B.C.	Before Christ
B.Ed.	Bachelor of Education
BOP	Balance of Payments
BOS	Beneficiary Oriented Scheme
BPL	Below Poverty Line
BRGF	Backward Regions Grant Fund

CACP	Commission for Agricultural costs and prices
CAD	Computer-Aided Design
CANSEA	Climate Action Network South East Asia
CART	Council for Advancement of Rural Technology
CBB	Commercial Bank Branch
CBD	Convention of Biological Diversity
CBRI	Central Building Research Institute
CCIS	Comprehensive Crop Insurance Development
CD	Compact Disc; Community Development
CD-I	Compact Disc Interactive
CD-R	Compact Disc Recordable
CD-ROM	Compact Disc Read Only Memory
CD-RW	Compact Disc Rewritable
C.E.	Common Era
CEA	Central Executive Officer
CETP	Common Effluent Treatment Plants
CIA	Central Intelligence Agency
CID	Central Investigation Department
CII	Confederation of Indian Industry
CIIL	Central Institute of India Languages
CIS	Central Institute of Tools Design
CNN	Cable News Network
CO	Commanding Officer
CPCB	Central Pollution Control Board
CPI	Consumer Price Index
CRAFICARD	Committee to Review Arrangement for Institutional Credit for Agriculture and Rural Development
CRR	Cash Reserve Ratio
CRSP	Central Rural Sanitation Programme
CSIR	Council of Scientific and Industrial Research
CSO	Central Statistical Organization
CSR	Corporate Social Responsibility
CST	Central Sales Tax
CSWB	Central Social Welfare Board
CVD	Countervailing Duty

CWC	Central Water Commission
CWDB	Central Wool Development Board
DAC	Department of Agriculture & Cooperation
DAT	Digital Audio Tape
DCS	Dairy Cooperative Societies
DDP	Desert Development Programme
DELTA	Diploma in English Language Teaching to Adults
DES	Directorate of Economics and Statistics
DFIs	Development Financial Institutions
DHDR	District Human Development Report
DPAP	Drought Prone Area Programme
DPEP	District Primary Education Programme
DRDA	District Rural Development Agency
DSS	Design Support System
DST	Department of Science and Technology
DWACRA	Development of Women and Children in Rural Areas
ECG	Electrocardiogram
EIA	Environment Impact Assessment
EMCBP	Environment Management Capacity Building Project
EMS	Enhanced Message Service
ENT	Ear, Nose and Throat
ENTs	Economic Needs Tests
EOUs	Export Oriented Units
EPCG	Export Promotion Capital Goods Scheme
EPZ	Export Processing Zone
EU	European Union
EXIM	Export Import Policy
FAO	Food and Agricultural Organization
FCI	Food Corporation of India
FDI	Foreign Direct Investment
FDS	Firka Development Scheme
FII	Foreign Institutional Investment
FPR	Flood Prone Rivers
FRBMA	Fiscal Reforms and Budget Management Act
FTAs	Free Trade Agreements

FTP	Foreign Trade Policy
FYM	Farm Yard Manure
GATT	General Agreement on Trade & Tariffs
GBS	Gross Budgetary Support
GCA	Gross Cropped Area
GCES	General Crop Estimate Survey
GCF	Gross Capital Formation
GDP	Gross Domestic Product
GHG	Green House Gases
GIC	General Insurance Corporation
GNP	Gross National Product
GSDP	Gross State Domestic Product
Ha	Hectares
HCR	Head Count Ratio
HE	Her/His Excellency; Higher Education
HDI	Human Development Index
HIV	Human Immunodeficiency Virus
HOPCOMS	Horticultural Producers Co-operative Marketing and Processing Society Ltd.,
HRD	Human Resource Development
HSD	High Speed Diesel
HSMD	Hazardous Substances Management Division
HYV	High Yielding Varieties
IAS	Improvement of Agricultural Statistics
IAY	Indira Awas Yojana
ICAI	Institute of Chartered Accountants of India
ICAR	Indian Council of Agricultural Research
ICDS	Integrated Child Development Service
ICOR	Incremental Capital Output Ratio
ICPS	Integrated Child Protection Service
ICS	Improvement of Crop Statistics
ICSI	Institute of Company Secretaries of India
ID	Identification, Identity
IFAD	International Fund for Agricultural Development
IGNOAPS	Indira Gandhi National Old Age Pension Scheme

IGNOU	Indira Gandhi National Open University
IIM	Indian Institute of Management
IISER	Indian Institute of Science, Education and Research
IIT	Indian Institute of Technology
ILO	International Labour Organization
IMF	International Monetary Fund
IMR	Infant Mortality Rate
IPCC	Inter government Panel On Climate Change
IPHS	Indian Public Health Standards
IPPP	Industrial Pollution Prevention Project
IRCTC	Indian Railway Catering and Tourism Corporation
IRDP	Integrated Rural Development Programme
IRFC	Indian Railway Finance Corporation
ISI	Indian Standards Institute
ISP	Internet Service Provider
IT	Information Technology
ITA	International Technology Agreement
ITK	Indigenous Technical Knowledge
JFM	Joint Forest Management
JNNURM	Jawaharlal Nehru National Urban Renewal Mission
JSY	Janani Suraksha Yojana
KCC	Kisan Credit Card
KCCS	Kisan Credit Card Scheme
KGBVS	Kasturba Gandhi Balika Vidyalaya Scheme
KMPH	Kilo Meters Per Hour
KSY	Kishori Shakthi Yojana
KVI	Khadi and Village Industries
KVK	Krishi Vigyan Kendra
KVP	Kisan Vikas Patra
LAF	Liquidity Adjustment Facility
LIC	Life Insurance Corporation
LPCPD	Litres Per Capita Per Day
LPG	Liquid Petroleum Gas

MD	Doctor of Medicine, Managing Director
MDMS	Mid-Day Meal Scheme
MIS	Market Intervention Scheme, Management Information System
MLA	Member of the Legislative Assembly
MLD	Million Litres per Day
MMR	Maternal Mortality Rate
MNCs	Multi-National Corporations
MSMED	Micro, Small and Medium Enterprises Development
MSP	Minimum Support Price; Mahila Samaikhya Programme
MSR	Marketed Surplus Ratio
MSSRF	M.S.Swaminathan's Research Foundation
NA	Not Available; Not Announced
NAAC	National Accreditation Assessment Council
NABARD	National Bank for Agriculture and Rural Development
NAEB	National Afforestration & Eco-development Board
NAIS	National Agricultural Insurance Scheme
NAP	National Afforestration Project
NC	Not Collected
NCA	National Commission on Agriculture
NCAER	National Council of Applied Economic Research
NCERT	National Council of Educational Research and Training
NCEUS	National Commission on Enterprises in the Unorganized Sector
NCMP	National Common Minimum Programme
NCPCR	National Commission for Protection of Child Rights
NDA	Net Domestic Asset
NDDB	National Dairy Development Board
NDMA	National Disaster Management Authority
NEP	New Economic Policy
NFBS	National Family Benefit Scheme
NFCG	National Foundation for Corporate Governance
NFFWP	National Food For Work Programme
NFSM	National Food Security Mission
NGO	Non-Governmental Organization
NHAI	National Highway Authority of India
NHB	National Housing Bank

NHM	National Horticulture Mission
NHPC	National Hydro-electric Power Corporation
NHRDP	National Human Resource Development Programme
NIDC	National Industrial Development Corporation
NISIET	National Institute of Small Industries Extension and Training
NLCP	National Lake Conservation Plan
NLSI	New Linguistic Survey of India
NMDFC	National Minorities Development and Finance Corporation
NNP	Net National Product
NNRMS	National Natural Resource Management System
NPAG	Nutrition Programme for Adolescent Girls
NPK	Nitrogen Phosphate Potassium
NRAA	National Rain fed Area Authority
NREGP	National Rural Employment Guarantee Programme
NRCD	National River Conservation Directorate
NRHM	National Rural Health Mission
NSAP	National Social Assistance Programme
NSA	Net Sown Area
NSIC	National Small Industries Corporation
NSS	National Sample Survey
NSSO	National Sample Survey Organization
NSTFDC	National Scheduled Tribes Finance Development Corporation
OECD	Organization of Economic Cooperation and Development
OHP	Overhead Projection
OU	Open University
PACS	Primary Agricultural Credit Society
PDEXIL	Power-loom Development and Export Promotion Council
PDS	Public Distribution System
PE	Provisional Estimates
PEM	Protein Energy Malnutrition
PGCE	Post Graduate Certificate in Education
PHM	Post-Harvest Management
Ph.D	Doctor of Philosophy
PIM	Participatory Irrigation Management
PLDB	Primary Land Development Bank

PMGRY	Prime Minister's Grameen Rozgar Yojana
PMGSY	Pradhan Manthri Gram Sadak Yojana
PMGY-RDW	Prime Minister's Gramodaya Yojana Rural Drinking Water
PMSSY	Pradhan Manthri Swasthya Suraksha Yojana
PNGRB	Petroleum and Natural Gas Regulatory Board
POA	Programme of Action
PPP	Public-Private Partnership
PRA	Participatory Rural Appraisal
PRIs	Panchayat Raj Institutions
PSEs	Public Sector Enterprises
PSS	Price Support Scheme
PURA	Providing Urban Amenities in Rural Areas
R&D	Research and Development
RBI	Reserve Bank of India
RCH	Reproductive Child Health
RE	Revised Estimates
RGGVY	Rajiv Gandhi Grameen Vidyutikaran Yojana
RIDF	Rural Infrastructure Development Fund
RKVY	Rashtriya Krishi Vikas Yojana
RRB	Regional Rural Bank
RSBY	Rashtriya Swasthya Bima Yojana
RSVY	Rashtriya Sam Vikas Yojana
RTI	Right to Information
RVNL	Rail Vikas Nigam Limited
RVP	River Valley Project
SAC	Space Application Centre
SASA	State Agricultural Statistical Authority
SAU	State Agricultural Universities
SCs	Scheduled Castes
SCB	State Cooperative Bank
SDP	State Domestic Product
SEBI	Securities and Exchange Board of India
SEZs	Special Economic Zones
SGRY	Sampoorna Grameen Razgar Yojana
SGSY	Swarnajayanthi Gram Swarozgar Yojana

SHDR	State Human Development Report
SHG	Self Help Group
SIDBI	Small Industries Development Bank of India
SISI	Small Industries Service Institute
SLDB	State Land Development Board
SME	Small and Medium Enterprises
SMS	Short Message Services
SRM	Sustainable Rural Marketing
SSA	Sarva Shiksha Abhiyan
SSI	Small Scale Industries
SSRy	Swarnajayanthi Shahari Rozgar Yojana
STs	Scheduled Tribes
STD	Subscriber Trunk Dialing; Sexually Transmitted Disease
STEP	Support to Training and Employment Programme for Women
STU	State Transmission Utility
TB	Treasury Bill
TPDS	Targeted Public Distribution System
TRIMS	Trade Related Investment Measures
TRIPs	Trade Related Intellectual Property Rights
TSC	Total Sanitation Campaign
UGS	User Groups
UGC	Universal Grant Commission
U.K.	United Kingdom
UEE	Universalization of Elementary Education
UI	Unirrigated
UN	United Nations
UNDP	United Nations Development Programme
UNESCO	United Nations Educational, Scientific and Cultural Organization
UNFCCC	United Nations Framework Convention on Climate Change
UNICEF	United Nations International Children's Education Fund
UNO	United Nations Organization
UNSC	United Nations Statistical Committee
USA	United States of America
USAID	United States Agency for International Development
USEP	Urban Self-Employment Programme
UT	Union Territory

VAT	Value Added Tax
VGB	Village Grain Banks Scheme
VHS	Video Home System
VIP	Very Important Person
VKGUY	Vishesh Krishi and Gram Udyog Yojana
VBPY	Varishta Bima Pension Yojana
VSAT	Very Small Aperture Terminal
VSI	Village and Small Scale Industries
VSS	Vana Samrakshana Samithi
VVV	Vikas Volunteer Vahini
WB	World Bank
WCF	Women's Credit Fund
WCP	Women's Component Plan
WDC	Women's Development Corporation
WHO	World Health Organization
WII	Wildlife Institute of India
WPI	Wholesale Price Index
WTO	World Trade Organization
YMCA	Young Men's Christian Association
YWCA	Young Women's Christian Association